Patient Assessment, Intervention, and Documentation for the VETERINARY TECHNICIAN

A Guide to Developing Care Plans and SOAPs

JODY ROCKETT, CYNTHIA LATTANZIO, KATIE ANDERSON

DELMAR
CENGAGE Learning

Australia • Brazil • Japan • Korea • Mexico • Singapore • Spain • United Kingdom • United States

Patient Assessment, Intervention, and Documentation for the Veterinary Technician: A Guide to Developing Care Plans and SOAPs, First Edition
Jody Rockett, Cynthia Lattanzio, and Katie Anderson

Vice President, Career and Professional Editorial:
Dave Garza

Director of Learning Solutions:
Matthew Kane

Acquisitions Editor:
David Rosenbaum

Managing Editor:
Marah Bellegarde

Product Manager:
Christina Gifford

Editorial Assistant: Scott Royael

Vice President, Career and Professional Marketing:
Jennifer McAvey

Marketing Director:
Deborah Yarnell

Marketing Coordinator:
Jonathan Sheehan

Production Director:
Carolyn Miller

Production Manager:
Mark Bernard

Content Project Manager:
Jeff Varecka

Art Director: David Arsenault

Technology Project Manager:
Sandy Charette

© 2009 Delmar, Cengage Learning

ALL RIGHTS RESERVED. No part of this work covered by the copyright herein may be reproduced, transmitted, stored or used in any form or by any means graphic, electronic, or mechanical, including but not limited to photocopying, recording, scanning, digitizing, taping, Web distribution, information networks, or information storage and retrieval systems, except as permitted under Section 107 or 108 of the 1976 United States Copyright Act, without the prior written permission of the publisher.

> For product information and technology assistance, contact us at
> **Professional & Career Group Customer Support, 1-800-648-7450**
> For permission to use material from this text or product, submit all requests online at **cengage.com/permissions**
> Further permissions questions can be e-mailed to
> **permissionrequest@cengage.com**

Library of Congress Control Number: 2008923412

ISBN-13: 978-1-4180-6749-6

ISBN-10: 1-4180-6749-0

Delmar
5 Maxwell Drive
Clifton Park, NY 12065-2919
USA

Cengage Learning products are represented in Canada by Nelson Education, Ltd.

For your lifelong learning solutions, visit **delmar.cengage.com**

Visit our corporate website at **cengage.com**

Notice to the Reader

Publisher does not warrant or guarantee any of the products described herein or perform any independent analysis in connection with any of the product information contained herein. Publisher does not assume, and expressly disclaims, any obligation to obtain and include information other than that provided to it by the manufacturer. The reader is expressly warned to consider and adopt all safety precautions that might be indicated by the activities described herein and to avoid all potential hazards. By following the instructions contained herein, the reader willingly assumes all risks in connection with such instructions. The publisher makes no representations or warranties of any kind, including but not limited to, the warranties of fitness for particular purpose or merchantability, nor are any such representations implied with respect to the material set forth herein, and the publisher takes no responsibility with respect to such material. The publisher shall not be liable for any special, consequential, or exemplary damages resulting, in whole or part, from the readers' use of, or reliance upon, this material.

Printed in Canada
1 2 3 4 5 6 7 12 11 10 09 08

Dedication

For the dogs who see me as I aspire to be. And for the cats who know better. Live well.
J. R.

To my long-suffering husband, who, in spite of allergies to animal fur, dander, dust, hay, and grass, has managed to coexist (I hope happily) with four children, nine cats, four dogs, an ever-changing number of horses, and assorted rodents.
C. L.

Thanks, Sally, Emmy, Frodo, and Redrock.
K. A.

Contents

Chapter 1
Veterinary Technician Practice Model and Documentation

Role of Veterinary Team Members 2
Veterinary Technician Practice Model 2
   Step 1: Gather Data 3
   Step 2: Identify and Prioritize Technician Evaluations 4
      Identify Technician Evaluations 4
      Prioritize Technician Evaluations 5
   Step 3: Develop Plan of Care and Implement Interventions 8
   Step 4: Evaluate Results 8
   Step 5: Add Data 8
Documentation 9
   SOAP Notes 9
      Steps of SOAP Composition **10**
   Case Scenario with Technician SOAP Note **11**
   Notations and Incorporations **13**
   Medical Administration/Order Record **13**
      MAOR Examples **18**
   Case Scenarios with SOAP Notes, Notations, and MAOR **27**
Bibliography **34**

Chapter 2
Generating the Database

Introduction **36**
Physical Examination **36**
   Performing the Physical Examination **37**

Diagnostic Tests and Procedures *46*
    Testing Category Classification *47*
        Blood Tests *47*
        Physical Tests *49*
        Fluid Collection/Evaluation *50*
        Pathology/Culture *51*
        Fecal/Urine. *51*
Bibliography. *51*

Chapter 3
Technician Evaluations with Suggested Interventions

Introduction. *54*
Abnormal Eating Behavior *54*
    Definition/Characteristics *54*
    Desired Resolution. *54*
    Interventions with Rationale/Amplification *54*
Acute Pain *58*
    Definition/Characteristics *58*
    Desired Resolution. *58*
    Interventions with Rationale/Amplification *58*
Aggression *60*
    Definition/Characteristics *60*
    Desired Resolution. *60*
    Interventions with Rationale/Amplification *60*
Altered Gas Diffusion *62*
    Definition/Characteristics *62*
    Desired Resolution. *62*
    Interventions with Rationale/Amplification *62*
Altered Mentation *63*
    Definition/Characteristics *63*
    Desired Resolution. *63*
    Interventions with Rationale/Amplification *63*
Altered Oral Health. *64*
    Definition/Characteristics *64*
    Desired Resolution. *64*
    Interventions with Rationale/Amplification *64*

Altered Sensory Perception **66**
   Definition/Characteristics **66**
   Desired Resolution.. **66**
   Interventions with Rationale/Amplification **66**
Altered Urinary Production **67**
   Definition/Characteristics **67**
   Desired Resolution.. **67**
   Interventions with Rationale/Amplification **67**
Altered Ventilation.. **68**
   Definition/Characteristics **68**
   Desired Resolution.. **68**
   Interventions with Rationale/Amplification **68**
Anxiety **69**
   Definition/Characteristics **69**
   Desired Resolution.. **69**
   Interventions with Rationale/Amplification **69**
Bowel Incontinence **69**
   Definition/Characteristics **69**
   Desired Resolution.. **70**
   Interventions with Rationale/Amplification **70**
Cardiac Insufficiency.. **71**
   Definition/Characteristics **71**
   Desired Resolution.. **71**
   Interventions with Rationale/Amplification **71**
Chronic Pain **72**
   Definition/Characteristics **72**
   Desired Resolution.. **72**
   Interventions with Rationale/Amplification **72**
Client Coping Deficit.. **73**
   Definition/Characteristics **73**
   Desired Resolution.. **73**
   Interventions with Rationale/Amplification **73**
Client Knowledge Deficit.. **75**
   Definition/Characteristics **75**
   Desired Resolution.. **75**
   Interventions with Rationale Amplification.................... **75**

Contents

Constipation **75**
    Definition/Characteristics **75**
    Desired Resolution **76**
    Interventions with Rationale/Amplification **76**

Decreased Perfusion **77**
    Definition/Characteristics **77**
    Desired Resolution **77**
    Interventions with Rationale/Amplification **78**
        Cerebral Perfusion **78**
        Cardiopulmonary Perfusion **78**
        Renal Perfusion **78**
        GI Perfusion **79**
        Peripheral Perfusion **79**

Diarrhea **80**
    Definition/Characteristics **80**
    Desired Resolution **80**
    Interventions with Rationale/Amplification **80**

Electrolyte Imbalance **82**
    Definition/Characteristics **82**
    Desired Resolution **82**
    Interventions with Rationale/Amplification **82**

Exercise Intolerance **83**
    Definition/Characteristics **83**
    Desired Resolution **83**
    Interventions with Rationale/Amplification **83**

Fear **84**
    Definition/Characteristics **84**
    Desired Resolution **84**
    Interventions with Rationale/Amplification **84**

Hyperthermia **85**
    Definition/Characteristics **85**
    Desired Resolution **85**
    Interventions with Rationale/Amplification **86**

Hypervolemia **87**
    Definition/Characteristics **87**
    Desired Resolution **87**
    Interventions with Rationale/Amplification **87**

Hypothermia **88**
   Definition/Characteristics **88**
   Desired Resolution.. **89**
   Interventions with Rationale/Amplification **89**
Hypovolemia **90**
   Definition/Characteristics **90**
   Desired Resolution.. **90**
   Interventions with Rationale/Amplification **90**
Impaired Tissue Integrity **91**
   Definition/Characteristics **91**
   Desired Resolution.. **91**
   Interventions with Rationale/Amplification **92**
Inappropriate Elimination **93**
   Definition/Characteristics **93**
      Inappropriate Elimination Feline Behavioral **93**
   Desired Resolution.. **93**
   Interventions with Rationale/Amplification **93**
      Inappropriate Elimination Feline Territorial **95**
   Desired Resolution.. **95**
   Interventions with Rationale/Amplification **95**
      Inappropriate Elimination Canine Behavioral. **96**
   Desired Resolution.. **96**
   Interventions with Rationale/Amplification **96**
      Inappropriate Elimination Canine Territorial **97**
   Desired Resolution.. **97**
   Interventions with Rationale/Amplification **98**
Ineffective Nursing **98**
   Definition/Characteristics **98**
   Desired Resolution.. **98**
   Interventions with Rationale/Amplification **98**
Noncompliant Owner **99**
   Definition/Characteristics **99**
   Desired Resolution.. **99**
   Interventions with Rationale/Amplification **100**
Obstructed Airway **100**
   Definition/Characteristics **100**

Contents

    Desired Resolution... *100*
    Interventions with Rationale/Amplification.................. *100*
Overweight *101*
    Definition/Characteristics *101*
    Desired Resolution... *101*
    Interventions with Rationale/Amplification.................. *102*
Postoperative Compliance *104*
    Definition/Characteristics *104*
    Desired Resolution... *104*
    Interventions with Rationale/Amplification.................. *104*
Preoperative Compliance *105*
    Definition/Characteristics *105*
    Desired Resolution... *105*
    Interventions with Rationale/Amplification.................. *105*
Reduced Mobility... *106*
    Definition/Characteristics *106*
    Desired Resolution... *107*
    Interventions with Rationale/Amplification.................. *107*
Reproductive Dysfunction... *108*
    Definition/Characteristics *108*
    Desired Resolution... *108*
    Interventions with Rationale/Amplification.................. *108*
Risk of Aspiration *108*
    Definition/Characteristics *108*
    Desired Resolution... *109*
    Interventions with Rationale/Amplification.................. *109*
Risk of Infection... *109*
    Definition/Characteristics *109*
    Desired Resolution... *109*
    Interventions with Rationale/Amplification.................. *109*
Risk of Infection Transmission..................................... *110*
    Definition/Characteristics *110*
    Desired Resolution... *110*
    Interventions with Rationale/Amplification.................. *110*
Self-Care Deficit... *111*
    Definition/Characteristics *111*
    Desired Resolution... *111*

Interventions with Rationale/Amplification *111*
    Feeding *111*
    Grooming *112*
    Toileting. *112*
    Owner Education. *112*
  Self-Inflicted Injury *112*
    Definition/Characteristics *112*
    Desired Resolution. *113*
    Interventions with Rationale/Amplification *113*
  Sleep Disturbance *113*
    Definition/Characteristics *113*
    Desired Resolution. *113*
    Interventions with Rationale/Amplification *113*
  Status Within Acceptable Parameters. *114*
    Definition/Characteristics *114*
    Desired Resolution. *114*
    Interventions with Rational/Amplification *114*
  Underweight *114*
    Definition/Characteristics *114*
    Desired Resolution. *114*
    Interventions with Rationale/Amplification *115*
  Urinary Incontinence *116*
    Definition/Characteristics *116*
    Desired Resolution. *116*
    Interventions with Rationale/Amplification *116*
  Vomiting/Nausea *119*
    Definition/Characteristics *119*
    Desired Resolution. *119*
    Interventions with Rationale/Amplification *119*
  Bibliography. *120*

Chapter 4
Medical Conditions and Associated Technician Evaluations

Introduction. *124*
Conditions Involving the Gastrointestinal (GI) System *124*

Contents

 Colitis **124**
 Dental Disease. **125**
 Dysphagia **126**
 Foreign Body Obstruction. **126**
 Gastric Dilatation Volvulus (GDV) **127**
 Gastritis **128**
 Intussusception. **129**
 Ileus **130**
 Malabsorption. **130**
 Maldigestion **131**
 Megaesophagus **132**
 Oral Trauma **133**
 Salivary Mucocele **134**
Conditions Involving the Liver, Gallbladder, and Pancreas **134**
 Cholangiohepatitis. **134**
 Hepatitis. **135**
 Hepatic Encephalopathy. **136**
 Hepatic Lipidosis **137**
 Jaundice. **139**
 Pancreatitis **139**
 Pancreatic Insufficiency. **141**
 Portosystemic Shunts (PSSs). **142**
Conditions Involving the Cardiovascular and Respiratory Systems **144**
 Cardiac Arrhythmia. **144**
 Cardiomyopathy **145**
 Congestive Heart Failure (CHF) **146**
 Heartworm Disease **147**
 Hypertension **148**
 Pericarditis. **149**
 Pneumothorax. **150**
 Pneumonia. **151**
 Septal Defects **152**
 Shock **153**
 Valvular Insufficiency **155**
Conditions Involving the Blood and Spleen
and the Lymphatic and Immune Systems **156**

- Allergic Reactions... **156**
- Anemia... **157**
- Atopy... **158**
- Coagulopathies... **159**
- Disseminated Intravascular Coagulation (DIC)... **160**
- Hemorrhage... **161**
- Immunodeficiencies... **162**
- Immune-Mediated Hemolytic Anemia (IMHA)... **163**
- Pemphigus... **164**
- Rheumatoid Arthritis... **164**
- Septicemia... **165**
- Splenomegaly... **165**
- Systemic Lupus Erythematosus (SLE)... **166**
- Thrombosis... **167**

Conditions Involving the Reproductive and Endocrine Systems... **168**
- Abortion... **168**
- Addison's Disease... **169**
- Agalactia... **170**
- Cushing's Disease... **170**
- Diabetes Insipidus... **171**
- Diabetes Mellitus... **172**
- Dystocia... **175**
- Eclampsia... **176**
- Hyperthyroidism... **177**
- Hypothyroidism... **178**
- Mastitis... **179**
- Paraphimosis... **180**
- Prostatitis... **180**
- Pseudopregnancy... **181**
- Pyometra... **182**
- Vaginitis... **182**

Conditions Involving the Renal and Urinary Systems... **183**
- Antifreeze Intoxication (Ethylene Glycol [EG] Intoxication)... **183**
- Cystitis... **184**
- Renal Failure (Acute and Chronic)... **185**

Urinary Bladder Perforation.................................. 187
Uroliths 187
Conditions Involving the Musculoskeletal and Neurologic Systems 189
Arthritis... 189
Ataxia.. 191
Brachial Plexus Avulsion....................................... 192
Coma 193
Degenerative Joint Disease (DJD) 194
Hip Dysplasia... 196
Intervertebral Disc Disease (IVDD)............................. 197
Meningitis 199
Osteomyelitis 200
Panosteitis 201
Seizures 201
Conditions Involving the Integument and Special Senses 202
Conjunctivitis 202
Corneal Ulceration.. 203
Entropion and Ectropion... 204
Flea Allergy Dermatitis 204
Glaucoma.. 206
Keratoconjunctivitis Sicca (KCS) 207
Ocular Proptosis.. 208
Otitis.. 208
Pyoderma.. 210
Retinal Detachment 210
Ringworm 211
Miscellaneous Neoplastic Conditions 212
Hemangioma.. 212
Hemangiosarcoma... 212
Insulinoma.. 213
Lymphoma 213
Mammary Tumors.. 214
Mast Cell Tumors.. 215
Osteosarcoma.. 216
Squamous Cell Carcinoma 217
Testicular Tumors... 217

Select Infectious Diseases *218*
   Rabies. *218*
   Feline Infectious Diseases *219*
      Feline Infectious Peritonitis (FIP) *219*
      Feline Immunodeficency Virus Infection (FIV) *220*
      Feline Leukemia (FeLV) *222*
      Feline Panleukopenia *223*
      Feline Upper Respiratory Disease (FURD) *224*
      Hemobartonella *226*
      Toxoplasmosis *227*
   Canine Infectious Diseases *228*
      Adenovirus *228*
      Bordetella Bronchiseptica *229*
      Canine Distemper *230*
      Hepatitis *232*
      Leptospirosis. *233*
      Parainfluenza *234*
      Parvovirus. *235*
Bibliography. *237*

Chapter 5
Surgical Procedures and Associated Technician Evaluations

Introduction *240*
Anal Gland Removal *240*
Amputation *241*
Arthroscopy *242*
Aural Hematoma *242*
Bladder Surgery (Cystotomy) *243*
Cesarean Section *244*
Declaw (Onychectomy) *245*
Dental Prophylaxis *246*
Diaphragmatic Hernia. *247*
Dislocations *248*
Ear Cropping *249*
Endoscopy *250*

Contents

Fractures *250*
    Pelvic Fractures *250*
    Limb Fractures *251*
    Facial Fractures *252*
Gastric Dilatation Volvulus (GDV) *253*
Hernias *254*
Intestinal Resection and Anastomosis *255*
Lacerations *256*
Ocular Injury/Surgery *256*
Prolapses *257*
Renal Surgery *258*
Spaying/Neutering *258*
Spinal Cord/Vertebrae Surgery *258*
Tail Docking *260*
Thoracic Surgery *261*
Tumor Removal *262*
Wounds *262*
Bibliography *264*

Chapter 6
Therapeutic Procedures and Associated Technician Evaluations

Introduction *266*
Behavioral Counseling *266*
Blood Transfusion *267*
Burn Therapy *268*
Cardiopulmonary Cerebrovascular Resuscitation (CPCR) *270*
Casts *272*
Chemotherapy *273*
Enteral Feeding Tubes *274*
Euthanasia *276*
Parenteral Nutrition (PN) *276*
Peritoneal Dialysis *277*
Radioactive Iodine Therapy *279*
Vascular Access via Catheterization *280*
Bibliography *281*

Chapter 7
Sample Cases with Documentation

Introduction **284**
    Case 1: Anal Gland Abscess **284**
    Case 2: Canine Neuter **285**
    Case 3: Cesarean Section **291**
    Case 4: Corneal Ulceration **298**
    Case 5: Feline Upper Respiratory Disease (FURD) **299**
    Case 6: Laceration **301**
    Case 7: New Puppy Exam **302**
    Case 8: Ringworm **304**
    Case 9: Tumor Removal **305**
    Case 10: Urethral Obstruction **310**
Bibliography **320**

Appendices
Canine and Feline Vaccinations **321**
Common Abbreviations/Acronyms Used in Documentation **321**
Weight and Liquid Conversions Commonly Used in Practice **322**
Temperature Conversions **323**
Normal TPR Values of the Dog and Cat **323**
Normal Hematology Values of the Dog and Cat **323**
Normal Blood Chemistry Values of the Dog and Cat **324**
Normal Urinalysis Values of the Dog and Cat **326**
Comparison of Common Crystalloid Fluids **327**
Conversion Formula for mm Hg to cm H_2O **328**
Normal Blood Pressure Values **328**
Normal Blood Gas Values of the Dog and Cat **328**
Directory of Pet Loss Hotlines **328**
Directory of Pet Loss Web Sites **329**
Bibliography **329**

Glossary **331**
Index **335**

Foreword

This book was written to address a perceived void in textbooks for the education of veterinary technicians. While many excellent texts provide substantive information to the student technician, none address the methodology used to apply this information to real-world problems. Nursing students have long benefited from instruction in the nursing process, which is a method that uses critical-thinking skills to apply theoretical knowledge to actual cases. This text is an attempt to develop a similar methodology for veterinary technicians, deemed the Veterinary Technician Practice Model by the authors. Development and acceptance of the nursing process took many years; although the profession of certified veterinary technician is relatively new, it is hoped that the proven success of the nursing process model will facilitate acceptance of a similar model in veterinary technology.

The first chapter of this book identifies and describes the various steps in the Veterinary Technician Practice Model and provides instruction and examples on documenting the results of that process. Particular emphasis is given to developing SOAP notes and Medication Administration/Order Records (MAORs).

Chapter 2 describes sources of subjective and objective data (the *S* and the *O* in a SOAP note), with a review of physical examination and commonly ordered tests. Chapter 3 is a list of technician evaluations (the *A* in a SOAP note). These evaluations are analogous to the Nursing Diagnosis portion of the nursing process. Included after each technician evaluation is a list of interventions commonly associated with the evaluation (the *P* in a SOAP note).

Chapters 4, 5, and 6 provide quick references to selected medical conditions, surgical procedures, and therapies, respectively, for canines and felines. To aid the technician in identifying proper technician evaluations, each identified condition, procedure, and therapy is followed by a list of appropriate technician evaluations.

Chapter 7 is a "demonstration" chapter that provides case scenarios with examples of documentation by the technician, including SOAP notes, notations, and MAORs.

The authors of *Patient Assessment, Intervention, and Documentation for the Veterinary Technician* are grateful for the opportunity to contribute to the profession of veterinary technology. It has been our sincere pleasure to develop this text. We hope the constructed model will serve to advance the profession and provide a framework for application of critical-thinking skills. As the development of any model is an ongoing process, the authors invite comments that will serve to further refine this first attempt.

Acknowledgments

The authors and Delmar Cengage Learning would like to thank the following individuals for reviewing this manuscript:

Bonnie Ballard
Gwinnett Technical College
Lawrenceville, GA

Sue Bosted, DVM
Animal Medical Clinic
Heyburn, ID

Eric R. Burrough, DVM
Kirkwood Community College
Cedar Rapids, IA

Anne Duffy
Kirkwood Community College
Cedar Rapids, IA

Sarah Hurley
Parkland Community College
Champaign, Illinois

Mary O'Horo Loomis, DVM
SUNY Canton
Canton, NY

Karl M. Peter, DVM
Foothill College
Los Altos Hills, CA

Janet Romich
Madison Area Technical College
Madison, WI

Lois Sargent, DVM
Miami Dade College
Miami, FL

About the **Authors**

After receiving her bachelor's degree in microbiology from the University of Wyoming, Jody Rockett, DVM, attended the University of Missouri, College of Veterinary Medicine. Following several years as an associate veterinarian in mixed animal practice, Dr. Rockett founded the Veterinary Technology program at the College of Southern Idaho (CSI). Dr. Rockett directs the program at CSI and is coauthor of *Veterinary Clinical Procedures in Large Animal Practice*. She served as the 2007 president of the Idaho Veterinary Medical Association. She, her husband, and their two children enjoy the rural lifestyle of southern Idaho and spend their free time raising and training a variety of animals.

Cynthia Lattanzio's AS, BA, and JD career choices reflect the diversity of her educational background. She received her BA in mathematics from the University of Connecticut, her JD from the University of Connecticut School of Law, and her AS in Nursing from Capital Community College. Her professional life has encompassed a spectrum of activities ranging from prosecuting attorney for the state juvenile courts to labor negotiator, medical-surgical nurse, and homemaker. Ms. Lattanzio currently resides in Connecticut with her husband, children, and numerous family pets.

Katie Anderson, CVT, received her associate degree in veterinary technology from the College of Southern Idaho and her BS in agricultural business from the University of Idaho.

Chapter 1
Veterinary Technician Practice Model and Documentation

ROLE OF VETERINARY TEAM MEMBERS

Most veterinary practices, clinics, and hospitals have a variety of professional and nonprofessional staff members that carry out all necessary functions. Veterinarians and veterinary technicians make up the professional staff, while the nonprofessional staff may include front office/clerical staff, kennel/barn workers, and veterinary assistants. In a small practice, the front office staff person serves as a receptionist and billing specialist. This individual may schedule appointments, greet clients, answer phones, send out bills, record payments, generate necessary paperwork (e.g., rabies vaccination certificates), and handle all general clerical duties. In larger practices and hospitals, these functions may be divided between several individuals. Kennel/barn workers provide housekeeping functions for the practice, such as cleaning cages/stalls and examination rooms and restocking supplies. The duties of veterinary assistants vary widely from practice to practice depending on training provided by the veterinarian and/or veterinary technician. Generally, veterinary assistants can perform duties such as restraining animals, feeding and watering patients, and gathering supplies for various procedures. However, since most assistants receive a limited amount of specialized college education, they should not perform procedures on animals or make independent judgments regarding patient care.

Together veterinary technicians and veterinarians make up the professional staff. Veterinarians are licensed by individual states to diagnose and treat diseases and health conditions of all types of animals. Veterinarians have the ultimate responsibility for any animal under their care. Veterinary technicians are college-prepared professionals credentialed by most states. They are trained to perform a wide variety of procedures (e.g., urinary catheter placement, venipuncture, and ECG), to execute orders of the veterinarian (e.g., administer fluids and medications), and to exercise independent judgment regarding patient care (e.g., administration of PRN medications and assessment of hydration status). For a complete list of technician skills, refer to the American Veterinary Medical Association's *Veterinary Technology Student Essential and Recommended Skills List*. You can obtain this information at the AVMA Web site <http://www.avma.org>.

VETERINARY TECHNICIAN PRACTICE MODEL

The veterinary technician has a role in veterinary medicine that is analogous to that of a nurse in human medicine. Both are trained professionals who assist the primary health caregiver (veterinarian or doctor) in providing optimum care to patients. The nursing profession has clearly identified the steps that a nurse takes to deliver care to a patient

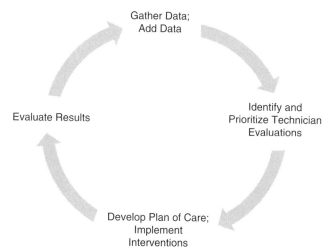

Figure 1-1: Steps of the Veterinary Technician Practice Model

and has identified these steps as the nursing process. As defined by Potter and Perry, "The nursing process is the traditional critical thinking competency that allows nurses to make clinical judgments and take actions based on reason" (Potter & Perry, 2001, p. 285). This text is an attempt to establish a similar orderly process for veterinary technicians. Throughout the text, this process will be referred to as the Veterinary Technician Practice Model. (See Figure 1-1.)

As Figure 1-1 shows, the Veterinary Technician Practice Model is a circular, or continuous, process that begins when the patient is first presented for care and ends only when the patient's medical issues are resolved or the animal is discharged from care. Each step in the process will be discussed separately.

Step 1: Gather Data

Information that is pertinent to the Veterinary Technician Practice Model comes from many sources, including but not limited to the technician's own examination, the veterinarian's examination, laboratory results, medical records, and the owner's observation.

Once data is collected, it may be divided into two separate classifications: objective and subjective. Objective data is data that can be measured or quantified. Some examples of objective data are temperature, blood pressure, respirations, wound size (which can be measured with a ruler), and laboratory results. Subjective data, on the other hand, is observable but not exactly measurable. Some examples of subjective data are decreased appetite, pain level, color of urine, and degree of edema. Data that is subjective in some situations may be objective in different circumstances. For example, urinary output is often subjective, based on observations such as urinary frequency and amount. However, when a urinary catheter is in place, the exact amount of urine can be measured; therefore, in this situation, urinary output is objective. Because a veterinary technician needs to be able to correctly classify data as either objective or subjective, it is important for a student technician to begin thinking in these terms.

All data that is collected during an exam should be recorded, even when the results are within normal limits (WNL). First, this indicates to subsequent caregivers that a thorough exam was done. More importantly, however, this information can serve as a baseline should the animal later develop any problems. Recording normal findings need not be time-consuming. The technician can use the acronym WNL for all systems that appear normal. For example, a cat is brought in with a laceration on its right front leg. An appropriate initial veterinary technician record of subjective data may read as follows: RF leg wound located above carpus. No active bleeding or swelling noted. Owner noticed wound when cat was let in house this morning. All other body systems WNL.

Many clinics and hospitals now have computerized databases that allow the technician to enter data on each body system. Usually, these databases provide a check-off box for WNL. Individualized paper databases also may be developed and preprinted by clinics.

Step 2: Identify and Prioritize Technician Evaluations
Identify Technician Evaluations

The technician evaluation is the veterinary technician's conclusion about the animal's or owner's *response* to physical and psychological challenges; it is based on the data that the technician collected in Step 1. The technician uses critical-thinking skills to cluster significant data in order to identify appropriate technician evaluations, which serve as an aid to the technician in determining proper interventions. This is the same process that the veterinarian uses to identify a diagnosis. Thus, the technician's independent judgment regarding technician evaluations is analogous to the veterinarian's judgment regarding diagnosis. However, the technician evaluation is separate and distinct from a veterinary diagnosis. Only a veterinarian is educated and licensed to make a medical diagnosis, which is the determination of the cause, course, and prognosis of a particular set of signs. For example, a dog is brought in for treatment exhibiting labored respirations, cyanosis in mucous membranes, and altered mentation. Only a veterinarian can diagnose the cause of these signs, which may be the result of a pulmonary embolism, cardiac failure, or a host of other conditions. However, regardless of the *cause* of the signs, the technician can make a judgment about the animal's *response* to the physiological change. In this case, the technician might make an evaluation of Decreased Perfusion, Cerebral and

Cardiopulmonary, and immediately institute appropriate interventions (e.g., provide supplemental oxygen).

In additional to making evaluations about actual or existing conditions, technicians also may make assessments about future or possible conditions (e.g., sequelae). For example, a 6-month-old puppy is brought in to a clinic when the owner observes the puppy gagging repeatedly. Upon examination, the veterinarian notices a sharp object lodged in the back of the puppy's throat. The object is easily removed with a Kelly clamp; and the puppy experiences immediate relief, with no apparent difficulties in breathing or swallowing. However, an experienced technician recognizes that the trauma to soft tissue in the pharynx may lead to swelling in the affected area, putting the puppy at risk for later difficulties in breathing or swallowing. Thus, the technician might make the evaluation Risk of Obstructed Airway. Virtually every technician evaluation can become a "risk for" evaluation if the technician anticipates the possible development of a problem.

It should also be noted that a few of the technician evaluations are already listed as "risk for" assessments. In these cases, it is not appropriate for the technician to make the assessment without the "risk for" modification, as this would result in diagnosing a condition, which is the sole province of the veterinarian. An example will serve to clarify: A dog is recovering from anesthesia following emergency surgery to repair a lacerated spleen. Because of the emergency nature of the surgery, the stomach was not empty at the time of surgery. The technician knows that because of the presence of stomach contents and because the gag reflex is suppressed by anesthesia, this animal is at risk for aspiration. Thus, the technician makes the evaluation Risk for Aspiration and initiates proper interventions. The following day the technician notices that the dog's breathing has become somewhat labored, there are rales in the lungs, and the dog's temperature is elevated. While the technician may suspect that the dog is suffering from aspiration pneumonia, this is a veterinary diagnosis that can be made only by a licensed veterinarian. A technician evaluation of aspiration would be a veterinary diagnosis that a technician is not qualified to make, as only a veterinarian can confirm that the dog did in fact aspirate.

Prioritize Technician Evaluations

In both human and animal medicine, patients often present with multiple problems. Prioritization is a methodical determination of the order in which each problem should be addressed. In human medicine, the nursing profession has used Maslow's Hierarchy of Needs as an aid in prioritization (Potter & Perry, 2001, p. 6). This model asserts that certain critical physiological needs (e.g., oxygenation) must be met before less critical physiological needs (e.g., nutrition) and higher human needs (e.g., love and self-esteem) are addressed. While this model is not precisely

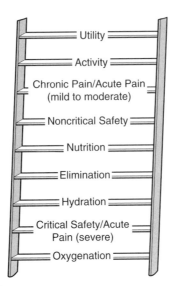

Figure 1-2: Needs Ladder

applicable to animals, the basic presumption that certain physiological needs must be met before other needs are addressed holds true for all living organisms. Keeping this fact in mind, the authors have developed a Needs Ladder that will aid the veterinary technician in prioritizing care; all the user needs to remember is to "go up the ladder." (See Figure 1-2.)

To ensure ease of use and to aid technicians in identifying and prioritizing appropriate technician evaluations, all technician evaluations listed in Chapter 3 have been grouped according to the Needs Ladder (see Figure 3-1) as well as by body system (see Figure 3-2) and in alphabetical order (see Figure 3-3).

Explanation of Needs Ladder

Oxygenation—Lack of oxygen will kill an animal more quickly than any other cause; thus, it is the first priority of care. An animal needs an unobstructed airway as well as adequate ventilation, respiration, and circulation for this need to be met. The technician can use the same ABC mnemonic that is employed in human medicine and taught in cardiopulmonary resuscitation. *A* represents airway, *B* represents breathing, and *C* represents circulation.

Critical Safety/Acute Pain (severe)—Critical safety comprises conditions that are or may be immediately life threatening to an animal. Acute pain (severe) is the sudden onset of pain so severe that it negatively impacts other physiological systems. These two needs are grouped together as the second priority of care; depending on the circumstances, each may take priority over the other when care is provided. For example, a stray dog has been hit by a car on a frigid winter day. It is not known when the accident occurred or how long the animal lay by the side of the road. The dog has fractured its hind leg and is barely responsive. The technician finds that the dog's temperature is 96.8°F, and concludes that the animal is dangerously hypothermic. In this case, priority must be given to the safety of the animal, and the hypothermia must be addressed before the pain from the fracture. Now suppose, in the same scenario, the dog is brought in with a temperature of 98.8°F. The dog would still be considered hypothermic; but in this case, relief of acute pain from the fracture would probably take precedence over the safety need.

Hydration—All living organisms need adequate water to maintain life. Although dehydration negatively impacts virtually every body system, the renal, cardiovascular, and neurosensory systems are especially vulnerable to damage and/or alteration. Severe dehydration can kill an animal in a short time.

Elimination—All animals need to excrete waste products through urine and feces. Failure to produce adequate urine leads to a buildup of toxins in the body that can ultimately prove fatal. Failure to pass urine (e.g., due to urinary obstruction) can lead to a ruptured bladder. Likewise, failure to pass feces can lead to an intestinal blockage, colic, and death.

Nutrition—Adequate nutrition is necessary for all body processes; it is especially important for an injured or sick animal to receive adequate nutrition to ensure timely healing. However, animals can tolerate suboptimal nutrition for substantial periods of time.

Noncritical Safety—This category includes factors and circumstances that affect the animal's well-being but are not immediately life threatening. Examples of noncritical safety include minor wounds that increase risk of infection, self-mutilation, and owner noncompliance with recommended medical regimen.

Chronic Pain/Acute Pain (mild to moderate)—Chronic pain impacts the animal's well-being and the animal's usefulness to its owner. In companion animals, chronic pain can lead to changes in behavior and withdrawal, which the owner may find upsetting. In work animals, chronic pain may reduce the animal's ability to perform. Food animals often experience slow weight gain when in distress from chronic pain. Acute pain (mild to moderate) is pain of sudden onset that does not immediately threaten the animal's physiological well-being.

Activity—All animals require adequate activity to maintain muscle mass, to prevent weight gain, to prevent joint stiffness and pressure ulcers, and to maintain psychological balance.

Utility—This category includes factors and circumstances that impact the animal's usefulness to its owner as a companion, a work animal, or a food producer. Examples include aggressiveness in a companion dog and reproductive dysfunction in a breeding animal.

Step 3: Develop Plan of Care and Implement Interventions

The plan of care for an animal is a listing of all interventions the technician intends to initiate to restore the animal to a state of well-being. Choosing appropriate interventions flows naturally from the identification of technician evaluations. For example, a dog is brought into the clinic exhibiting sunken orbits, dry mucous membranes, and skin tenting. The owner states that the animal had been inadvertently locked in a garage for 2 days. The technician selects Hypovolemia as the first-priority technician evaluation; this selection naturally leads the technician to plan certain interventions, such as administering fluids as ordered by the veterinarian, monitoring vital signs and lab results, recording fluid intake and output, and weighing the animal daily. As a second technician evaluation, the technician selects Client Knowledge Deficit. The technician then institutes a plan to educate the owner about the dangers of dehydration and the need to monitor the animal's location and safety.

In emergency situations, planning and implementing interventions may be almost simultaneous and the technician may be required to take action without any written plan. Nevertheless, identifying technician evaluations aids the technician in providing appropriate care, just as diagnosing a condition aids the veterinarian in providing optimum care.

Step 4: Evaluate Results

Technicians must continuously evaluate an animal's response to interventions. The animal's condition may improve, remain unchanged, or even worsen, in which case new interventions may be required. Continual reevaluation during the entire time the animal is in the technician's care ensures a timely response to the animal's changing condition.

Step 5: Add Data

Step 5 is essentially a repeat of Step 1. The technician examines the animal to obtain additional data. With this new data, it is possible to determine whether the existing plan of care is meeting the patient's needs and may be continued or whether the care needs to be modified in any way.

DOCUMENTATION

The veterinary medical record is a compilation of all written information, reports, and communications regarding a particular animal's veterinary care. The record may include intake information (e.g., the owner's name and the party responsible for payment), a database, consent forms, estimates, vaccination sheets, flowsheets (e.g., the anesthesia log used during surgery and the medical administration record), radiographs, lab results, referrals, discharge summaries, and progress notes (communications about the animal's response to treatment). Subsequent caregivers should be able to consult the medical record and develop a full understanding of the animal's health status.

The veterinary medical record is a legal document, so certain conventions should be followed when writing in or adding to the record. First, the animal's and owner's names as well as the animal's species should be clearly written on every piece of paper in the record, including the back of each sheet if any writing appears there. Second, all handwritten entries should be made in black ink for the simple reason that black ink produces more legible copies if the record needs to be duplicated. Third, the date and time of the entry should be clearly noted, preferably at the beginning of the entry; the name and position of the person making the entry should be noted at the end. (When using initials only, a master signature sheet should be included on which each individual who is making an entry writes his or her signature next to his or her initials. This form can be used to identify the author of every entry.) Finally, no part of the record should be written over, "erased" with correction fluid, or otherwise obliterated. If an incorrect entry is made, a single line should be drawn through the incorrect information and the word "error" as well as the name of the person making the change should be written next to the entry.

The technician should document every aspect of care that is provided to an animal. From a legal standpoint, it is difficult or impossible to prove that a certain action was taken when that action is not documented in the record. Nurses have an adage that veterinary technicians would do well to follow: "Care not documented is care not given."

SOAP Notes

One of the primary means of communication between members of the veterinary medical team is the progress note, which, as stated previously, is a communication regarding the animal's condition and response to treatment. The SOAP (SOAP is an acronym for subjective, objective, assessment, plan) is a standardized format for writing a progress note. (See Figure 1-3.) Documentation in the SOAP note should

> S — Subjective Data: Pertinent signs or symptoms that are observable but not exactly measurable.
>
> O — Objective Data: Pertinent signs or symptoms that are measurable or quantifiable.
>
> A — Assessment: Technician evaluations are listed in order of priority. Assessment of progress may also be recorded in a second or subsequent SOAP note.
>
> P — Plan of Care for the Animal. This may be the original plan or a revised plan.

Figure 1-3: SOAP Description

not be confused with a database or flowsheet. A database records all findings from the original examination, including findings that are WNL. Flowsheets document one item of information over time (e.g., a flowsheet documenting anesthesia administration during surgery). The information contained in the database and/or flowsheets may or may not be included in the SOAP note depending on the relevance of the data. The focus of the SOAP note is the specific problem or problems for which the animal is being treated.

A SOAP note is written when an animal is first seen and then at least daily throughout the animal's hospitalization. The frequency of SOAP notes depends on the severity of the animal's condition. Generally, notes are more frequent when the animal's condition is critical. The first SOAP may seem redundant, as the subjective and objective data were already recorded on the database. However, the use of daily (or more frequent) SOAP notes gives all health care personnel an easy way to follow an animal's progress. This text addresses SOAPs written by the technician only; it is expected that the attending DVM will document his or her findings and interventions as well.

Steps of SOAP Composition

Inexperienced students or technicians may require assistance in the "mechanics" of SOAP composition. The following quick "steps of composition" can facilitate this process.

- **Step 1:** Categorize data as subjective or objective. This provides the *S* and *O* components of the SOAP.
    - Data is collected through physical examination or review of laboratory results.
    - Refer to Chapters 1 and 2 for categorization and collection of data.

- **Step 2:** Based on collected data, identify and rank appropriate technician evaluations. This provides the *A* component of the SOAP.
  - Refer to Chapter 3 for a complete list of evaluations.
  - Refer to Chapters 4, 5, and 6 for suggestions on appropriate evaluations. These chapters provide evaluations commonly associated with the majority of medical, surgical, diagnostic, and treatment procedures.
- **Step 3:** Based on identified technician evaluations, list appropriate interventions for each evaluation and initiate a MAOR. This provides the *P* component of the SOAP.
  - Refer to Chapter 3 for interventions commonly associated with each technician evaluation. It is extremely important to refer to Chapter 3 at this point as Chapters 4, 5, and 6 provide only interventions unique to the identified condition and technician evaluation.
  - Refer to Chapters 4, 5, and 6 for interventions that may be unique to the medical condition or procedure.
  - Refer to Chapter 1 for rules regarding MAOR construction.

CASE SCENARIOS

Case Scenario with Technician SOAP Note

A 5-year-old castrated male Labrador retriever is brought in with a history of head shaking, ear scratching, and whining for 2 weeks' duration. The owner describes these signs as gradually worsening during the 2 weeks. Abnormalities found on examination include a pain response of 1/5, pinnal hyperemia, vertical and horizontal ear canal inflammation, and nonvisible tympanic membrane due to the presence of a waxy brown exudate with a foul/sour odor. All noted abnormalities are bilateral. TPR 101°F, 120 bpm, panting, all other physical parameters WNL.

The veterinarian requests a swab and cytology to be completed by the technician. Cytology confirms a yeast infection. The veterinarian requests that the technician clean and flush both ears with dilute chlorhexidine. Prescriptions for Conofite, Rimadyl, and Epi-Otic are ordered by the veterinarian and filled by the technician. The veterinarian then explains the diagnosis and prognosis to the owner. The technician demonstrates ear-cleaning technique, at-home care, and application of medication to the owner. (See Figure 1-4.)

| Patient ID: | Progress Notes |

10/1/08 10 AM

S- Owner states dog shaking head, scratching ears, whining. PE: Bilateral: pinna reddened, ear canals inflamed, waxy brown exudate, foul odor, 1/5 pain. All other parameters WNL.

Comment: These are all subjective signs because none are measurable. Use of clear abbreviations streamlines SOAP note.

O- TPR: 101°F, 120, panting

Comment: Vital signs are objective data because they are measurable.

A- 1. Knowledge Deficit 2. Acute Pain 3. Impaired Tissue Integrity 4. Risk for Self-Inflicted Injury

P- 1. Ear swab, stain per order. Results: positive yeast.
2. Fill order Rimadyl, Conofite, Epi-Otic.
3. Clean/flush ears with dilute chlorhexidine. Apply Conofite. Give Rimadyl per order.
4. Demonstrate cleaning, med adm to owner. Explain purpose of meds and prevention techniques.
5. Schedule recheck in 7 days (10-8-08) per order.
C. Smith, CVT

Comment: Document that owner understands interventions/prescriptions and can perform necessary procedures. Notice that interventions were implemented that addressed all four, technician evaluations: demonstration and education regarding medication purpose/use addressed the evaluation of client knowledge deficit; prescription for Rimadyl addressed acute pain; use of chlorhexidine and Conofite will resolve impaired tissue integrity; owner knowledge of prevention techniques reduces risk of self-inflicted injury.

Figure 1-4: Sample SOAP with Comments

Notations and Incorporations

Occasionally, the technician may need to enter information in the record; but a full SOAP is not required. These entries can be handled as a notation to the record. A notation should contain only data (factual information) that has become available since the last data was recorded. For example, a cat is admitted to the clinic and as part of the diagnostic workup, the veterinarian orders the technician to perform a FeLV/FIV test. The results of that test are available 1 hour after admission. The technician would make the following entry:

> 10/1/08 1 PM. Results of FeLV/FIV test negative, Dr. Cook notified. C. Smith, CVT.

Many test results can be entered in the record as a notation. However, if the test results are produced on a separate document, there is no need to transcribe the results into the record. Documents that are full-sized can be referenced in a notation and then incorporated into the record. The technician should ensure that a date and time appear on the results; if they are not present, the technician should add them. All entries added to the record should be signed.

Test results that are printed on smaller sheets of paper can be taped inside the record. To ensure authenticity, the technician should place his or her signature such that it begins on the paper showing the test results and crosses over onto the paper to which the results are taped. This entry also should be dated, timed, and signed. (See Figure 1-5.)

Medical Administration/Order Record

A Medical Administration/Order Record, usually referred to as an MAOR, provides documentation of every medication and intravenous fluid that is given to an animal during its stay in the hospital/clinic, as well as ordered treatments, tests, and diets. (An MAOR is not used for an in/out visit. All information regarding medications, treatments, tests, etc., is written in the progress notes as a SOAP note or a notation. Usually, an MAOR is initiated for any animal admitted for an overnight stay.) An MAOR provides at-a-glance information to all members of the health care team. For example, the veterinarian can look at the MAOR to determine new dosing when an animal is not responding to current dose levels. The veterinary technician can use the MAOR to quickly determine when medications are due. All team members can use the MAOR to verify that ordered mediations and/or fluids have been administered in a timely manner. Although use of MAORs is not widespread in veterinary medicine, they are an established method of documentation in human medicine because of their usefulness; and the

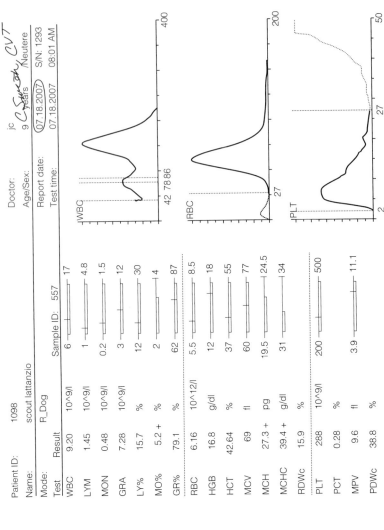
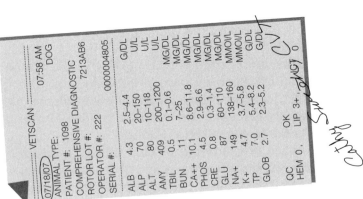

Figure 1-5: Example of Incorporation into Record

benefits of use transfer directly to veterinary medicine. Hospitals that have access to computer records may have a computerized MAOR; however, a paper MAOR is just as easy to use. Individual clinics often draft MAORs to suit their particular needs. (See Figure 1-6.)

Medication Administration/Order Record

Patient ID	Allergies	Initials	Signature	Title

Medication Administration

Date of Order	Medication	Time						

Fluids and IV Drips

Figure 1-6: MAOR

Patient ID

Allergies

Order Record

Treatments									
Date	Treatment	Time							
Tests									
Date	Test	Time							
Diet									
Date	Diet	Time							

Figure 1-6: Continued

The following rules apply to paper MAORs.

1. Every medication should be listed by its entire name (i.e., no abbreviations), along with its dose or dosage, route of administration (e.g., PO, IV, or topical), and frequency of administration. The technician should write the order on the MAOR *exactly* as it is written by the

veterinarian. If the order is not clear or is incomplete, the technician must verify the order with the veterinarian. The technician should never make assumptions about an order. (Note: Medications and fluids that are administered during surgery are normally entered on the anesthesia or surgery record and do not appear on the MAOR.)

2. The times of administration are written next to the order. The person who administers a dose initials the appropriate space to indicate that the dose was given.

   When the time of dosage differs significantly from the written time, the actual time of administration should be written next to the person's initials. (Usually, medications given 1 hour before or after the stated time are considered "on time." However, since this rule can vary from clinic to clinic, the technician should ascertain the parameters at his or her clinic.)

3. When a dose is ordered for a predetermined period of time, an "X" should be placed under the dates for which the medication is *not* ordered. The word "discontinued" should not be used.

4. When a dose is not given for any reason, the technician should write "not given" (or NG) in the appropriate space. Again, this is an easy way for subsequent caregivers to ascertain which medications have been given. (The technician must notify the veterinarian when a medication is withheld.)

5. When an error is made, one line should be drawn through the incorrect information and the word "error" written next to the initials of the provider who made the incorrect entry.

6. Orders for fluids should include the exact fluid ordered and the rate of administration. The time each new bag of fluid is hung and the time fluids are discontinued should be noted next to the provider's initials. This section of the MAOR also can be used to document any medications that are administered via IV drip.

7. When any medication is discontinued, the word "discontinued" should be noted. The record of discontinued medications should never be crossed out, "erased" with correction fluid, or otherwise obliterated.

8. When an animal is discharged, boxes corresponding to future times/dates should be left blank.

9. Orders for treatments, tests, and diets are entered on the MAOR, applying the same rules.

10. The patient's name is always written on the MAOR, as are any known allergies to medication that the animal has.

11. The full signature of each provider who enters information on the MAOR should be written at the bottom of the appropriate box on the MAOR next to his or her initials. This facilitates tracking the identity of the individuals who administer medications and perform treatments, tests, etc.

MAOR Examples

The following examples are random orders chosen to illustrate the rules for using an MAOR; they are not applicable to any particular case or animal.

Example 1

The veterinarian writes the following: Meloxicam 0.1 mg/kg p.o. s.i.d. for 3 days. Date of order is 10-1-08. (See Figure 1-7.)

Since the veterinarian wrote "meloxicam," this is what the technician should write in the MAOR. The technician should not write "Metacam" (the proprietary name for meloxicam).

The clinic has meloxicam available in a concentration of 1.5 mg/mL. The technician does the appropriate calculations and determines that the 8.8 lb cat for which the order was written should receive 0.27 mL of meloxicam at each administration. However, the technician does not write "0.27" mL on the MAOR because this is not how the veterinarian wrote the order. The technician places an "X" in the boxes for 10/4, 10/5, 10/6, and 10/7 because meloxicam was not ordered for those days. The technician writes "8 AM" as the time of administration for this once-a-day medication. The time chosen for administration may vary from clinic to clinic.

Medication Administration/Order Record

Patient ID	Allergies	Initials	Signature	Title
Any Animal	none	CS	C Smith	CVT

Medication Administration

Date of Order	Medication	Time	10/1	10/2	10/3	10/4	10/5	10/6	10/7
10/1	Meloxicam 0.1 mg/kg p.o. s.i.d. for 3 days	8A	CS	CS	CS	X	X	X	X

Figure 1-7

Example 2

The veterinarian writes the following: Baytril tablet 5 mg/kg p.o. b.i.d. for 7 days. The animal is discharged on Day 5. Date of order is 10-1-08. (See Figure 1-8.)

The technician does not substitute the word "enrofloxicin" (the nonproprietary name for Baytril) on the MAOR. The clinic has Baytril tablets in 22 mg, 68 mg, and 136 mg form. The technician does the appropriate calculations and determines that the 44 lb dog for which the order is given should receive 100 mg of Bytril at each administration and decides to administer three fourths of the 136 mg pill. The technician should not write "Give ¾ of 136 mg tablet" on the MAOR. The technician leaves Days 6 and 7 blank since the animal was discharged.

Medication Administration/Order Record

Patient ID	Allergies	Initials	Signature	Title
Any Animal	none	CS	Cathy Smith	CVT
		KP	Karan Pelletier	CVT

Medication Administration

Date of Order	Medication	Time	10/1	10/2	10/3	10/4	10/5	10/6	10/7
10/1	Baytril tablet 5 mg/kg p.o. b.i.d. for 7 days	8A	CS	KP	KP	CS	CS		
		6P	CS	KP	KP	CS	CS		

Figure 1-8

Example 3

The veterinarian writes the following: Baytril tablet: Administer ¾ of 136 mg tablet p.o. b.i.d. for 7 days. Date of order is 10-1-08. (See Figure 1-9.)

The technician transcribes the order onto the MAOR exactly as the veterinarian wrote it.

Medication Administration/Order Record

Patient ID	Allergies	Initials	Signature	Title
Any Animal	none	CS	Cathy Smith	CVT
		KP	Katu Pelletier	CVT

Medication Administration

Date of Order	Medication	Time	10/1	10/2	10/3	10/4	10/5	10/6	10/7
10/1	Baytril tablet: Administer 3/4 of 136 mg tablet p.o.b.i.d. for 7 days	8A	CS	KP	KP	CS	CS	CS	CS
		6P	CS	KP	KP	CS	CS	CS	CS

Figure 1-9

Example 4

The veterinarian writes the following: Amoxi-drops 10 mg/kg p.o. b.i.d. for 7 days. The second dose of this medication is not given. Date of order is 10-1-08. (See Figure 1-10.)

The technician does not substitute the word "amoxicillin" for the term "Amoxi-drops." Amoxi-drops are available in a concentration of 50 mg/mL. The technician determines that the 8.8 lb cat for which this order was written should receive 0.8 mL at each administration. The technician does not write "0.8 mL" on the MAOR.

The technician writes "not given" on the box for the second dose.

Veterinary Technician Practice Model and Documentation

Medication Administration/Order Record

Patient ID	Allergies	Initials	Signature	Title
Any Animal	none	CS	Cathy Smith	CVT
		KP	Karan Pelletier	CVT

Medication Administration

Date of Order	Medication	Time	10/1	10/2	10/3	10/4	10/5	10/6	10/7
10/1	Amoxi-drops 10 mg/kg p.o. b.i.d. for 7 days	8A	CS	KP	KP	CS	CS	CS	CS
		6P	not given	KP	KP	CS	CS	CS	CS

Figure 1-10

Example 5

The veterinarian writes the following: Buprenorphine 0.1 mL p.o. t.i.d. (written at 4 PM on the day of admission). Date of order is 10-1-08. (See Figure 1-11.)

The technician writes the order in the MAOR; and per clinic procedure, he or she sets times of t.i.d. administration as 6 AM, 2 PM, and 10 PM.

Medication Administration/Order Record

Patient ID	Allergies	Initials	Signature	Title
Any Animal	none	CS	Cathy Smith	CVT
		KP	Karan Pelletier	CVT
		DD	Donald Diamond	CVT
		NR	Nancy Reed	CVT

Medication Administration

Date of Order	Medication	Time	10/1	10/2	10/3	10/4	10/5	10/6	10/7
10/1	Buprenorphine 0.1 ml p.o. t.i.d.	6A		DD	DD	DD	DD	DD	DD
		2P	4P CS	KP	KP	CS	CS	NR	NR
		10P	DD	KP	KP	DD	DD	DD	KP

Figure 1-11

The first dose is not given until 4 PM. Since the first dose is administered late, the technician notes the actual time of administration next to his or her initials. This information is important in determining when the next dose should be given. Since buprenorphine is an opioid, proper spacing of doses is an issue. The next dose was being administered by a different person; however, that person knew to hold the dose past 10 PM because the entry in the MAOR indicated that the first dose was late. Without the MAOR, that information might not have been passed along to the second health care provider.

The technician places an "X" in the box for the 6 AM dose on Day 1 because the medication was not ordered at that time.

Example 6

On the day of discharge following a 4-day hospitalization, the veterinarian writes the following: Revolution 120 mg tube, apply topically prior to discharge. The dog is admitted on 10/1 and is discharged on 10/4. Date of order is 10-1-08. (See Figure 1-12.)

The technician writes the order and places an "X" in the boxes for 10/1, 10/2, and 10/3 because no order for this medication was given on those days. The technician writes "4 PM" as the time of administration and places his or her initials in the appropriate box. Since the animal was then discharged, the remaining boxes were left blank.

Medication Administration/Order Record

Patient ID	Allergies	Initials	Signature	Title
Any Animal	none	CS	Cathy Smith	CVT

Medication Administration

Date of Order	Medication	Time	10/1	10/2	10/3	10/4	10/5	10/6	10/7
10/1	Revolution 120 mg tube, apply topically prior to discharge	4P	X	X	X	CS			

Figure 1-12

Example 7

The veterinarian writes the following: Flush and clean ears using Epi-Otic. Date of order is 10-1-08. (See Figure 1-13.)

The technician writes the order on the MAOR under the treatment section. When the treatment is completed, the technician notes the time and includes his or her initials. Since this is a one-time order, the technician places an "X" in the remaining boxes.

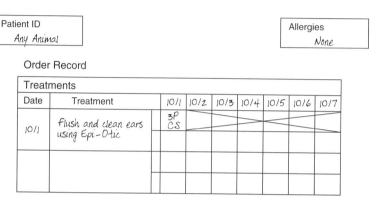

Figure 1-13

Example 8

The veterinarian writes the following: Flush jugular catheter q 4 hours with heparinized saline. (The catheter was placed at 1 PM on 10/1 and removed at 7 AM on 10/3.) Date of order is 10-1-08. (See Figure 1-14.)

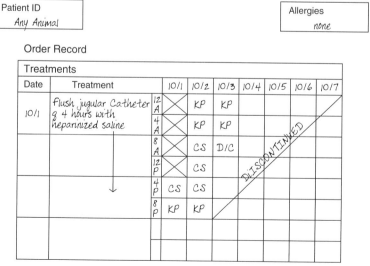

Figure 1-14

The technician writes the order on the MAOR. Since six entries are needed for the q 4-hour treatments, the technician draws a line down through the succeeding boxes. Times for treatment are written next to the order. Since the first treatment is not due until 4 PM on 10/1, the technician places an "X" in the boxes for 12 AM, 4 AM, 8 AM, and 12 PM. (This indicates that no order was given for these times.) After the catheter is removed, the technician writes "discontinued" in the succeeding boxes.

Example 9

The veterinarian writes the following: FeLV/FIV test Date of order is 10-1-08. (See Figure 1-15.)

The technician writes in the order and notes the time the test was performed. (The results of the test can be entered in the record as a notation or included in a SOAP.)

Patient ID	Allergies
Any Animal	none

Order Record

Treatments									
Date	Treatment								

Tests									
Date	Test		10/1	10/2	10/3	10/4	10/5	10/6	10/7
10/1	FeLV/FIV		4p	CS					

Diet									
Date	Diet								

Figure 1-15

Veterinary Technician Practice Model and Documentation

Example 10

The veterinarian writes the following: Administer 0.9% NaCl at 5cc/lb/h for 4 hours, then change to Normosol-R at 20 cc/h. Date of order is 10-1-08. The animal is discharged on Day 3. (See Figure 1-16.)

Medication Administration/Order Record

Patient ID	Allergies	Initials	Signature	Title
Any Animal	none	CS	Cathy Smith	CVT

Medication Administration

Date of Order	Medication	Time							

Fluids and IV Drips

Date	Medication	Time		10/1	10/2	10/3	10/4	10/5	10/6	10/7
10/1	Administer 0.9% NaCl at 5cc/lb/h for 4 hours, Then	Start 8A Stop 12P		CS CS						
10/1	Change to Normosol-R @ 20 cc/h.	Start 12P	#1 CS	#2 CS						

Figure 1-16

The technician writes the order for normal saline in the first box with a time marked "start" and a time marked "stop" and initials the appropriate boxes when the fluid is actually started and stopped. The order is written exactly as the veterinarian wrote it; the technician's calculation of a cc/h rate does not appear on the MAOR. Since this order does not continue for the remaining days on the MAOR, an X is placed in the box for each succeeding day. The order for Normosol-R is written exactly as the veterinarian wrote it. Next to his or her initials, the technician notes the time fluid administration was started. He or she writes "#1" to indicate that this is the first bag of fluid hung. Each succeeding bag also should be numbered.

Since this is an ongoing order, the remaining boxes are not crossed out.

Example 11

At 7 AM, on the first day of hospitalization, the veterinarian writes the following: n.p.o. Date of order is 10-1-08. At 6 PM that day, the veterinarian writes the following: Provide H_2O. At 8 AM on Day 2, the veterinarian writes the following: Provide H_2O and moist food. The animal is discharged at 5 PM on Day 4. (See Figure 1-17.)

The technician notes the time the animal is made n.p.o. When the order for water is written, the technician writes "discontinued" across the remaining boxes of the n.p.o. order. The technician notes the time water is provided to the animal. When the order for food and water is given, the technician writes "discontinued" across the remaining boxes of the order for water only. The technician notes the time food and water are provided. The remaining boxes are left blank after the animal is discharged.

Patient ID		Allergies
Any Animal		none

Order Record

Treatments									
Date	Treatment								

Tests									
Date	Test								

Diet									
Date	Diet		10/1	10/2	10/3	10/4	10/5	10/6	10/7
10/1	n.p.o		7A CS	DISCONTINUED					
10/1	Provide H2O		6P CM	DISCONTINUED					
10/2	Provide H2O and moist food		✕	8A CM	8A CM	8A CM			

Figure 1-17

🐾 CASE SCENARIO

Case Scenario with SOAP Notes, Notations, and MAOR

Day 1
10/1/08

8 AM: A 4-year-old intact male DSH outdoor cat presented for a large swelling "at the top of the tail" that had appeared 2 days prior. The owner stated that the cat would not allow her to touch the area. Although normally very friendly, the cat tried to bite when picked up.

The cat had not received any vaccinations, and the owner regularly dewormed the animal with an over-the-counter product.

Physical exam results included TPR 103.8°F, 125 bpm, 28 bpm, a 3/5 pain response, soft flocculent 2 × 3 in swelling with small scab in center located 3 in cranial to the tail head. All other physical findings were WNL.

The veterinarian aspirated a purulent bloody material from the swelling to confirm an abscess. The veterinarian asked the technician to perform a FeLV/FIV test. The cat tested negative for FeLV and FIV. The veterinarian recommended general anesthesia to drain and flush the abscess. The technician provided the client with a cost estimate and consent forms. The veterinarian ordered Amoxi-drops 0.8 mL p.o. b.i.d., Metacam 0.4 mg p.o. s.i.d. for 3 days, post-op wound flushing b.i.d. starting post-op Day 1, and a pre-surgical n.p.o. status. The medication order was filled by the technician, who then administered the first doses. Presurgical blood work was declined by the owner. The cat was placed in a hospital ward with an n.p.o. cage card.

1 PM: The cat was premedicated with atropine and acepromazine, then anesthetized using ketamine. The wound was opened; drained of a moderate amount of purulent, bloody fluid; and flushed with dilute chlorhexidine. The incision was not sutured and was allowed to close via second intention healing. The cat recovered uneventfully from anesthesia.

5 PM: The cat was alert and responsive; a pain response of 2/5 was noted. A second dose of Amoxi-drops was administered. A small amount of serosanguinous fluid was noted draining from the wound. The veterinarian discontinued the n.p.o. status and ordered H_2O in PM followed by normal feeding in AM. A litter box was placed in the cage.

Day 2
10/2/08

7 AM: A physical exam was performed. Results included TPR 102.5°F, 120 bpm, 25 bpm. Minimal swelling with a small amount of serosanguinous discharge was noted. The cat still resented touching of area and was assigned a 2/5 pain response. Other physical parameters were WNL. A moderate amount of urine was noted in the litter box. The technician administered Amoxi-drops and Metacam and then flushed the wound with dilute chlorhexidine per order. Food and water were offered.

5 PM: A brief examination revealed no changes in physical status. The cat had eaten during the day, and urine was noted in the litter box. The cat was discharged to the owner. Dispensed medications included Amoxicillin, Metacam, and dilute chlorhexidine flushing solution. The owner was educated regarding purpose and administration of medications, procedure for wound flushing, signs of returning/increasing infection, importance of additional vaccinations, and benefits of castration. (See Figures 1-18 and 1-19.)

Veterinary Technician Practice Model and Documentation

Sample Documentation

Patient ID: Lightning Friedman

Progress Notes

10/1/08, 8 AM

S- Soft swelling, cranial to tail head; small scab in center. Pus, bloody aspirate. Painful along dorsum. Pain response 3/5. Owner states change in behavior (biting) when picked up. Not UTD on vaccines.

> Comment: These are subjective findings because they are not measurable. Use of abbreviations (e.g., UTD) shortens SOAP.

O- TPR: 103.8, 125, 28. Swelling 2x3 inches, 3 inches cranial to tail head.

> Comment: The dimensions of the swollen area are included here because they can be measured. Vital signs are always objective data.

A- 1. Preoperative Compliance 2. Acute Pain 3. Impaired Tissue Integrity 4. Client Knowledge Deficit 5. Risk of Infection Transmission.

> Comment: Risk of infection transmission is included because the animal is not vaccinated and therefore may spread disease to other animals. Some states require a rabies vaccination when an animal presents with a wound of unknown origin.

P- 1. Initiate MAOR to include Amoxi-drops, Metacam, NPO, post-op flushing, FeLV/FIV test.
2. Provide client with estimate; obtain consent. Presurgical blood work declined.
3. Monitor pain status.
4. Educate owner prior to discharge.

C.Smith, CVT

Figure 1-18a

Patient ID:
Lightning Friedman

Progress Notes

10/1/08, 9 AM

Notation: FeLV/FIV negative. DVM informed.

C. Smith, CVT

10/1/08, 5 PM

S- BAR. Small amount serosanguinous drainage. Gag reflex present. Pain response 2/5.

O- No data.

A- Preoperative compliance resolved. Continue TE 2,3,4,5. Cat recovered well from surgery, no post-op complications noted.

P- 1. Modify MAOR. change diet.
2. Continue plan.

C. Smith, CVT

Figure 1-18b

Comment: Note that it is not necessary to rewrite continuing technician evaluations for each SOAP. Technicians should use techniques to streamline use of SOAP notes as ease of use increases likelihood that proper and timely notes will be written. However, technician evaluations should be rewritten in the first SOAP of each day; this prevents having to refer back through multiple entries.

Comment: The plan does not have to be rewritten in every SOAP when the existing plan is being continued. However, the plan should be rewritten in the first SOAP note of each day; this prevents having to refer back through multiple entries.

Patient ID: Lightning Friedman

Progress Notes

10/2/08, 7 AM

S- PE WNL except wound. Small amount serosanguinous drainage; minimal swelling. Pain response 2/5. Patient urinated moderate amount during night.

O- T 102.5; P 120 bpm; R 25 bpm

A- 1. Acute Pain 2. Impaired Tissue Integrity 3. Risk of Infection Transmission 4. Client Knowledge Deficit

P- 1. Complete MAOR.
2. Monitor pain status.
3. Educate owner.

C. Smith, CVT

10/2/08 5 PM

S- PE WNL. Minimal discharge.

O- PR WNL.

A- 1. Risk for Owner Noncompliance. 2. Risk of Infection 3. Risk for Infection Transmission.

P- 1. Educate owner:
a. purpose and administration of medications
b. wound flushing
c. signs/symptoms of returning infection
d. importance of vaccinations; benefits of castration
2. Dispense Amoxi-drops, Metacam, and dilute chlorhexadine.
3. Discharge cat.
4. Call owner in 2 days for update on cat's condition.

C. Smith, CVT

Figure 1-18c

Medication Administration/Order Record

Patient ID	Allergies	Initials	Signature	Title
Lightning Friedman	none	CS	Cathy Smith	CVT

Medication Administration

Date of Order	Medication	Time	10/1	10/2	10/3	10/4	10/5	10/6	10/7
10/1	Amoxi-drops 0.8 ml p.o.b.i.d	7A	CS	CS					
		5P	CS						
10/1	Metacam 0.4 mg p.o. s.i.d. for 3 days	7A	CS	CS		X	X		

Fluids and IV Drips

Figure 1-19a

Patient ID	Allergies
Lightning Friedman	none

Order Record

Treatments

Date	Treatment		10/1	10/2	10/3	10/4	10/5	10/6	10/7
10/1	flush wound BID starting post-op day 1	7A		CS					
		5P							

Tests

Date	Test	10/1	10/2	10/3	10/4	10/5	10/6	10/7
10/1	FeLV/FIV	9A CS						

Diet

Date	Diet	10/1	10/2	10/3	10/4	10/5	10/6	10/7
10/1	NPO	8A CM						
10/1	Provide H2O	5P CS						
10/1	Provide H2O + food in AM 10/2		7A CS					

Figure 1-19b

BIBLIOGRAPHY

Doenges, M. E., Moorhouse, M. F., & Geissler-Murr, A. C. (2002). *Nurse's pocket guide* (8th ed.). Philadelphia: F. A. Davis.

McCurnin, D. M., & Bassert, J. M. (1998). *Clinical textbook for veterinary technicians* (4th ed.). Philadelphia: W. B. Saunders.

Phipps, W. J., Monahan, F., Sands, J. K., Marek, J. F., & Neighbors, M. (2003). *Medical-surgical nursing: Health and illness perspectives* (7th ed.). St. Louis, MO: Mosby.

Potter, P. A., & Perry, A. G. (2001). *Fundamentals of nursing* (5th ed.). St. Louis, MO: Mosby.

Chapter 2
Generating the Database

INTRODUCTION

Data provides the information fundamental to making the evidence-based, analytical decisions required of medicine. Obtained through history, physical examination, observation, and testing, a database is used to generate appropriate technician evaluations and interventions. Databases effectively serve as the cornerstone from which assessments and plans of care are built.

The intent of this chapter is to refresh the veterinary technician's understanding of the components of physical examination and to outline the various diagnostic tests/procedures used in data collection. Technicians wanting a broader, more detailed description of these components are referred to standard veterinary technician and veterinary medical textbooks.

PHYSICAL EXAMINATION

Physical examination is the most valuable diagnostic tool available in veterinary medicine/technology. Information obtained through a complete history and examination guides the selection of additional diagnostic tests and assessments. The results of the examination and the tests serve as the subjective and objective data upon which the veterinarian bases a diagnosis and a veterinary technician bases a technician evaluation.

A variety of approaches have been used to ensure adequate examination of the animal. Regardless of the approach or technique selected, the most important criterion of examination remains consistency. All technicians must strive to develop a technique that is concise, consistent, and complete.

> **Reader Note**
>
> Both subjective (painful, alert) and objective information (respiratory rate 20 bpm) are obtained during physical examination. Whenever possible, provide numerical, quantifiable (objective) information.

> **Reader Note**
>
> Use of a standardized hospital physical examination form is highly recommended.

Performing the Physical Examination

The following sequence can be used to ensure adequate examination.

1. **Identify Patient**
   - *Refer to the patient and owner by name during the examination.*
   - *Record any tattoo or microchip identification numbers.*
   - *Confirm signalment: age, breed, sex, and reproductive status.*
2. **Obtain History**
   Confirmation of Presenting Complaint
   - *Confirm presenting complaint (why the client is at the office) and time elapsed since problem was initially noted. Ask owner to describe the progression of the problem, including clinical signs, time of onset, duration, severity, and attempted home treatments.*

   History of Presenting Complaint
   - *Ask a specific set of history questions relevant to presenting complaint. Standardized clinic questionnaires are recommended to ensure thoroughness and consistency in obtaining information.*

   Sample: Vomiting Dog History Questionnaire
   - When did the vomiting start?
   - How many times per day is the dog vomiting now? Has the vomiting increased or decreased in frequency since you first noticed the problem?
   - When does the vomiting occur? For example, is it continual or does it happen only after large meals?
   - Can you describe what the vomit looks like (blood, bile, digested versus undigested)?
   - What do you normally feed your dog?
   - How often do you feed your dog?
   - Have you fed your dog anything different from the regular diet?
   - Has the dog had access to table scraps, trash, or other ingestible items?
   - Is the dog known to chew on plants, rugs, furniture, or other items around the house?
   - To your knowledge, has the dog been exposed recently to chemicals such as fertilizers, pesticides, prescription medications, or household cleaners?
   - Is the dog able to drink and keep water down?
   - Does the dog have diarrhea or other changes in stool?
   - Are any other animals in the home affected?

   Body Systems History
   - *Ask a series of general history questions to obtain information regarding various body systems, including cardiovascular, pulmonary, hepatic, renal, gastrointestinal, integumentary, musculoskeletal, nervous and special senses, and endocrine.*

Table 2-1: General History Questions

QUESTION: "HAVE YOU SEEN/NOTICED ANY:"	SYSTEMS EVALUATED
Vomiting?	Renal, GI
Coughing?	Cardiovascular, Pulmonary
Changes in activity level?	Pulmonary, Cardiovascular, Musculoskeletal
Changes in water consumption or urination?	Renal, Endocrine
Episodes of fainting or seizures?	Nervous, Hepatic, Cardiovascular
Itching or hair loss?	Integumentary, Endocrine
Diarrhea or blood in stool?	GI, Hepatic
Changes in your animal that you would like to tell us about?	All systems

A standardized form is recommended. Owners typically can complete history forms of this nature while they are in the waiting room. (See Table 2-1.)

Pertinent Past History
– Determine whether any changes have occurred in the environment (e.g., housing, travel, and additional pets) or the animal's diet. Assess vaccination, heartworm prevention, and deworming status. Confirm any previous major medical issues such as past surgeries and ongoing diseases.
– History depth reflects the system or disease involved, time elapsed since previous visit, and number of presenting complaints.

Historical Summary
– Summarize pertinent history in three to five sentences.

3. **Obtain Weight and Assign Body Condition Score**
   – *Confirm validity of previous diet recommendation based on current weight and body condition score (BCS).* (See Table 2-2.)
4. **Obtain Vital Signs**
   – *Vital signs include heart rate, pulse rate and quality, respiratory rate and quality, mucous membrane color, capillary refill time (CRT), temperature, level of consciousness (LOC), and a pain assessment.* (See Tables 2-3, 2-4, 2-5, and 2-6.)

Table 2-2: Body Condition Score (BCS) Assignment

BCS	DESCRIPTION	DEFINING PHYSICAL CHARACTERISTICS
1	Very Thin	Ribs clearly visible, severe waist, pelvic bones clearly visible and protruding, severe abdominal tuck
2	Underweight	Minimal fat covering ribs, waist easily visible, minimal fat around base of tail, prominent abdominal tuck
3	Ideal Weight	Ribs palpable with moderate fat cover, waist visible, tail base smooth but pelvic bones palpable, abdominal tuck
4	Overweight	Ribs difficult to palpate, very slight waist, no abdominal tuck, back slightly broadened
5	Obese	Ribs extremely difficult to palpate with extreme amount of fat covering, no visible waist, no abdominal tuck with adipose hanging below abdomen, very broad back

Table 2-3: Normal Vital Sign Values

ANIMAL SPECIES	TEMPERATURE (°F)	PULSE RATE (BPM)	RESPIRATORY RATE (BPM)	CAPILLARY REFILL TIMES (SECONDS)
Adult Large Breed Dog	99.5–102.5	70–100	8–30	1–2
Adult Small Breed Dog	99.5–102.5	90–160	8–30	1–2
Adult Cat	100–102.5	110–200	8–20	1–2
Neonate Cat	Birth–1 week: 96–97 2+ weeks: 100	220–260	15–35	1–2

Table 2-4: Mucous Membrane Color and Capillary Refill Time Assessment

NORMAL	ABNORMAL
Pink	Pale/White (anemia, decreased perfusion, shock)
	Yellow (icterus)
	Blue (inadequate oxygenation)
	Dark red (sepsis, hypercapnia, shock)
	Hemorrhages (petechial, ecchymotic)
CRT < 2 seconds	CRT > 2 seconds (decreased perfusion)

Table 2-5: Levels of Consciousness

NORMAL DESCRIPTIVE TERMS	ABNORMAL DESCRIPTIVE TERMS
Bright, Alert, Responsive (BAR)	Depressed (awake but slow to arouse)
Quiet, Alert, Responsive (QAR)	Stupor (asleep except when aroused by strong stimuli)
	Coma (unresponsive to all stimuli)

Table 2-6: Pain Assessment Scale

NUMERIC SCALE	DESCRIPTION	EXAMPLE OF CONDITION ASSOCIATED WITH CORRESPONDING NUMERIC ASSESSMENT
0	No Pain	Healthy dog, no interventions
1	Mild, Intermittent Discomfort	Vaccine injection
2	Mild Pain	Dental prophylactic with no extractions or underlying dental disease
3	Moderate Pain	Ovariohysterectomy
4	Severe Pain	Pelvic/femoral fracture
5	Extreme Pain	Thoracotomy with rib resection

5. **Auscultate Thoracic Cavity (Heart and Lungs)**
   – *This is often performed at the same time heart and respiratory rates are checked.* **The animal should be auscultated for respiratory depth, cardiac rhythm, all valve areas, and any abnormal sounds. All four thoracic quadrants on each side of the animal should be auscultated (craniodorsal, cranioventral, caudodorsal, and caudoventral).** (See Tables 2-7 and 2-8 and Figures 2-1 and 2-2.)

Table 2-7: Descriptive Terms Used for Cardiopulmonary Auscultation

SYSTEM	NORMAL AUSCULTATION DESCRIPTIVE TERMS	ABNORMAL AUSCULTATION DESCRIPTIVE TERMS
Cardiac	Normal sinus rhythm Normal sinus arrhythmia (normal in dogs, change in rate associated with respiration)	Muffled heart sounds Murmur Irregular rhythm (arrhythmia) • Regular • Irregular
Respiratory	Normal bronchiovesicular (soft rustling noises) Clear	Wheeze (musical) Crackles (fluid, air popping) Gurgling Friction rubs (squeaking) Moist Increased bronchiovesicular, rhonchi (harsh, rales) Absence of lung sounds Drum sound (tympanic resonance)

Table 2-8: Abnormal Respiratory Patterns and Descriptions

Shallow (decreased depth)
Abdominal effort required
Open mouth breathing
Dyspnea (labored, difficult breathing)
Deep

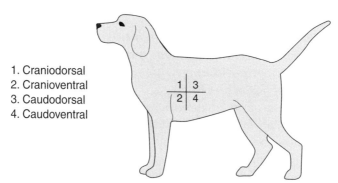

1. Craniodorsal
2. Cranioventral
3. Caudodorsal
4. Caudoventral

Figure 2-1: Thoracic Quadrants

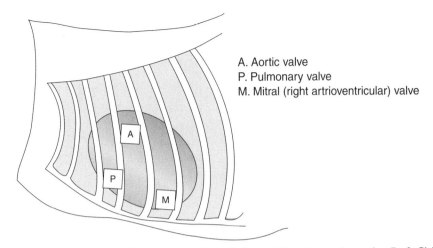

Figure 2-2A: Cardiac Valve Auscultation: Points of Maximum Intensity (Left Side)

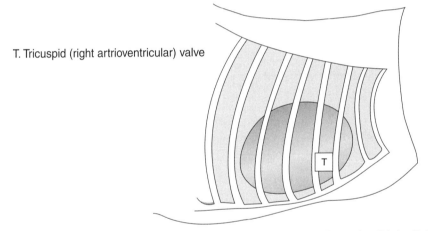

Figure 2-2B: Cardiac Valve Auscultation: Points of Maximum Intensity (Right Side)

6. Examine Head and Neck
   - *Assess for symmetry of head, eye position, ocular discharge or redness, pupil size, corneal integrity, eyelid conformation, scleral and conjunctival color, nasal discharge, otic odor and cleanliness, otic discharge, pinna thickness and malleability, lip swelling, or ulceration.*
   - *Retract lips or open mouth for oral examination. Assess for halitosis, tooth or periodontal abnormalities, tongue size and integrity, excessive salivation, oral ulcerations, oral swellings, or foreign bodies.*
   - *Palpate salivary glands (parotid and mandibular) and lymph nodes (cervical and submandibular) for pain or swelling.*

- *Palpate larynx, trachea, and thyroid gland for any abnormalities. Palpation should not elicit a coughing episode in a normal animal.*
7. **Examine Trunk and Extremities**
   - *Visually examine and/or palpate trunk for symmetry, vertebral column alignment or pain, swellings, alopecia, ectoparasites, hair coat quality, and skin abnormalities. Observe for abdominal distension or herniation.*
   - *Visually examine and/or palpate limbs for uniformity, abnormal swellings, changes in weight bearing, muscle tone, angular deformities, patella position, condition of feet and nail beds, joint range of motion, and crepitus. Palpate prescapular, axillary, and popliteal lymph nodes.*
   - *Limbs should be palpated when the animal is weight-bearing and non-weight-bearing.*
8. **Palpate Abdomen**
   - *Palpate the three abdominal quadrants: cranial—stomach, duodenum, liver, biliary structures, and pancreas; middle—spleen, kidneys, and intestines; caudal—bladder, prostate, colon, and uterus. Palpate mammary glands.* Organs most commonly identified/recognized during palpation include the bladder, spleen, small intestines, kidney, and liver. (See Figure 2-3.)

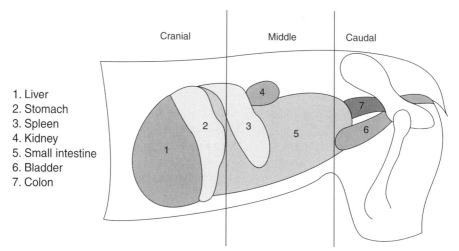

Figure 2-3A: Abdominal Quadrants (Left Side)

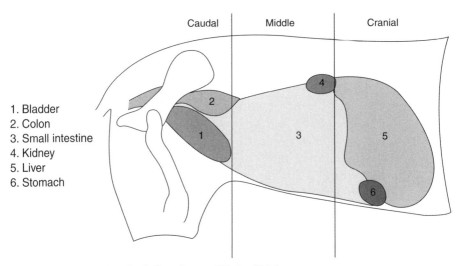

Figure 2-3B: Abdominal Quadrants (Right Side)

1. Bladder
2. Colon
3. Small intestine
4. Kidney
5. Liver
6. Stomach

9. Examine Peritoneum and External Genitalia
   - *Visually examine anus and perineal area for swellings, ulceration, fecal matting, and evidence of diarrhea.*
   - *Palpate scrotum for uniformity of testicular size, shape, and consistency. Visually examine penis/prepuce or vulva for swelling, ulceration, abnormal odor, and discharge.*
10. Record All Findings
    - *Documentation of abnormal and normal findings is fundamental to maintaining accurate medical records. In addition to recording all abnormal findings, body systems assessed within normal limits also should be documented as such. Use of designated/standardized within normal limits (WNL) parameters is advised to ensure consistency and appropriateness of the WNL designation. (See Table 2-9.)*

Table 2-9: Within Normal Limits (WNL) Parameters

SYSTEM/AREA OF EVALUATION	EVALUATION PARAMETERS
Cardiovascular	• Strong, regular pulse palpable with normal rate and quality • Absence of abnormal heart sounds and rhythm • Heart rate within established parameters • Mucous membranes pink, CRT < 2 seconds • Absence of systemic edema
Pulmonary	• Respirations regular and unlabored • Respiratory rate within established parameters • Visible bilateral equal chest expansion • Clear bilateral lung sounds • Airway patent with absence of abnormal secretions
Gastrointestinal	• Absence of vomiting, diarrhea, and constipation • Normal abdominal palpation with absence of tenderness, pain, and distension • Normal bowel movements • Oral cavity normal with mucous membranes pink and intact • Eating and drinking
Renal/Urinary	• Voiding clear, yellow urine with appropriate frequency and volume • Nondistended, nonpainful bladder upon palpation • Absence of stranguria • Absence of discharge from penis or vulva
Musculoskeletal	• Equal strength and weight bearing bilaterally • Absence of muscle weakness • Normal gait • Nontender muscle and limb palpation with full functional range of motion • Symmetrical muscle mass
Reproductive	• Absence of discharge from penis, vulva, or teats • Nonpainful, normal mammary or testicular tissue upon palpation • Mucous membranes pink and intact • Ability to urinate with uninterrupted stream and to retract penis into prepuce (males) • Production of appropriate milk volume and consistency (lactating females) • Estrus cycle within appropriate parameters and sperm production within acceptable limits (intact animals)

(Continued)

Table 2-9: Continued

SYSTEM/AREA OF EVALUATION	EVALUATION PARAMETERS
Integumentary/Lymph	• Skin warm, dry, clean, and intact • Normal skin turgor and mobility • Mucous membranes pink and intact • Hair coat symmetrical, clean, and nonbrittle • Absence of external parasites • Absence of lymph node enlargement upon palpation
Neurologic/Sensory	• Alert with appropriate LOC and orientated to place and time • Exhibiting behaviors typical/appropriate for species • Eyes open spontaneously and responds to external stimuli (visual, tactile, and auditory) • Symmetrical movement with intact proprioception and appropriate gait • Intact pain response
Incisions/Wounds	• Intact suture line • Absence of hyperemia, swelling, and discharge • Absence of abnormal odors • Bandages/dressings clean, dry, and intact
Pain	• Pain assessment of less than 1/5 • Absence of clinical signs of pain
Self-Care	• Ability to eat, drink, defecate, urinate, and groom without assistance
Peripheral/Invasive Lines (Catheters, Drains)	• Absence of edema, pain, heat, and discharge at site of entry • Extremity distal to IV catheter site warm, with no swelling • Fluid line/catheter (urinary, IV)/drain system patent/unobstructed • Bandages/dressings clean, dry, and intact

DIAGNOSTIC TESTS AND PROCEDURES

Data regarding the health status of an animal can come from a variety of sources. Tests and procedures ordered by the veterinarian assist in diagnosis, establish prognosis, and facilitate appropriate treatment. These same tests/procedures also assist the technician in developing technician evaluations, appropriate interventions, and plans of care. Timely completion of diagnostic procedures, sample collection, and specimen submission facilitates accurate results and the best possible care for patients.

> **Reader Note**
> Laboratory results are considered objective data.

The majority of tests/procedures performed in a small animal practice have been identified in this section. Tests are listed alphabetically in accordance with the five categories classification system and are followed by a brief description. Technicians requiring in-depth or specialized testing information are referred to traditional laboratory texts.

Testing Category Classification
- Blood Tests (general, hormonal, coagulation)
- Physical Tests (imaging, pressures, ocular, other)
- Fluid Collection/Evaluation
- Pathology/Culture
- Fecal/Urine

Blood Tests
General
- **Allergy testing (serum):** Detects antibodies to specific allergens. Used to diagnose allergies.
- **Ammonia tolerance test:** Measures blood ammonia levels at baseline and following administration of exogenous ammonia. Used to diagnose portosystemic shunts.
- **Antinuclear antibody (ANA) test:** Immunofluorescent antibody test (IFA) detects the presence of ANAs in the blood. Commonly used in the diagnosis of immune-mediated diseases, especially systemic lupus erythematosus (SLE).
- **Bile acid measurement:** Measures the concentration of bile acid, both fasting and postprandial. Used to assess cholestasis, liver disease, and portosystemic shunts.
- **Blood gas analysis:** Evaluates blood pH, pCO_2, pO_2, and bicarbonate. Used to assess acid-base status.
- **Blood glucose curves (also called the glucose tolerance test):** A series of tests that measure blood glucose levels at baseline following administration of glucose. Most commonly used to diagnose diabetes mellitus.

- **Complete blood count:** A group of tests that assesses red blood cells, white blood cells (WBC), and platelets. Typically includes packed cell volume (PCV), WBC count and differential, plasma protein concentration, platelet count, hemoglobin concentration, mean corpuscular volume (MCV), and mean corpuscular hemoglobin concentration (MCHC). Used to diagnose a wide variety of conditions and diseases.
- **Coombs' test:** Detects the presence of autoantibodies to erythrocytes. Commonly used to diagnose immune-mediated hemolytic anemia.
- **Heartworm test:** Detects adult heartworm antigens or antibodies to adult or microfilaria antigens. Used to diagnose heartworm infection (*Dirofilaria immitis*).
- **Knott's test:** Detects the presence of microfilaria in the blood. Used to diagnose heartworm (*Dirofilaria immitis*) infection.
- **LE cell preparation:** Detects the presence of ANAs. Used in the diagnosis of systemic lupus erythematosus (SLE).
- **Serum antibody titers:** Detects the presence of antibodies to specific microorganisms. Used in the diagnosis of various diseases.
- **Serum chemistry:** Measures enzymes, metabolites, or metabolic by-products of various organs to assess function. Typically includes total protein, albumin, bilirubin, alkaline phosphatase (ALP), alanine aminotransferase (ALT), aspartate aminotransferase (AST), lactate dehydrogenase (LDH), blood urea nitrogen (BUN), creatinine, amylase, lipase, and glucose.
- **Toxicology panels:** Detects the presence of toxic substances in the bloodstream. Used to diagnose poisoning.
- **Triglyceride measurement:** Measures circulating blood levels of triglycerides (long-chain fatty acids).

Hormonal

- **ACTH stimulation test:** Measures plasma concentration of cortisol pre- and postadministration of adrenocorticotropic hormone (ACTH). Used in the diagnosis of hyperadrenocorticism and hypoadrenocorticism.
- **Dexamethasone suppression test:** Measures plasma concentration of cortisol pre- and postadministration of dexamethasone. Used to distinguish pituitary-dependent hyperadrenocorticism from adrenal-dependent hyperadrenocorticism.
- **Endogenous ACTH:** Measures plasma concentration of ACTH. Used in the diagnosis of hyperadrenocorticism or hypoadrenocorticism.
- **Serum cortisol:** Measures plasma concentration of cortisol. Used in the diagnosis of hyperadrenocorticism or hypoadrenocorticism.
- **Thyroglobulin antibody test:** Detects the presence of autoantibodies. Used in the diagnosis of immune-mediated thyroiditis.

- **Thyroxine (T4):** Measures plasma concentration of T4. Used in the diagnosis of hypothyroidism or hyperthyroidism.
- **Thyroid-stimulating hormone (TSH):** Measures plasma concentration of TSH. Used in the diagnosis of hypothyroidism.

Coagulation
- **Activated clotting time (ACT):** Measures the time it takes for whole blood to clot once it is exposed to an activating agent. Used in the diagnosis of clotting deficiencies.
- **Activated partial thromboplastin time (APTT):** Screening test for intrinsic and common coagulation pathways. Used in the diagnosis of clotting deficiencies.
- **Fibrinogen:** Measures plasma concentration of fibrinogen. Used in the diagnosis of clotting disorders and inflammatory diseases.
- **Prothrombin time (PT):** Screening test for extrinsic and common coagulation pathways. Used in the diagnosis of clotting disorders.
- **von Willebrand factor (vWF) antigen assay:** Measures plasma concentration of vWF. Used in the diagnosis of prolonged bleeding times caused by von Willebrand's disease.

Physical Tests
Imaging
- **Computed tomography (CT):** An x-ray technique that produces a cross-sectional image (slices) of the area examined.
- **Doppler:** An ultrasound technique that detects blood flow velocity. Typically used to diagnose cardiac and vascular abnormalities.
- **Echocardiography:** An ultrasound technique that examines position and motion of the heart walls and internal structures. Produces a "moving picture" of the heart.
- **Magnetic resonance imaging (MRI):** An imaging technique that uses a magnetic field and radiofrequency signals to produce cross-sectional images of selected soft tissues.
- **Myelogram:** An x-ray image of the spinal cord taken after a radiopaque medium is injected into the subarachnoid space.
- **Nuclear scintigraphy:** A photographic recording of radioactivity distribution and intensity in a patient, postinjection of a radiopharmaceutical.
- **Orthopedic Foundation Association evaluation:** An evaluation of a ventrodorsal radiograph of the extended coxofemoral joints for hip dysplasia.
- **PennHIP evaluation (Pennsylvania Hip Improvement Program):** An evaluation of a series of radiographs of the coxofemoral joint to examine hip conformation and laxity (distraction versus compression). Used to evaluate for hip dysplasia.

- **Radiographic contrast studies:** The radiograph produced after the patient has ingested or been injected with a radiopaque medium.
- **Radiographs:** An x-ray image captured on radiographic film.
- **Ultrasound examination:** An imaging technique that uses high-frequency sound waves.

Blood Pressure/O_2 Measurement

- **Direct blood pressure (arterial catheter with pressure transducer):** A pressure transducer records the mean, systolic, and diastolic arterial pressures.
- **Indirect blood pressure (Doppler and Dinamap):** An indirect method of monitoring the patient's arterial blood pressure.
- **Pulse oximetry:** A noninvasive method of monitoring the patient's pulse rate and arterial blood oxygen saturation.

Ocular Tests

- **Fluorescein stain:** A fluorescein stain is applied to the eye to detect the presence of corneal ulcers and to evaluate patency of the nasolacrimal duct.
- **IOP measurement through indentation tonometry (Schiotz tonometer) or applanation tonometry (Tono-Pen):** A Tono-Pen or tonometer measure intraocular pressure. The presence of glaucoma is evaluated through this test.
- **Schirmer tear test:** Measures tear/lacrimal secretory capacity of the eye. Used to diagnose keratoconjunctivitis sicca.

Other Tests and Measurements

- **Electrocardiogram (ECG):** A tracing that represents the electrical activity of the heart. Used to diagnose arrhythmias and other cardiac abnormalities.
- **Endoscopy:** Visualization of the interior organs and cavities of the patient's body by using an endoscope.
- **Intradermal testing:** A method of testing for allergies by injection of suspected allergens.

Fluid Collection/Evaluation

- **Bronchoalveolar lavage:** Use of a saline wash to collect fluid/cellular specimens from the respiratory tract for the purpose of culture or cytology.
- **Centesis (arthrocentesis, abdominocentesis, and thoracocentesis):** The perforation or puncture of a cavity for the purpose of collecting specimens for further laboratory testing.

- **Prostate wash:** A procedure in which prostate fluid is collected for further laboratory testing.

Pathology/Culture
- **Biopsy:** The removal of tissue for testing. Typically includes histology and culture.
- **Cytology:** The study of cells, including their formation, origin, structure, function, and pathology.
- **Fine needle aspiration:** The process of removing a small amount of tissue through a narrow gauged needle via negative pressure. Samples are typically submitted for cytology and culture.
- **Histology:** The microscopic study of cells and tissues.
- **Swab culture (nasal, pharyngeal, and cervical):** A procedure that obtains samples via the use of a cotton swab, which is then submitted for culture.
- **Wound culture:** Culture of contaminated wounds for identification of contaminating agents such as bacteria and fungi.

Fecal/Urine
- **Fecal culture:** Evaluation of feces for enteric pathogenic bacteria.
- **Fecal flotation:** Test that identifies parasite eggs, oocysts, and various life cycle stages present in feces.
- **Urinalysis:** A series of tests performed on urine. Includes physical, chemical, and microscopic examination of urine. Common evaluation parameters include urine color; clarity; pH; specific gravity; and presence of protein, glucose, ketones, bilirubin, urobilinogen, nitrite, cells (rbc, wbc, epithelial, or renal), crystals, casts, parasites, yeast, bacteria, and sperm.

BIBLIOGRAPHY

Anderson, K. N., Anderson, L. E., & Glanze, W. L. (1994). *Mosby's medical, nursing, & allied health dictionary* (4th ed.). St. Louis, MO: Mosby.

Cote, E. (2007). *Clinical veterinary advisor: Dogs and cats.* St. Louis, MO: Mosby.

Crow, S. E., & Walshaw, S. O. (1997). *Manual of clinical procedures in the dog, cat, & rabbit* (2nd ed.). Philadelphia: Lippincott Williams & Wilkins.

Doxey, D. L., & Nathan, M. B. F. (1995). *Manual of laboratory techniques.* Cheltenham, UK: British Small Animal Veterinary Association.

McCurnin, D. M., & Bassert, J. A. (1998). *Clinical textbook for veterinary technicians* (4th ed.). Philadelphia: W. B. Saunders.

McCurnin, D. M., & Poffenbarger, E. M. (1991). *Small animal physical diagnosis and clinical procedures*. Philadelphia: W. B. Saunders.

Pratt, P. W. (1997). *Laboratory procedures for veterinary technicians* (3rd ed.). St. Louis, MO: Mosby.

Sirois, M. (2004). *Principles and practice of veterinary technology* (2nd ed.). St. Louis, MO: Mosby.

Chapter 3
Technician Evaluations with Suggested Interventions

INTRODUCTION

This chapter contains a list of all technician evaluations with each evaluation defined and the desired resolution identified. Evaluations are followed by suggested or common interventions and rationales. However, the suggested interventions are not exhaustive; that is, in certain cases, the veterinarian or the veterinary technician may decide that additional interventions are appropriate. Likewise, the use of all suggested interventions may not be warranted in every case. Ultimately, the selection of appropriate interventions and plans of care reflect individual patient needs and the judgment of the medical team.

The various technician evaluations are printed alphabetically. However, for easy reference, they have been grouped according to the Needs Ladder (Figure 3-1) as well as by body system (Figure 3-2) and in alphabetical order (Figure 3-3).

ABNORMAL EATING BEHAVIOR

Definition/Characteristics

Abnormal eating behavior is consumption of food in a manner that is atypical of normal species behavior. Atypical behaviors can include rapid ingestion of food (wolfing), food regurgitation followed by repeated consumption, pica, grass eating, begging, and coprophagy. Secondary signs of abnormal eating patterns can include bloating, nausea and/or vomiting, and other signs of gastrointestinal (GI) distress such as anorexia, rolling, posturing, and restlessness.

Desired Resolution

Consumption of appropriate feeds in a manner that is acceptable to owners and minimizes adverse medically associated conditions.

Interventions with Rationale/Amplification

1. Assess the animal for signs of acute GI distress. Notify the veterinarian immediately if such signs are noted. *All animals, but particularly large breeds of dogs such as Great Danes and English mastiffs, may suffer serious sequelae from bolting large meals. Pica (ingestion of aberrant food or garbage) can result in diarrhea and vomiting with resulting dehydration.*
2. Collect samples and complete diagnostic tests as directed by the veterinarian. *Disease processes can induce numerous eating abnormalities. Medical causes should be ruled out prior to initiating behavioral approaches to treatment.*

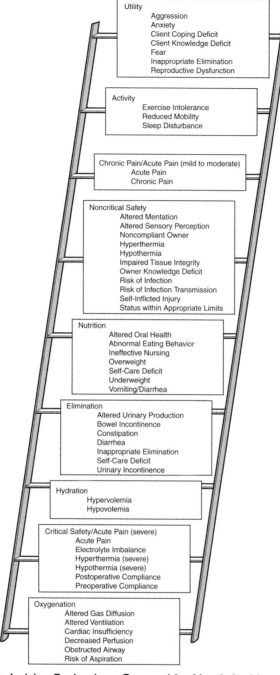

Figure 3-1: Technician Evaluations Grouped by Needs Ladder

Gastrointestinal
Abnormal Eating Behavior
Altered Oral Health
Bowel Incontinence
Constipation
Decreased Perfusion, Gastrointestinal
Diarrhea
Ineffective Nursing
Overweight
Underweight
Vomiting/Nausea

Genitourinary
Altered Urinary Production
Inappropriate Elimination
Decreased Perfusion, Renal
Reproductive Dysfunction
Urinary Incontinence

Other
Client Coping Deficit
Client Knowledge Deficit
Electrolyte Imbalance
Hyperthermia
Hypothermia
Noncompliant Owner
Preoperative Compliance
Postoperative Compliance
Self-Care Deficit
Status within Acceptable Parameters

Neurologic/Psychologic
Acute Pain
Altered Mentation
Aggression
Altered Sensory Perception
Anxiety
Chronic Pain
Decreased Perfusion, Cerebral
Fear
Self-Inflicted Injury
Sleep Disturbance

Musculoskeletal
Exercise Intolerance
Reduced Mobility

Respiratory
Altered Gas Diffusion
Altered Ventilation
Decreased Perfusion, Cardiopulmonary
Obstructed Airway
Risk of Aspiration

Cardiovascular
Cardiac Insufficiency
Decreased Perfusion, Cardiopulmonary
Hypervolemia
Hypovolemia

Skin
Decreased Perfusion, Peripheral
Impaired Tissue Integrity

Endocrine/Immune
Risk of Infection
Risk of Infection Transmission

Figure 3-2: Technician Evaluations Grouped by Body System

Abnormal Eating Behavior	Hypervolemia
Acute Pain	Hypothermia
Aggression	Hypovolemia
Altered Gas Diffusion	Impaired Tissue Integrity
Altered Mentation	Inappropriate Elimination
Altered Oral Health	Ineffective Nursing
Altered Sensory Perception	Noncompliant Owner
Altered Urinary Production	Obstructed Airway
Altered Ventilation	Overweight
Anxiety	Postoperative Compliance
Bowel Incontinence	Preoperative Compliance
Cardiac Insufficiency	Reduced Mobility
Chronic Pain	Reproductive Dysfunction
Client Coping Deficit	Risk of Aspiration
Client Knowledge Deficit	Risk of Infection
Constipation	Risk of Infection Transmission
Decreased Perfusion	Self-Care Deficit
Diarrhea	Self-Inflicted Injury
Electrolyte Imbalance	Sleep Disturbance
Exercise Intolerance	Status Within Appropriate Limits
Fear	Underweight
Hyperthermia	Urinary Incontinence
	Vomiting/Nausea

Figure 3-3: Technician Evaluations Listed in Alphabetical Order

3. Provide client education:
   - *Wolfing food.* Divide daily ration into three or four small meals per day. Provide a quiet, stress-free location for consumption of food. Separate dog from other animals during mealtimes. Particular kibble shapes may help to encourage chewing and reduce this wolfing behavior.
   - *Pica.* Treatment is often unrewarding; aversion therapy can be attempted. Most cases are treated using avoidance.
   - *Begging.* Approximately 30 percent of dogs beg for food. Begging behaviors typically lead to excessive caloric intake and obesity. Treatment includes ignoring begging behavior and removing the animal from area during mealtimes and food preparation. Feed dog in separate area of home while owners are eating. Recommend using low-calorie snacks if owner insists on feeding the animal treats.
   - *Coprophagy.* Treatment is often unrewarding. Avoidance techniques include application of For-Bid, fresh pineapple, or meat tenderizer (sprinkle over meals at 1 tsp/25 lb) to all meals. Covering cat litter box or immediately removing feces from yard is highly recommended. Free-choice feedings or diets high in fiber may have some beneficial effect. Routine prophylactic deworming is recommended for all coprophagic animals.
   - *Grass eating.* This is considered normal behavior for dogs.

ACUTE PAIN

Definition/Characteristics

Acute pain is the sudden onset of an unpleasant sensory experience, the intensity of which can be mild to severe. Acute pain is associated with changes in behavior (aggression, fear, restlessness, withdrawal, and/or vocalizations), guarding and/or posturing, repetitive behavior (pacing and/or repeatedly lying down and arising), excessive grooming, anorexia, diaphoresis, pupillary dilation, and changes in vital signs (blood pressure [BP], heart rate, and respirations).

Desired Resolution

Absence of signs of acute pain; the animal appears comfortable; behavior has returned to normal or baseline.

Interventions with Rationale/Amplification

1. Rank pain response on a level of 0–5. Record findings and alert the veterinarian to the level of pain.

Pain may be quantified with the following scale:

- *0 = no pain*
- *1 = mild intermittent discomfort*
- *2 = mild pain*
- *3 = moderate pain*
- *4 = severe pain*
- *5 = extreme pain*

The pain response in animals is frequently identified through behavioral changes such as alteration in normal vocalization patterns (dogs will whine; cats may growl or purr), withdrawal (cats will attempt to hide; dogs will suddenly appear timid) or attention-seeking behaviors, aggression, excessive grooming or licking (or a sudden lack of any grooming), repetitive behaviors, fixed unresponsive staring, guarding of affected area, tendency to refuse to move, and anorexia. Physiological signs of pain can include increased BP, heart rate, and respiratory rate. Pale mucous membranes caused by peripheral vasoconstriction also are common.

Pain is most successfully treated via prevention (e.g., administering pain medication prior to surgical interventions) or early intervention. Once the pain response has been activated, a higher dose of medication is typically required to provide relief. Tolerance to pain is variable and individualistic; it is affected by factors such as species, gender, health, age, and breed. In general, younger animals exhibit more signs of pain than do geriatric animals. Prey species such as birds, rabbits, horses, and sheep take active measures to hide a pain response as long as possible.

Ranking pain response during each examination of the animal will assist the veterinarian in determining whether the animal has responded to treatment, as evidenced by progressively decreasing scores. Veterinarians should be notified if pain is unrelieved by treatment or appears to be increasing.

2. Monitor and record vital signs. *Worsening vital signs may indicate that the animal is still experiencing pain. As pain resolves, these parameters will trend toward normal.*
3. Use nonpharmacological approaches to control pain when appropriate. *Application of heat (heating pads or hot packs), cold (ice packs or hydrotherapy), gentle stroking, or environmental modifications (deep bedding or soothing sounds) have been shown to decrease the pain response. If possible, position the animal to decrease sensation of pain. Occasionally, positioning can partially relieve acute pain. Although most animals instinctively assume the position that provides the most relief, some animals may be too weak to position themselves. Be very careful about positioning an animal who has suffered blunt trauma (e.g., was hit by automobile or fell) as spinal injury is possible and movement of the animal may exacerbate the initial injury.*
4. Administer pain medications as directed by the veterinarian. *Analgesics frequently used in veterinary facilities include the nonsteroidal anti-inflammatories (NSAIDs) or opioids (narcotics). Recent advances in identification and control of pain have led to increased use of preemptive and multimodal analgesia. In addition, local, transdermal, and topical modes of administration are becoming increasingly popular. Never apply topical heat (e.g., heating pads) over topical patches as this can increase rate of delivery of drug to toxic levels. Monitor for adverse side effects of any pain medications.*

Adverse side effects noted with NSAID use include:
- *Vomiting or changes in appetite*
- *Changes in bowel movements such as black, tarry, or bloody stool*
- *Changes in behavior*
- *Jaundice*
- *Changes in drinking or urination habits*
- *Changes in skin, such as redness, itching, or scabs*

Adverse effects associated with narcotics include:
- *Constipation*
- *Respiratory depression*
- *Weakness or ataxia*
- *Nausea*
- *Mania (cats)*
- *Skin reactions to topical patches*
- *Euphoria*
- *Changes in appetite*

5. Ensure safety of owner and veterinary personnel through the use of appropriate restraint techniques. *Animals in acute pain may become aggressive, particularly as they are being handled and examined by a stranger. Restraint and muzzling provides increased safety for the animal's caregivers. Use extreme caution when applying a muzzle to an animal that shows signs of respiratory compromise or shock, perhaps using alternative measures such as an E-collar, which won't interfere with respiratory efforts.*
6. Educate owner about signs of pain as well as methods of administering prescribed analgesic medications, necessary precautions when administering medications, and side effects of medications. Instruct owner to call the veterinarian if pain reoccurs or appears to be unrelieved by treatment. *Providing both verbal and written instruction facilitates client understanding and compliance. Since many animals do not vocalize pain, owners must be given specific instructions on what signs to look for regarding pain.*

AGGRESSION

Definition/Characteristics

Aggression is defined as hostile behavior exhibited toward other animals and/or people. Aggression can be characterized by increased vocalizations (barking, growling, or hissing), raised hackles, baring of teeth, eye-to-eye contact, biting, kicking, and scratching.

Desired Resolution

Animal exhibits no hostility or challenging behavior toward other animals and/or people unless so instructed by handler.

Interventions with Rationale/Amplification

1. Restrain the animal using a technique that eliminates or minimizes potential of injury to the handler or animal. *Aggressive animals pose a threat to caregivers and/or other animals. The type of equipment used for restraint varies with each species. Restraint can induce undesirable physiological changes in the animal.*
2. Interview owner carefully to identify situations in which the animal exhibits aggression as well as the duration of the problem. Classify the type of aggression. *Identification of triggers may enable the owner, veterinarian, and technician to develop proper management strategies. Although highly undesirable, most incidences of aggression are not considered abnormal. Aggressive behavior or actions are typically a normal species behavioral pattern that is incompatible with owner's lifestyle or needs. Aggression can be classified as follows:*

- *Dominance aggression:* Usually males that attempt to dominate family members. Often become aggressive when approached while eating, disturbed in sleep, restrained, pulled, pushed or lifted, groomed, or medicated.
- *Possessive aggression:* Related to dominance aggression. Occurs when an object or food is withdrawn from the animal.
- *Play aggression:* Usually in young/juvenile dogs and involves growling and nipping that doesn't break the skin. Animals often simultaneously exhibit play behaviors such as bowing, wagging their tails, and running in circles.
- *Fear-induced aggression:* Occurs in either sex; animals often bite when reached for or cornered. Animals often simultaneously exhibit fearful body language.
- *Pain-induced aggression:* Occurs in either sex when humans attempts to medicate or examine a painful area.
- *Territorial aggression:* Occurs in either sex and usually involves strangers (animals or humans) approaching an area that dog perceives as its territory.
- *Predatory aggression:* Typical hunting behavior exhibited. Dog is attacking what he or she believes to be prey.
- *Parental aggression:* Occurs in females protecting offspring.

3. Discuss various coping techniques with owner to identify best options for the particular animal. Inform owners of dangers associated with animal aggression; address the need to protect vulnerable individuals such as children and visitors.
    - *Dominance aggression:* Consider castration. Initiate behavior modification therapy and general obedience training. Long-term prognosis for this form of aggression is poor.
    - *Possessive aggression and territorial aggression:* Physical separation such as kennels and fenced-in yard. Initiate behavioral modification therapy.
    - *Play aggression:* Terminate inappropriate play (such as children running and playing tug-of-war). Use appropriate toys and decrease stimuli during play periods.
    - *Fear-induced aggression:* Identify eliciting stimuli and initiate counterconditioning.
    - *Pain-induced aggression:* If possible, identify source and eliminate pain. Initiate counterconditioning. Avoid punishment.
    - *Predatory aggression:* Extremely difficult to eliminate; confinement is the only reliable method of control.
    - *Parental aggression:* Limit number of humans interacting with female until weaning has occurred. Separate female when handling offspring.

4. Identify community resources that are available to owner. *Training facilities, private trainers, therapists, and humane societies may be able to provide help to owner.*

5. Check the animal's vaccination history to ensure that all vaccinations, especially rabies, are up to date. *Vaccination of an aggressive animal provides protection to victims of aggression by decreasing the likelihood of disease transmission.*

ALTERED GAS DIFFUSION

Definition/Characteristics

Altered gas diffusion is an imbalance in normal oxygen and carbon dioxide exchange at the alveolar level. Causes include changes in the alveoli (e.g., chronic obstructive pulmonary disease, pneumonia, and pulmonary edema), changes in blood flow (e.g., congestive heart failure and pulmonary embolism), and changes in oxygenation of the blood (e.g., oxygen deprivation, carbon monoxide poisoning, and hypoventilation). Altered gas diffusion is associated with dyspnea; tachycardia; cyanotic mucous membranes; altered arterial blood gases (ABGs) and oxygen saturation; and/or changes in behavior and LOC, including confusion, restlessness, lethargy, and coma.

Desired Resolution

Adequate oxygenation as demonstrated by improved breathing pattern, normal heart rate, pink mucous membranes, ABGs and oxygen saturation WNL and improved level of consciousness.

Interventions with Rationale/Amplification

1. Provide supplemental oxygen as ordered. Monitor and record oxygen saturation levels. *In many cases, supplemental oxygen will increase oxygen saturation in the blood. Monitoring of oxygen saturation levels over time will indicate whether the animal is responding to treatment of underlying cause.*
2. If possible, position the animal for maximum airway clearance. *Extension of the head allows for maximum airflow.*
3. Assemble all necessary emergency equipment, including endotracheal tubes, tracheotomy sets, and suction catheters. *In emergency situations, easy access to emergency equipment improves delivery of care to the animal.*
4. Monitor and record all signs and symptoms of respiratory distress. Notify the veterinarian *immediately* if signs and symptoms increase or do not improve following treatment. *Acute respiratory distress can quickly escalate to an emergency situation. Failure to improve after treatment indicates other treatment options need to be considered.*
5. Monitor pertinent lab values such as ABGs. Report all lab values to the veterinarian. *Following successful treatment, laboratory values will normalize or return to patient baseline.*

6. Provide fluids at rate and route ordered by the veterinarian. *Adequate hydration allows the animal to mobilize and excrete secretions.*
7. Administer medications as ordered by the veterinarian. *Various medications such as broncodilators, diuretics, and corticosteroids may aid in increasing oxygenation. Antibiotics may be ordered to treat or prevent infection. Heparin or coumadin will aid in preventing emboli.*

ALTERED MENTATION

Definition/Characteristics

Altered Mentation is a change in cognition, emotional state, and/or level of consciousness. Altered mentation may be of chronic or acute onset. It is characterized by changes in behavior (e.g., being withdrawn, displaying aggression, and walking in circles), regression in training (e.g., housebroken dog soiling in the house), confusion (e.g., not recognizing owner), decreased response to external and internal stimuli (e.g., decreased pain response), excessive sleeping, and/or coma. Altered mentation of a chronic nature may be caused by the aging process or by various neurological diseases. Changes in mentation with an acute onset may be caused by trauma, infection, electrolyte imbalances, or any physiological process that alters brain chemistry and oxygenation.

Desired Resolution

Return to baseline cognition, emotional state, and level of consciousness. (Chronic changes in mentation may be progressive and irreversible.)

Interventions with Rationale/Amplification

1. Obtain a complete history from owner regarding the animal's baseline level of mental functioning and any changes that have been observed. The veterinarian should be notified immediately of any animal that exhibits a decreased level of consciousness as this may indicate an emergency situation. *A complete history will assist the veterinarian in determining whether the changes in mentation are chronic or acute and will provide information important in diagnosing the cause.*
2. Obtain samples of urine and blood as ordered by the veterinarian; perform laboratory tests of samples. *These tests may yield important information regarding the cause of changes in mentation. For example, altered liver function enzymes may indicate hepatic dysfunction, which can lead to portosystemic encephalopathy. Electrolyte imbalances and hypoglycemia also are possible causes of altered mentation.*
3. Administer fluids and medication as ordered by the veterinarian. *The use of fluids and choice of medications will be dependent upon the veterinarian's conclusions regarding the underlying cause of altered mentation. Note that any animal with a decreased level of consciousness should not be given*

anything by mouth unless it has been determined that the animal has an intact gag reflex.

ALTERED ORAL HEALTH

Definition/Characteristics
Altered oral health is a decline in the physical condition of the oral cavity that can be characterized by halitosis, gingivitis, periodontitis, glossitis, dental tartar, oral masses, dysphagia, anorexia, ulceration, drooling, dryness, and tooth loss.

Desired Resolution
Restoration of normal tooth, tongue, and gingival surfaces, accompanied by pain resolution. Initiation of appropriate oral prophylactic care program.

Interventions with Rationale/Amplification

1. Collect samples and perform diagnostic tests as directed by the veterinarian. *A complete blood count (CBC) and chemistry profiles are frequently performed prior to a routine dental cleaning. Various tests such as cite tests for FeLV/FIV or biopsies can be performed when oral health reflects systemic or neoplastic disease.*
2. Perform dental cleaning as directed under the supervision of the veterinarian. *Dental cleaning requires the use of general anesthesia. Routine dental cleaning and prophylactic procedures should be performed by the technician. Numerous texts providing a step-by-step description of dental prophylaxis and cleaning are available.*
3. Provide medication as directed by the veterinarian. *The type of medication selected is determined by the underlying cause of oral disease. Antibiotics, anti-inflammatories (both nonsteroidal and steroidal), and occasionally immunosuppressive agents may be utilized.*
4. Provide nutrition support as necessary. *Animals suffering from oral masses, trauma, or severe dental disease may be unable to prehend, masticate, or swallow. Nutritional assistance in the form of altered diet texture (balanced, syringeable prepared critical care diets or gruels), pharyngostomy or JPEG tubes, or total parenteral nutrition (TPN) may be required.*
5. Moisten oral cavity as necessary. *Comatose animals and those unable to keep their tongue within the oral cavity will experience excessive drying of tongue and gingival surfaces. Application of a moistening agent such as glycerin is warranted.*
6. Provide client education. *Technician involvement is fundamental to establishing an appropriate preventative oral health care program. Technicians should educate/counsel clients on the following:*
    - *Explain risk factors associated with oral disease. Liver, renal, and cardiovascular diseases have been linked to poor dental health. Numerous*

pamphlets, handouts, videos, and models are available for client education purposes.
- Explain the pet's current dental problems and treatments/procedures recommended by the veterinarian. Using the pet's dental chart when discussing problems is very helpful. Clients are more likely to treat their pet's medical condition when they fully understand the medical significance of the problem.
- Develop an individualized preventative home care plan that addresses plaque control. There are effectively three methods for accomplishing plaque control. These include mechanical (tooth brushing, dental diets, and chews), chemical (chlorhexidine solutions) and barrier (oravet) methods.
  - Tooth brushing. This is the gold standard of plaque control. Human or pet toothbrushes that have soft bristles can be used. Toddler-size brushes are appropriate for small dogs and cats; compact head brushes can be used for larger animals. Brush length and curvature is a matter of preference. Do not use human toothpaste as it can cause emesis, gastric irritation, or fluoride intoxication. Flavored pet toothpaste is recommended. Recommended products should have the Veterinary Oral Health Council (VOHC) endorsement; this is the veterinary equivalent to the American Dental Association (ADA).
    - Instruct client to hold brush at a 45° angle to tooth surface. Using a circular motion, brush the buccal surface of the tooth. The lingual surface does not need to be brushed. Brush for 1 minute on each side. Ideally, the teeth should be brushed daily; however, clients should be encouraged to develop a frequency that will fit their schedule.
    - Problem animals can be acclimated to the procedure slowly through positive reinforcement. Owners can start by simply allowing the animal to lick the toothpaste from the brush. It is helpful initially to place dogs in a corner to prevent backing away. Minimal use of restraint is recommended on cats.
  - Dental-specific diets. Most dental diets are formulated such that they "scrape" the tooth surface and remove plaque when the food is eaten or they employ a calcium chelating agent that makes calcium unavailable for inclusion in tartar (calculus). Dental-specific diets are available through Hill's, Royal Canin, Iams, Purina, and Eukanuba.
  - Chews and toys. These items also remove plaque mechanically, although their importance is somewhat limited. Rawhide items should be avoided as they are choking hazards. Chew ropes or Kongs can be utilized.
  - Chlorhexidine. Numerous types of solutions and gels are available to remove plaque chemically. Zinc-containing solutions also are used.

- Barriers. Inert polymers can be applied to the tooth surface weekly to prevent bacterial adherence, thereby minimizing tartar buildup.

ALTERED SENSORY PERCEPTION

Definition/Characteristics
Altered sensory perception is an impaired ability to assimilate environmental information. Alterations commonly involve ocular and auditory systems.

Desired Resolution
Restoration of sense/perception or patient's and owner's adjustment for long-term care.

Interventions with Rationale/Amplification
Visual
1. Provide a safe environment. *Remove unnecessary items from kennel (toys and empty bowls) and secure water/food bowls to prevent tipping. Provide a quiet environment.*
2. Always have larger animals on a leash/harness when outside the kennel. *Be aware of potential obstacles for the patient. Maintaining slight pressure on the leash will provide the animal with a degree of guidance. Carry smaller animals.*
3. Always alert the patient to your presence by speaking softly. Touch can be used as a reassuring measure if the animal is already aware of your presence. *It is easy to frighten a visually impaired animal.*
4. Monitor food and water consumption. Animals should be shown the location of food and water bowls. *Bowls that are not secured should be removed after feeding.*
5. Place animals on a PVC grate if needed. *This will help keep the patient dry and clean from spilled food and waste.*
6. Educate owners regarding safe home environments. *Remove obstacles, allow the animal to become familiar with the environment, and provide an enclosed area for outside exercise. If needed, assist owners in locating trainers who work with impaired animals.*

Auditory
1. Always place the patient on a leash/harness when outside the kennel. *Animals with auditory impairment are less able to sense danger (e.g., an approaching car); leashing the animal provides safety.*
2. Prior to touching the animal, make visual contact to ensure that the animal is cognizant of your presence. *Animals with auditory impairment may be startled by a sudden touch and may react aggressively.*

ALTERED URINARY PRODUCTION

Definition/Characteristics

Altered urinary production is a variance from normal levels of urinary elimination or a change in urine characteristics. Changes in urinary output may originate at the renal level (e.g., kidney failure and kidney infection) or in the ureters, bladder, or urethra (e.g., urethral obstruction). Altered urinary output may be characterized by anuria, oliguria, polyuria, dysuria, hematuria, or other changes in the color/consistency of urine (e.g., cloudy, presence of sediment, and foul odor).

Desired Resolution

Production and elimination of normal amounts of clear, yellow urine. No pain with urination.

Interventions with Rationale/Amplification

1. Obtain a detailed history from owner regarding the animal's normal urinary pattern as well as observed changes in urination. Other changes (e.g., behavior, eating, resting, and presence of pain) also should be noted. *A good history provides the veterinarian with baseline data for the particular animal and aids in diagnosing the condition.*
2. Monitor and record the animal's vital signs at intervals ordered by the veterinarian. *Since the kidneys are intimately involved in regulating BP, renal abnormalities can impact vital signs. A drop in BP results in a decreased pulse pressure (weak pulse) but an increased rate. An increase in BP results in a pounding pulse and possibly vein distension.*
3. Obtain a urine sample and perform urinalysis as ordered by the veterinarian. Note and record any deviations from normal in urine color/characteristics. *There are various methods for obtaining urine samples. Dogs may be taken outside on a leash and allowed to explore areas where other animals have urinated; this may prompt the animal to mark the area, and the urine can be captured. Urine can sometimes be expressed from cats; if expression is not possible, a urinary centesis may be performed. A urinalysis provides valuable information to the veterinarian and serves as an aid in diagnosing the condition.*
4. Record the animal's urinary output every 8 hours or more frequently when ordered by the veterinarian. Notify the veterinarian when output is greater or less than normal for the species of animal. *Prolonged anuria or oliguria can result in permanent damage to the kidneys; prompt intervention can save renal function. Urinary output is usually estimated; however, when a urinary catheter is in place, the amount can be measured exactly.*

5. Administer fluids and medications as ordered by the veterinarian. *Animals whose urinary output is low secondary to dehydration need to be rehydrated. Certain conditions (e.g., urethral obstruction) are treated with superhydration (hydration that exceeds the body's metabolic needs) to flush the renal system. When an animal is superhydrated, expect to see higher-than-normal levels of urine production. Certain medications such as diuretics, antibiotics, and analgesics may be ordered by the veterinarian.*
6. Place urinary catheter when ordered by the veterinarian. *Animals that are producing adequate amounts of urine but are unable to pass the urine risk bladder perforation. A catheter also may be placed if the veterinarian needs to know the exact amount of urinary production.*

ALTERED VENTILATION

Definition/Characteristics

Altered ventilation is a change in inspiration/expiration that leads to abnormal oxygenation. This state is associated with dyspnea, orthopnea, adventitious lung sounds, use of accessory muscles for breathing, nasal flaring, mucous membrane cyanosis, decreased oxygen saturation, and altered ABGs.

Desired Resolution

Reestablish a normal pattern of respiration. Absence of signs of respiratory distress such as dyspnea, orthopnea, nasal flaring, use of accessory muscles, and cyanosis. Oxygen saturation and ABGs are normal or are returned to patient baseline.

Interventions with Rationale/Amplification

1. Provide supplemental oxygen as ordered for hypoventilation. Note and record patient's oxygen saturation level. *Supplemental oxygen can help to increase oxygen saturation of the blood.*
2. Monitor and record vital signs and signs of respiratory distress, including dyspnea, orthopnea, nasal flaring, and use of accessory muscles. Notify the veterinarian immediately if an increase in respiratory distress is noted. *Trending of these signs over time will assist the veterinarian in determining whether treatment is effective. Acute/increased respiratory distress is a veterinary emergency.*
3. Auscultate lungs for adventitious lung sounds, which may include rales, rhonchi, crackles, and/or wheezing. Record results. *Adventitious lung sounds should dissipate with effective treatment.*
4. Assemble all necessary emergency equipment, including endotracheal tubes, tracheotomy sets and suction catheters. *In emergency situations, easy access to emergency equipment improves delivery of care to the patient.*

ANXIETY

Definition/Characteristics

Anxiety is a distressing change in cognitive/sensory/emotional baseline characterized by purposeless or repetitive movements (pacing, constant lying down and arising, or walking in circles); heightened or decreased attention to external stimuli; panting and/or drooling; changes in facial expression; vocalizations; physiological changes such as increased BP, heart rate, and respiration; and dilated pupils. Anxiety is part of the fight-or-flight reaction; and it ranges from mild (as the animal perceives a possible threat) through moderate, severe, and panic (as the animal perceives imminent injury or death).

Desired Resolution

Absence of anxiety as evidenced by return to normal or baseline behavior; normal physiological parameters.

Interventions with Rationale/Amplification

1. If possible, identify and remove cause of the animal's anxiety. *Owner information is especially critical in assessing anxiety as behavior that prompts a visit to the veterinarian may not be evident at the time of the visit.*
2. If it is not possible to remove the cause of anxiety, discuss various coping strategies with owner. *Strategies include removing the animal from source of distressing stimuli (e.g., bringing the animal inside during thunderstorms) and isolating (crating) the animal.*
3. Administer/dispense medication as ordered by the veterinarian. *Certain medications such as anxiolytics and tranquilizers may be used by owner on an as-needed basis.*
4. Educate owner about medication, including time for onset, peak, and duration of effect. Discuss use of medication prior to anticipated anxiety-producing experience. *Owners who can anticipate the animal's need may be able to prevent anxiety from occurring through proper use of medication and coping techniques.*

BOWEL INCONTINENCE

Definition/Characteristics

Bowel incontinence is the inability to voluntarily control bowel elimination. Total incontinence is associated with unpredictable loss of stool or constant oozing of soft stool. The animal may exhibit red or excoriated skin in the anal area or fecal matter clinging to the coat in the anal area. Total incontinence may be an acute and short-lived phenomena (e.g., an animal suffering from acute diarrhea) or may be chronic and of long duration. Partial bowel incontinence is characterized by loss of stool at somewhat predictable times (e.g., following meals and during the overnight period when the animal is not taken outside for several hours).

Desired Resolution

Animal achieves voluntary control of bowel elimination or achieves a pattern of bowel elimination that is acceptable to owner.

Interventions with Rationale/Amplification

1. Obtain a complete history from owner, including times and amounts of bowel movements, length of time problem has been occurring, and circumstances surrounding incontinence. *A complete history will assist the veterinarian in selecting appropriate diagnostic tests, possible treatments, and management techniques.*
2. Carefully inspect the skin for signs of irritation or scald caused by repeated exposure to feces. *Animals that have diarrhea are particularly prone to fecal scald in the perianal area. Application of protective barriers such as petroleum jelly may be warranted. Clipping the perianal area can help to minimize fecal/hair matting.*
3. Provide appropriate environmental modifications. *Placing animals on a PVC-coated cage grate/floor assists in keeping them clean and dry, especially in cases associated with diarrhea. Always remove the collar when an animal is placed on the grate. Check bedding throughout the day for fecal contamination.*
4. Collect samples and run diagnostic tests as directed by the veterinarian. *Fecal samples are frequently examined in animals suffering from acute incontinence secondary to diarrhea. Other diagnostic tests can include CBCs and serum chemistry profiles.*
5. Administer medications and diets as ordered by the veterinarian. *Certain medications such as antibiotics/antimicrobials and antihelmentics may be ordered by the veterinarian to treat causes of diarrhea. Topical ointments may be ordered to treat skin irritation. Feeding a low-residue diet can decrease amount of stool produced. Several low-residue diets are on the market, including Hill's i/d, Royal Canin DIGESTIVE LOW FAT LF, and Eukanuba Low-Residue. When inflammatory bowel disease (IBD) is the diagnosis, a limited-antigen diet may be the veterinarian's choice. Limited ingredient diets are produced by Royal Canin, Hill's, Purina, and Iams. Some cases may be fiber-responsive and necessitate the use of a high-fiber diet such as Royal Canin CALORIE CONTROL CC HIGH FIBER and Hill's w/d or r/d.*
6. Discuss possible management techniques with owner. *New onset incontinence may resolve quickly when underlying causes are treated. Partial incontinence (short- or long-term) may be treated by toileting the animal more frequently and more regularly, particularly immediately following meals.*

Total incontinence presents more of a management problem, with confinement, crating, or outdoor living being the only viable alternative for the animal. The use of pet diapers helps to minimize soiling of the living areas but increases the frequency of perianal scald. Unfortunately, because of the associated management problems, many owners elect to euthanize animals that have become totally incontinent.

It is important to clarify with the owner that inappropriate elimination or house soiling is not always synonymous with incontinence and previously house-trained animals may exhibit sudden onset of incontinence for a variety of medical reasons.

CARDIAC INSUFFICIENCY

Definition/Characteristics

Cardiac insufficiency is the inability of the heart to provide adequate blood flow to meet metabolic requirements. Cardiac output may be altered by reduced preload, increased afterload, reduced contractility, or alterations in heart rate and/or rhythm; signs and symptoms may vary according to the cause. Signs of reduced preload include jugular vein distention (JVD), peripheral edema, weight gain, fatigue, and dyspnea. Signs of increased afterload include increased BP, shortness of breath, mucous membrane cyanosis, increased capillary refill time (CRT), and/or skin temperature changes. Signs of reduced contractility include generalized, peripheral and/or dependent edema, sudden weight gain, shortness of breath and dyspnea, adventitious lung sounds, and cough. Signs of altered heart rate/rhythm include electrocardiogram (ECG) changes, respiratory distress, fatigue, and anxiety. Reduced urine output may be present whenever cardiac output is decreased as the kidneys are not properly perfused.

Desired Resolution

Normal or stable ECG; vital signs within normal limits; no signs/symptoms of respiratory distress; absence of edema; return to baseline weight; animal is able to return to normal or baseline activity level.

Interventions with Rationale/Amplification

1. Provide supplemental oxygen as ordered. *Supplemental oxygen increases available oxygen.*
2. Weigh patient daily. *One kilogram (2.2 lb) of weight increase indicates a liter of retained fluid. Trending of this parameter over time will indicate whether the animal's condition is stable, improving, or deteriorating.*
3. Monitor and record vital signs, oxygen saturation, and CRT every 4 hours or more frequently as ordered by the veterinarian. *Generally, more critical*

patients receive more frequent monitoring. Trending of these parameters will assist the veterinarian in determining whether treatment is effective.
4. Provide or limit fluids as ordered by the veterinarian. *Hemodynamic status is negatively impacted by decreased cardiac output. Excess fluids may contribute to massive edema; conversely, since fluid that has third-spaced is not available for intravascular use, adequate fluids must be provided to maintain blood volume.*
5. Monitor and record ECG results. *Trending of this parameter over time will assist the veterinarian in determining whether treatment is effective.*
6. Administer medications and therapeutic diets as ordered by the veterinarian. *Monitor and record the animal's response to treatment. Various medications such as diuretics, vasodilators, antihypertensives, antidysrhythmics, antibiotics, and inotropics may be ordered. Diets that may be selected include Royal Canin EARLY CARDIAC EC, Hill's k/d, Hill's h/d, and Purina CV Cardiovascular.*

CHRONIC PAIN

Definition/Characteristics

Chronic pain is a distressing sensory experience that has existed for more than 1 or 2 months. Onset may be gradual (e.g., osteoarthritis) or sudden (e.g., tendon injury). Chronic pain is associated with changes in behavior (e.g., withdrawal, vocalization, and guarding), decreased appetite, general appearance of unthriftiness, "dull" facial expression, and/or musculoskeletal changes (e.g., limping and muscle atrophy). Vital signs are often normal as the animal's body has acclimated to the presence of pain.

Desired Resolution

Absence of signs of chronic pain; animal appears comfortable; behavior has returned to normal or baseline.

Interventions with Rationale/Amplification

1. Restrain and muzzle the animal if necessary. *Animals in pain may become aggressive, particularly when being handled and examined by a stranger. Restraint and/or muzzling provides increased safety for the animal's caregivers.*
2. Observe the animal for signs of chronic pain; record results. Question owner regarding onset, duration, and patterns of chronic pain; record results. *This information will aid the veterinarian in determining the cause of pain.*

3. Administer/dispense medications as ordered by the veterinarian. *Certain medications such as analgesics, corticosteroids, and muscle relaxants may be ordered to reduce the animal's pain.*
4. Educate owner regarding signs of continued pain. *Instruct owner to call the veterinarian if pain is unrelieved by treatment. Since many animals do not vocalize pain, owners must be instructed about methods for assessing pain in the animal.*

CLIENT COPING DEFICIT

Definition/Characteristics

Client coping deficit is the inability of the client to cope with, understand, or accept a pet's medical condition. Coping deficits are frequently accompanied by feelings of fear, anxiety, anger, denial, apathy, frustration, or grief. Responses can be directed at self, pet, or medical staff.

Desired Resolution

Client adapts to, accepts, and appropriately responds to the pet's medical needs.

Interventions with Rationale/Amplification

1. Provide an appropriate environment for the client to receive potentially disturbing information. *Busy reception areas and noisy treatment rooms are inappropriate sites for giving clients medical information. Clients should be brought into an examination or consultation room to minimize distractions and to ensure privacy during medical discussions.*

   Remove potentially interruptive items such as cell phones and pagers from the room. As pets can be distracting when moving around, whining, or seeking comfort from their owner, technicians are advised to discuss medical issues prior to bringing the pet to the client.

2. Provide client education regarding the diagnosis and treatment of the identified medical condition. *Clients often require detailed information to understand and accept a pet's medical condition. Information can be supplied in the form of demonstrations, printed materials, videos, Web sites, or verbal instruction. Remember that written information should supplement, not substitute for, verbal communication.*

   Comparison to similar human medical conditions, such as diabetes, can be made when discussing long-term care issues. However, care must be taken to avoid anthropomorphisizing, or placing human emotions

on an animal. For example, dogs would be unlikely to suffer the same emotional distress that humans would experience when informed of a limb amputation.
3. Provide support. *Clients may require emotional support when making medical or end-of-life decisions regarding a pet. Nonverbal communication is a very important component of emotional support. The use of attending behaviors is recommended. These can include direct eye contact, nonjudgmental expressions, open body postures, and speech directed to the client at eye-to-eye level.*

    It is very important to listen actively and avoid the tendency to interrupt. Ultimately, the goal of the veterinary medical team is to facilitate and support the client during the interaction, not to make the medical decisions for the client.
4. Provide client education regarding euthanasia if warranted. *Humane euthanasia is a stressful yet sometimes necessary procedure. Technicians must recognize the grief and difficulties associated with euthanasia and strive to provide the compassion and tact warranted. The following are general recommendations regarding client education/involvement in humane euthanasia:*
    - *Schedule euthanasia appointments for more quiet times of the day.*
    - *Schedule a quiet room if the client elects to be present for the euthanasia.*
    - *Complete all paperwork, including consent forms and body disposal options, prior to admitting the animal. Ideally, these details are arranged the day prior to the euthanasia.*
    - *If the client wants to know, describe the drugs used and protocol for euthanasia. Provide educational materials such as pamphlets if needed.*
    - *If the client elects to be present during the procedure, prepare the client for what he or she might see during and after death. Be sure to inform the client that the procedure is painless.*
    - *If the client is present during the euthanasia, medical personnel should "pronounce the pet dead." Clients may wrongly believe that the animal is still alive if the eyes are open.*
    - *Allow the client quiet time with the pet after the euthanasia. Assist the client in clipping a lock of fur, removing a collar, or performing any other symbolic ritual that is meaningful to him or her.*
    - *Prepare the body appropriately if the client wants to take the animal home for burial. (Note that some areas do not allow home burial of euthanized animals.) Carry the body to the client's car (use a side or back door if possible) and ensure that the client is able to drive.*
    - *Update the patient chart as deceased.*
    - *Send a condolence card.*
    - *Refer client to appropriate support groups if warranted. (Most veterinary colleges have hotlines and support groups.)*

CLIENT KNOWLEDGE DEFICIT

Definition/Characteristics

Client knowledge deficit is a lack of information or understanding by the owner regarding the animal's diagnosis/prognosis, recommended procedures, or at-home care. Owners may or may not verbalize a need for education.

Desired Resolution

Owner verbalizes understanding of all aspects of the animal's condition, including instructions for at-home care. Owner is able to demonstrate any procedures (e.g., medication administration) required for at-home care.

Interventions with Rationale Amplification

1. Assess client's current level of knowledge. *Owners vary greatly in their familiarity with animals. A first-time owner requires more education than a long-time pet owner.*
2. Educate owner regarding all aspects of the animal's condition, including but not limited to diagnosis, prognosis, care provided or to be provided by veterinary team, and care to be provided at home. *An educated owner is more likely to follow discharge instructions and provide proper care for the animal.*
3. Demonstrate any procedures that client will need to perform at home (e.g., medication administration, ear cleaning, and wound care). If necessary, before leaving the hospital, have owner demonstrate procedure to ensure proper technique. *Many owners are unfamiliar with certain aspects of the aftercare required by the animal.*
4. Provide written discharge instructions to owner; if available, provide written information regarding the animal's diagnosis/condition. *Owners picking up an animal may feel stressed by the thought of assuming care for the animal. Generally, an individual's ability to learn and process new information is decreased by stress. Providing written information gives the owner an opportunity to absorb new material in the nonstressful environment of his or her home.*
5. Inform owner that the staff is available to answer any questions about the animal's condition and care after discharge. *Emphasizing availability will decrease the owner's anxiety.*

CONSTIPATION

Definition/Characteristics

Constipation is the inability to pass feces or infrequent passage of small, hard feces. Signs of constipation include tenesmus, anorexia, hypoactive bowel sounds, abdominal bloating, abdominal tenderness, and the animal's dragging the anal area on the ground. Severe cases of constipation can develop into a bowel obstruction.

Desired Resolution

Passage of normal amount of softly formed stool with no signs of abdominal distress.

Interventions with Rationale/Amplification

1. Obtain a complete history from owner, including description of normal bowel habits, regular diet, signs of constipation, and date of onset. *A complete history will aid the veterinarian in diagnosing the underlying cause.*
2. Monitor and record frequency, volume, and consistency of any bowel movements. Note the presence of blood, mucus, or foreign material. *Monitoring fecal output will help to assess the degree of constipation and will influence the need for and type of interventions.*
3. Assess the animal for signs of dehydration, including dry mucous membranes and poor skin turgor ("tenting"). Administer fluids at rate and route ordered by the veterinarian. *As an animal becomes dehydrated, the body pulls fluid from the gut, leading to hard, dry stools. Proper hydration is an important consideration in relieving and preventing constipation.*
4. Record fluid intake and output volumes. *Constipation is commonly associated with dehydration.*
5. Monitor and record signs of tenesmus or abdominal distress, including abdominal bloating/tenderness, anorexia, and abnormal bowel sounds. *Specifically note whether tenesmus occurs before or after defecation. Tenesmus prior to defecation suggests that constipation may be the major problem. Postdefecation tenesmus can be suggestive of colitis or prostate problems. Signs tend to dissipate as successful treatment progresses.*
6. Administer medications and treatments as ordered by the veterinarian. *Rectal suppositories and/or enemas may aid the animal in passing hard stool. Digital disimpaction may be possible in a larger animal. Laxatives and stool softeners may be administered orally.*
7. Alter diet as directed by the veterinarian. *A high-fiber, high-moisture diet may be recommended for animals that are prone to constipation. High-fiber diets include Royal Canin CC HIGH FIBER, Hill's w/d and r/d, and Purina DCO Diabetes Colitis and OM Overweight Management.*
8. When hospitalized, allow ample opportunity for the animal to defecate. *Some animals require longer and more frequent opportunities to defecate. Most animals are hesitant to defecate in front of strangers. Some cats require one litter box for defecation and one litter box for urination.*
9. Encourage physical activity if appropriate. *Activity helps to stimulate defecation. Animals that are physically able should be encouraged to exercise. This movement stimulates GI blood flow and motility.*

10. Educate owner regarding proper diet and exercise program for the animal. *An educated owner is more likely to comply with the suggested treatment program.*

DECREASED PERFUSION

Definition/Characteristics

Decreased perfusion is a reduction of blood flow through the capillaries causing insufficient oxygenation of cells and tissues. Any system in the body may be subject to altered tissue perfusion; the affected system should be referenced. Most commonly, reference is made to cerebral, cardiopulmonary, renal, GI, or peripheral perfusion. Defining characteristics vary depending on which body system is affected.

- Altered cerebral perfusion may result in changes in behavior and/or LOC; changes in vital signs, reflexes, and pupillary reaction; and changes in motor skills.
- Altered cardiopulmonary perfusion may result in dyspnea, dysrhythmias, increased CRT, and changes in behavior (anxiety and restlessness).
- Altered renal perfusion may result in oliguria, anuria, and/or hematuria; increased specific gravity (i.e., increased concentration) of urine; increases in BUN and creatinine (Cr); and changes in BP.
- Altered GI perfusion may result in anorexia, nausea and/or vomiting, abdominal distention, pain/tenderness in the affected area, melena, and altered bowel sounds.
- Altered peripheral circulation may result in localized erythema, changes in skin temperature, tenderness/pain in the affected extremity, edema, changes in pulses, and necrosis.

Desired Resolution

Reestablish adequate oxygenation at the cellular level.

- Cerebral perfusion: return to normal or baseline behavior and LOC; normal vital signs, reflexes, and pupillary reactions; return of normal or baseline motor skills.
- Cardiopulmonary perfusion: vital signs normal; respirations unlabored; CRT < 2 seconds; return to baseline behavior.
- Renal perfusion: normal volume of clear, yellow urine; normal lab values; normal BP.
- GI perfusion: appetite normal with no signs of GI upset; stool normal volume, color, and consistency; normal bowel sounds.
- Peripheral perfusion: skin normal color and temperature; absence of pain and edema; normal pulses.

Interventions with Rationale/Amplification
Cerebral Perfusion
1. Provide supplemental oxygen as ordered. Monitor and record oxygen saturation levels.
2. Monitor and record patient's vital signs every 4 hours or more frequently as ordered by the veterinarian. Notify the veterinarian immediately if vital signs deteriorate. *Inadequate perfusion to cerebral tissues can affect all body systems.*
3. Monitor and record patient's behavior, LOC, reflexes, pupillary reactions, and motor skill levels. Notify the veterinarian immediately of any sudden change in status. *Trending of these parameters over time will assist the veterinarian in determining whether the patient is responding to therapy.*
4. Administer medications and fluids as ordered by the veterinarian. *Various medications such as corticosteroids and embolytics may increase cerebral perfusion. Animals with decreased cerebral perfusion may be unable to self-hydrate due to changes in LOC.*

Cardiopulmonary Perfusion
1. Provide supplemental oxygen as ordered. *Supplemental oxygen can increase oxygen saturation of the blood, providing optimum oxygenation to the heart and lungs.*
2. Monitor and record vital signs, including CRT and oxygen saturation, every 4 hours or more frequently as needed. Notify the veterinarian immediately if vital signs deteriorate. *Inadequate perfusion to the heart or lung tissue may induce changes in vital signs.*
3. Note and record signs of respiratory distress, including flaring nares, use of accessory muscles, orthopnea, and hemoptysis. Notify the veterinarian immediately of signs of increased respiratory distress. Hemoptysis signifies an extreme emergency. *Inadequate perfusion to the lungs often results in signs of respiratory distress.*
4. Observe and record patient's behavior every 4 hours or more frequently as ordered by the veterinarian. *Patients experiencing decreased cardiopulmonary perfusion often exhibit marked restlessness and anxiety.*
5. Administer medications and fluids as ordered by the veterinarian. *Medications such as nitroglycerin, beta blockers, ACE inhibitors, and antiarrhythmics may increase cardiac perfusion. Antiembolytics and diuretics may be used to increase pulmonary perfusion.*

Renal Perfusion
1. Monitor and record patient's intake and output every 8 hours or more often as directed by the veterinarian. Also note color and clarity of urine. *Volumes less than 0.27 mL/kg/h are suggestive of oliguria.*

Normal urine production volumes are as follows: cat, 10–20 mL/kg/day; dog, 20–80 mL/kg/day.
2. Monitor and record patient's vital signs every 4 hours. *Decreased kidney perfusion can result in significant alterations in BP, which in turn impacts heart rate and respiratory rate.*
3. Monitor lab values daily. *High BUN and Creatinine indicate current or impending renal failure. As perfusion increases, these values should trend toward normal.*
4. Administer medications and fluids as ordered by the veterinarian. *Medications that increase cardiac output may result in a concurrent increase in renal perfusion. Diuretics also may be ordered to increase urine production. Fluid volume is regulated to provide maximum renal perfusion.*

GI Perfusion

1. Monitor and record patient's stool output every 8 hours. Also note color and consistency of stool. *Decreased volume of stool, melena, constipation/diarrhea, and alterations in frequency of bowel movements can be signs of decreased GI tissue perfusion.*
2. Monitor for and record signs of abdominal distress, including change in eating habits, abdominal distention, or pain in the affected area. *These are all signs of decreased GI perfusion.*
3. Monitor and record patient's bowel sounds every 8 hours. *Decreased GI tissue perfusion may result in hypoactive or absent bowel sounds.*
4. Administer medications and fluids as ordered by the veterinarian. *Various medications such as stool softeners and laxatives may aid in re-establishment of normal bowel activity. Antibiotics and analgesics also may be considered. Adequate hydration is necessary for normal stool consistency.*

Peripheral Perfusion

1. Monitor and record skin color and temperature, edema, pulse strength, and tenderness/pain in the extremities.

Edema may be quantified using the following scale:
- *0 = no edema*
- *1 = mild*
- *2 = moderate*
- *3 = severe*

Pulse strength may be quantified using the following scale:
- *0 = no pulse*
- *1 = weak pulse*
- *2 = normal pulse*
- *3 = pounding pulse*
- *D = pulse present but determined by Doppler*

Pain may be quantified with the following scale:
- *0 = no pain*
- *1 = mild, intermittent discomfort*
- *2 = mild pain*
- *3 = moderate pain*
- *4 = severe pain*
- *5 = extreme pain*

2. Administer medications as ordered by the veterinarian. *Analgesics may be ordered for pain in the affected extremities.*

DIARRHEA

Definition/Characteristics

Diarrhea is the presence of loose, unformed stool. It is characterized by abdominal pain, bloating, increased defecation frequency and urgency, foul fecal odor, abnormal color of feces, and changes in behavior. Bowel sounds may be hypo- or hyperactive. Frequent passage of small amounts of stool is common, and previously continent animals may exhibit bowel incontinence during episodes of diarrhea. Causes of diarrhea are numerous and can include microorganisms, parasites, irritation to the digestive tract (e.g., certain medications, chronic inflammatory processes, and sudden change in diet), or stress. Diarrhea may be accompanied by dehydration, electrolyte imbalances, and metabolic acidosis.

Desired Resolution

Production of a normal amount of softly formed stool, free from abnormal odors, mucus, or blood. Animal is continent of stool.

Interventions with Rationale/Amplification

1. Monitor and record the animal's vital signs, including heart rate, respiratory rate, CRT, mucous membrane color, skin turgor, and temperature. *Frequency of vital monitoring will depend on the severity of the diarrhea. Dehydration may result in decreased BP, increased heart rate, and increased respiratory rate. Metabolic acidosis may result in increased respiratory rate as the body attempts to compensate by "blowing off" excess carbon dioxide. The animal may have an increased temperature when diarrhea is caused by an infectious microorganism.*
2. Monitor and record frequency, volume, and consistency of diarrhea. Specifically note the presence of blood, mucus, or foreign materials. *Characteristics associated with small intestine diarrhea can include watery, profuse stool and presence of melena. Characteristics associated more with large intestine diarrhea include hematochezia, mucus, and increased frequency of defecation.*

3. Monitor and record presence of tenesmus. *Straining to defecate is typically associated with large intestine issues.*
4. Observe for signs of electrolyte imbalances. *Sodium and potassium loss occurs secondary to severe diarrhea. Clinical signs can include weakness or dysrhythmias.*
5. Weigh the animal daily. *Severe diarrhea can result in rapid fluid loss and dehydration. A 500 cc fluid loss is equivalent to a 1 lb loss of body weight.*
6. Examine perianal area for evidence of fecal scalding or irritation at least 2 times per day. *Frequent defecation or diarrhea can irritate the perianal tissues. Always keep the perianal area clean and free from feces. Appropriate cleaning and protection of perianal tissues is imperative. (Apply petroleum jelly or another scald barrier.)*
7. Administer fluids as directed. Record rate, route, and volume of fluids administered. *Fluids can be administered orally, subcutaneously, or intravenously. The route is influenced by the degree of dehydration and presence of vomiting. Diarrheas can be associated with increased fluid loss and dehydration. Dehydration secondary to prolonged or severe diarrhea is especially dangerous for very young or elderly animals.*
8. Obtain laboratory specimens for diagnostics tests as ordered by the veterinarian. *Commonly performed tests include fecal floatation, direct smears, parvovirus cite test, and bacterial cultures of fecal specimens. Blood samples for CBCs, electrolyte levels, and serum chemistries also may be warranted. Hemoglobin and hematocrit (H&H) may be increased due to concentration of blood; WBC may be elevated if microorganisms are present.*
9. Administer medications as ordered by the veterinarian. *Various medications may be ordered depending on the cause of the diarrhea. Diarrhea caused by microorganisms may be treated with antibiotics/antimicrobials. Parasite infestation may be treated with antihelmentics. Inflammatory processes may be treated with corticosteroids. Pancreatic enzyme supplementation may be used in cases of maldigestion. Diarrhea of any origin may be treated with antidiarrheals.*
10. Provide appropriate diet. *Animals experiencing diarrhea may initially have food withheld to rest the GI tract. Vomiting animals should be placed on an n.p.o. status. Appropriate diets for diarrheal conditions are bland and contain highly digestible protein, low fiber, and relatively low fat. Examples of suitable commercial diets include Hill's i/d, Purina EN, and Royal Canin DIGESTIVE LOW FAT LF. Food should be provided in small, frequent meals with no restriction on water. The animal's normal diet can be gradually reintroduced 2–3 days after the diarrhea has resolved.*
11. Use appropriate precautions to prevent spread of infectious or zoonotic diarrheas. *Hands should be thoroughly washed between contact with all veterinary patients. Use of designated laboratory coats or other*

garments in cases of suspected infectious diarrhea is highly recommended. Housing areas and bedding should be meticulously disinfected between patients. Dogs brought outside to defecate should be restricted to designated areas away from healthy animals.
12. During patient admission, carefully question owner regarding onset, duration, and frequency of bowel movements and changes in the animal's diet, routine, or environment. *This information will aid the veterinarian in determining the cause of the diarrhea.*
13. Educate owner regarding medications dispensed, diet, and follow-up care. Ensure that owner understands all instructions. *Providing both written and oral instructions will increase the likelihood of owner compliance.*

ELECTROLYTE IMBALANCE

Definition/Characteristics
Electrolyte imbalance is an alteration in normal electrolyte values. Imbalances can result from an increase or decrease in an electrolyte level.

Desired Resolution
Electrolyte values within normal limits.

Interventions with Rationale/Amplification
1. Collect samples and conduct tests as directed by the veterinarian to monitor electrolyte levels. Notify the veterinarian of results and document in the patient chart. *Many conditions warrant electrolyte sampling every 8 hours. Conditions frequently associated with electrolyte changes include the following:*
   - *Hyponatremia: diarrhea, heart failure, and vomiting*
   - *Hypernatremia: renal failure and diarrhea*
   - *Hyperkalemia: decreased urine production, renal failure, urinary obstruction, and Cushing's disease*
   - *Hypokalemia: anorexia, administration of fluids low in potassium, vomiting, diarrhea, renal disease, diuresis post urinary obstruction*
   - *Hyperphosphatemia: diabetes, insulin therapy, and parenteral nutrition*
   - *Hypomagnesemia: administration of fluids low in magnesium and diarrhea*
2. Monitor for, record, and report any clinical signs associated with electrolyte abnormalities. *Clinical signs can include the following:*
   - *Hypernatremia: anorexia, apparent thirst, vomiting, muscular weakness, behavioral changes, and seizures or coma*
   - *Hyponatremia: lethargy, nausea, and weight changes*

- *Hyperkalemia: muscle weakness, ECG changes (peaked narrowed T waves, shortened QT interval, and widened QRS interval), and arrythmias*
- *Hypokalemia: muscle weakness, ventroflexion of head, forelimb hypermetria, arrythmias, and PU/PD*
- *Hypocalcemia: muscle twitching, face rubbing, stiff gate, behavioral changes, ataxia, and seizures*

3. Administer medications as directed by the veterinarian and record in patient chart. *Treatments ordered by the veterinarian can include the following:*
   - *Hypophosphatemia: intravenous (IV) potassium chloride*
   - *Hypernatremia: IV glucose solutions, sodium bicarbonate, calcium gluconate, and diuretics*
   - *Hyponatremia: isotonic or hypertonic (extreme cases) saline*
   - *Hypocalcemia: calcium gluconate and vitamin D*

EXERCISE INTOLERANCE

Definition/Characteristics

Exercise intolerance is the physical inability to engage in normal or desired activities. Exercise intolerance is characterized by changes in vital signs (increased heart rate, respiratory rate, and BP) and shortness of breath with exertion, fatigue, muscle weakness, and/or changes in daily behavior (e.g., increased sleeping/resting and decreased play).

Desired Resolution

Animal exhibits sufficient physical energy to engage in normal or baseline activity.

Interventions with Rationale/Amplification

1. Obtain complete history of exercise intolerance from owner, including onset (sudden versus progressive), duration, and symptoms noticed by owner. Also confirm heartworm status (i.e., whether the animal is on a preventative program). *This information will aid the veterinarian in making a medical diagnosis regarding the underlying cause of exercise intolerance.*
2. Monitor and record the animal's vital signs. *This information will assist the veterinarian in developing a treatment plan for the animal.*
3. Provide supplemental oxygen as ordered by the veterinarian. *Supplemental oxygen may increase oxygen saturation in the blood, thus decreasing shortness of breath.*
4. Administer medications as ordered by the veterinarian. *Depending on the cause of activity intolerance, many different medications may be ordered by the veterinarian.*

5. Educate owner regarding the animal's physical limitations. Explain exercise plan to owner. *An owner who is educated about the daily treatment plan is more likely to be compliant.*

FEAR

Definition/Characteristics

Fear is a distressing emotional response to a threat. The threat may be real (e.g., a predator) or imagined (e.g., blowing paper). Fear response may be innate, such as fear of a predator, or may be a learned response, such as fear of a veterinarian. Fearful animals can exhibit changes in facial expression, decreased response to environmental stimuli other than the threat, and heightened response to all stimuli associated with the threat. The fight-or-flight response to fear induces physiological changes that include increased heart rate, respiratory rate, and BP; dilated pupils; panting; diaphoresis; and/or nausea/vomiting.

Desired Resolution

Animal controls fear response to the extent that physiological changes are minimized and attending personnel are not harmed.

Interventions with Rationale/Amplification

1. Identify and remove the source of fear if possible. *Common inciters of fear within the veterinary facility include loud equipment such as clippers and vacuums, unfamiliar people, and barking or noisy animals. The following techniques can be used to minimize these issues.*
   - *Board or house cats separate from dogs. Separate birds from all other species.*
   - *Use the animal's name whenever possible to establish rapport/familiarity with the animal.*
   - *Wait an appropriate length of time for the drug to take effect if a premedication or sedative is used.*
   - *Allow animals to visualize and hear loud vibrating equipment such as clippers prior to applying to the animal.*
   - *Work slowly and deliberately around the animal. Do not attempt to rush the animal into the procedure.*
   - *Limit the animal's exposure to high-traffic, noisy areas of the clinic.*
   - *Some animals may benefit from "nonmedical treat trips" to the clinic. Puppies and kittens in particular can benefit from friendly, nonthreatening trips as a means of lessening the incidence of conditioned fear responses.*
2. Note in the animal's medical record any extremely fearful responses. *Fear can artificially alter laboratory values such as blood glucose and BP.*

If an animal is very distressed during blood draws or other procedures, that fact should be noted as it may affect the interpretation of test values. Medical notation also is important for purposes of alerting staff to modify future procedures or protocols, thus ensuring staff safety when an animal responds with fear-induced aggression.

3. Use appropriate restraint techniques. *Safety of owner and personnel is of utmost concern. Fear-induced aggression is frequently manifested by biting and scratching. The restraint technique selected should be based on the animal's behavior. Many animals respond to a calm, reassuring voice, while others may require being muzzled, placed in a cat bag, wrapped in a towel, or restrained using another measure. Animals that become highly aggressive due to fear may require tranquilization or sedation.*
4. Discuss long- and short-term management options with owner. *Some animals may require intervention to minimize the disruption of fear-based behaviors. Learned fear responses can sometimes be unlearned with proper training and relearning. Innate fear responses are more difficult to eradicate but may be lessened with training. Frequently cited causes include thunderstorms, fireworks, car rides, and certain family members or strangers. Treatment options include one or more of the following: avoidance, behavior modification therapy, or medication. Owners electing to pursue behavioral modifications should be referred to a qualified animal trainer/behaviorist.*
5. Administer medications as ordered by the veterinarian. *Certain medications such as sedatives, tranquilizers, and anxiolytics may reduce fear response.*

HYPERTHERMIA

Definition/Characteristics

Hyperthermia is an increase in normal body temperature. Hyperthermia may be associated with increased respiratory rate, tachycardia, panting, warm skin, dry mucous membranes, altered laboratory values (electrolyte imbalance, increased PCV due to dehydration, or metabolic acidosis), seizures/convulsions, and changes in LOC. Hyperthermia can range from a low-grade fever to a life-threatening condition. (Fever is an increase in body temperature in response to infection, an immune-mediated disease, drug reactions, or adverse environmental conditions.)

Desired Resolution

Body temperature within normal range.

Interventions with Rationale/Amplification

1. Monitor and record temperature, pulse rate and character, respiratory rate and character, CRT, mucous membrane color, and LOC. *The frequency of monitoring will be ordered by the veterinarian and is dependent upon the severity of the hyperthermia. Trending of these parameters over time will indicate whether the animal is responding to treatment. Normal temperature values (°F): dogs, 99.5–102; cats, 100–102; ferrets, 100–102; horses, 99–101; cattle, 101–102. Animals exhibiting a change in LOC or seizures should be under constant supervision.*

   Extreme hyperthermia is a veterinary emergency; the veterinarian should be notified immediately if the animal's temperature is above 104°F.

2. Provide a cool environment for the animal; use fans to increase air circulation and bathe the animal in tepid or slightly cool (never cold) water. If bathing is not possible; apply ice packs or cold water to the groin, base of neck, and foot pads. *Cold water and/or ice should never be applied to the entire body of the animal as this may result in constriction of surface blood vessels, which causes an increase in body temperature. Tepid/cool rinses prevent vasoconstriction while promoting cooling. Ice packs and cold water may be applied only to areas of high blood flow and areas of cooling, such the groin (between the hind legs), base of neck, and foot pads.*

   In cases of life-threatening hyperthermia, a cool water enema can be administered to decrease core body temperature. However, this procedure will prevent further use of rectal thermometers for obtaining the animal's temperature.

   Measures should be continued until the temperature is less than 103°F (all species).

3. Monitor fluid intake and provide fluids at rate and route ordered by the veterinarian. *Hyperthermia is often accompanied by dehydration.*

4. Provide supplemental oxygen as ordered by the veterinarian. *Increased respiratory rate results in increased oxygen demand. Animals requiring more than ½ hour of oxygen therapy should receive humidified oxygen.*

5. Monitor fluid loss, noting color of urine. *Hyperthermia may cause dehydration, which results in anuria or oliguria. Animals suffering from dehydration often have concentrated, very dark urine or may not produce urine for a few hours.*

6. Obtain laboratory samples as requested by the veterinarian. *Common tests performed to assess causes and effects of hyperthermia include CBC, urinalysis, chemistry profiles, electrolyte levels, urinalysis, and serum titers. As the cause of hyperthermia resolves, these values should trend toward normal.*

7. Administer medications as ordered by the veterinarian. *Common medications used to treat hyperthermia include antipyretics, corticosteroids (prednisone), electrolytes, and antibiotics. Antibiotic response fevers should drop within 48 hours of administration of medication.*

HYPERVOLEMIA

Definition/Characteristics

Hypervolemia is an increase in intravascular, interstitial, or intracellular fluid. It is associated with dyspnea, increased respiratory rate, coughing, nasal discharge, adventitious lung sounds (crackles or rales), edema, rapid weight gain, JVD, abdominal distention, and/or increased BP (pounding pulse). Laboratory values show a decrease in urinary concentration (i.e., a decrease in specific gravity), and PCV and may show concurrent electrolyte imbalances. Changes in mental status, such as anxiety and/or restlessness, may occur. Hypervolemia can be associated with congestive heart failure, renal failure, hepatic cirrhosis, steroid use, psychogenic polydipsia, and iatrogenic water loading.

Desired Resolution

Reestablish normal fluid volume with accompanying improvements in lung sounds, return to normal BP and urine concentration, decrease in edema, stable weight, and return to baseline mental status.

Interventions with Rationale/Amplification

1. Obtain vital signs and assess for specific indicators of fluid overload every 4 hours or at intervals ordered by the veterinarian. *Critical conditions can warrant more frequent monitoring. Common signs of fluid overload include serous nasal discharge, increased breathing sounds, increased respiratory rate, distended jugular veins, moist coughing, frequent urination, edema, ascites, and exophthalmos. If equipment is available, monitoring of BP is warranted. Lungs should be auscultated every 2–4 hours when animals are placed on IV fluids.*
2. Weigh the animal daily. *Rapid increases in body weight are associated with fluid retention. Weight loss is an indication that the animal is mobilizing and excreting excess fluids. Each pound of weight loss indicates the reductions of approximately 500 cc of fluid.*
3. Monitor and record fluid intake and output daily. *As the body mobilizes excess fluid volume, daily output will exceed input. Output is typically estimated. Fluid loss occurs through urination, defecation, vomiting, and insensible loss such as respiration.*

4. Obtain samples and conduct tests as directed by the veterinarian. *Tests frequently run to assess fluid volume include PCV, total protein (TP), CBCs, urine specific gravity, and electrolytes. Laboratory values will trend toward normal as excess fluid is removed.*
5. Institute sodium and fluid restrictions as ordered by the veterinarian. *Low-sodium diets are often instituted because "water follows sodium" (i.e., an increase in sodium levels results in an increase in fluid volume levels; conversely, a decrease in sodium levels results in a decrease in fluid volume levels). Suitable diets include Hill's k/d.*
6. Administer medications as directed by the veterinarian and record in patient chart. *Diuretics such as furosemide are frequently administered. Diuretics help the animal mobilize and excrete excess fluid.*
7. Evaluate degree of edema daily using numeric scale. *Edema may be rated on the following scale:*
   - *0 = no edema*
   - *1 = mild edema*
   - *2 = moderate edema*
   - *3 = severe edema*

   Edema results when excess fluid moves out of the vascular system into interstitial spaces. As fluid volume excess is mobilized and excreted, edema decreases.
8. Evaluate edematous skin for signs of abrasion or trauma and protect from injury if necessary. *Edematous skin is very fragile and injures easily. Bandaging high pressure areas or providing additional padding may be warranted.*
9. Change body position of recumbent animals every 1–2 hours. *Fluid gathers in dependent portions of the body. Frequent changes in body position can help minimize dependent edema.*
10. Encourage walking or mild exercise if physically appropriate. *Skeletal muscle movement increases lymphatic return, thereby decreasing peripheral edema.*

HYPOTHERMIA

Definition/Characteristics

Hypothermia is a decrease in normal body temperature. It is associated with shivering, piloerection, decreased CRT, mucous membrane cyanosis, altered heart rhythm and rate (initially tachycardia, progressing to bradycardia as body temperature falls), altered respirations (initially hyperventilation, progressing to hypoventilation as body temperature falls), altered laboratory values (metabolic acidosis and altered ABGs), and changes in LOC.

Desired Resolution

Body temperature within normal range.

Interventions with Rationale/Amplification

1. Monitor and record temperature, pulse rate, respiratory rate, CRT, mucous membrane color, and LOC. *The frequency of monitoring will depend on the severity of the hypothermia. Normal temperature values (°F): dogs, 99.5–102.5; cats, 100–103. Trending of these parameters over time will indicate whether the animal is responding to treatment.*

   Severely hypothermic animals will have a decreased LOC.

   Animals suffering from postsurgical hypothermia should be monitored very closely. Procedures necessitating prolonged anesthetic time or surgeries requiring opening of the abdominal cavity have a higher likelihood of postsurgical hypothermia. Hypothermia increases the time it takes for anesthetics to be removed from the body.

   Neonatal animals such as those born via cesarean section are especially prone to hypothermia. These animals should be maintained in a warm environment (incubator) until the mother is fully recovered from anesthesia.

2. Provide a warm environment for the animal, dry the animal with warm towels if wet, and wrap in warm blankets if possible. *Exercise extreme caution when using heat lamps, hot water bottles and/or heating pads on a hypothermic animal. The resulting surface vasodilation may lead to hypotension and consequent shock. Thermal burns also are a possible consequence. These rewarming methods should not be used on "critically hypothermic" animals. Warm blankets are much safer.*

   Recumbent animals or animals with decreased LOC are unable to move away from excessive heat. All animals in the immediate postoperative period should be placed on warm blankets or provided an alternative heat source until fully recovered.

3. Provide warmed fluids at rate and route ordered by the veterinarian. *The animal should be given warmed fluids to drink if swallowing and gag reflex is intact. Intravenous fluids must be given judiciously as the hypothermic heart cannot respond rapidly to increased fluid volume. Never warm IV fluids in the microwave. Microwaving fluids can cause "hot pockets" to develop in the fluids, which may cause vascular damage.*

4. Provide supplemental oxygen as ordered by the veterinarian. *Respiratory effort may decrease with hypothermia. Supplemental oxygen provides optimum oxygen saturation of the blood. Animals requiring more than ½ hour of oxygen therapy should receive humidified oxygen.*

5. Monitor fluid intake and output, especially noting volume of urine produced. *Hypothermia causes decreased cardiac output, which may result in anuria or oliguria.*
6. Administer medication as ordered by the veterinarian. *Various drugs such as antiarrythmics and cardiac stimulants may be ordered to counter the effects of hypothermia.*

HYPOVOLEMIA

Definition/Characteristics

Hypovolemia is a decrease in intravascular, interstitial, or intracellular fluid. This state is associated with decreases in urinary output, pulse strength, and BP. Concurrent increases in urine concentration, CRT, PCV, and heart rate also are noted. Nonquantifiable changes include sunken orbits; dry, pale, or "muddy" mucous membranes; malaise; skin tenting; cool extremities; and changes in LOC.

Desired Resolution

Reestablish normal hydration and fluid balance with accompanying improvements in urine output, pink moist muocus membranes, CRT less than 2 seconds, good skin turgor, normal BP, and return to baseline LOC.

Interventions with Rationale/Amplification

1. Monitor and record fluid intake and output, paying close attention to color and approximate volume of urine produced. *Dark-colored urine is suggestive of continued dehydration. Input includes fluid provided through oral, subcutaneous, intravenous, intraosseus, or intraperitoneal routes of administration.*
2. Monitor and record ongoing fluid loss due to vomiting or diarrhea. *These volumes are typically estimated. The presence of blood, mucus, or foreign material in the fluids should be noted.*
3. Provide fluids at the rate and route directed by the veterinarian. *Common routes include oral, subcutaneous, and intravenous. The route selected reflects the severity of dehydration and presence of vomiting. Fluids can be classified as crystalloids (substances having small diffusible particles) and colloids (particles with high molecular weights that do not diffuse across the capillary membrane and therefore exert intravascular persistence). Crystalloids frequently used in veterinary medicine for fluid replacement include lactated Ringer's solution (LRS), normal saline, and Normosol-R. Common colloids include whole blood, packed red blood cells (RBCs), plasma, Hetastarch (hydroxyethyl starch), and dextrans (Dextran 70 and Dextran 40).*

Fluid requirements are typically calculated using the following formula:

$$\text{Total fluids (mL)} = \text{Rehydration fluids (mL)} + \text{Maintenance fluids (mL)} + \text{Ongoing losses (mL)}$$

$$\text{Rehydration fluid (mL)} = \% \text{ dehydration} \times \text{body weight (kg)} \times 1{,}000 \text{ mL/L}$$

$$\text{Maintenance fluid (mL)} = 60 \text{ mL/kg/day}$$

$$\text{Ongoing losses (mL)} = \text{Estimate of fluid loss through vomiting, diarrhea, etc. (mL)}$$

4. Weigh the animal daily. *Sudden weight changes reflect fluid loss. A pound of weight loss is typically associated with a 500 cc fluid loss.*
5. Monitor temperature, pulse, respiration, CRT, mucous membrane color, and skin turgor every 4 hours or at intervals ordered by the veterinarian. Specifically assess for tachycardia, a decrease in palpable pulse pressure, and urinary output or increases in CRT, skin tenting, or respiratory rate.

Dehydration can induce changes in many physical parameters.

The percent of dehydration is often estimated based on the following physical indicators:

- *Less than 5 percent dehydrated: no visible changes*
- *5 percent dehydration: dry mucous membranes*
- *6–8 percent dehydration: increased skin turgor and dry mucous membranes*
- *10–12 percent dehydration: increased skin turgor, weak and rapid pulse, increased CRT, dry mucous membranes, and depression*

6. Administer specific treatments and medications prescribed by the attending veterinarian. Notify the veterinarian if signs of dehydration persist or worsen after prescribed treatment. *Most animals appear noticeably brighter after rehydration.*
7. Record fluid intake and output every 8 hours or more frequently as ordered by the veterinarian.

IMPAIRED TISSUE INTEGRITY

Definition/Characteristics

Impaired tissue integrity is the loss of or damage to tissues constituting primary defenses against microorganisms, including integumentary, mucosal, and corneal tissues. This state is characterized by lesions, inflammation with erythema, purulent discharge, localized edema, pain at the affected location, and altered laboratory results (e.g., increased WBCs and positive wound cultures).

Desired Resolution

Intact integumentary, mucosal, and corneal tissues.

Interventions with Rationale/Amplification

1. Note and record characteristics including location, size, color, amount and type of discharge, edema, and temperature of surrounding tissues. *The initial assessment provides a baseline for monitoring the progression of the lesion.*
2. Assist the veterinarian in cleaning and dressing wounds. *Open lesions are a pathway for microorganisms to enter the animal's body. Upon initial presentation to the veterinarian, the wound should be cleaned and dressed if necessary.*
3. Change dressings, clean wounds (e.g., flushing and irrigation), and medicate (e.g., topical ointments and sprays) as ordered by the veterinarian. Note and record characteristics of lesion during dressing changes or on a daily basis when there is no dressing on the wound. Notify the veterinarian of any adverse changes in wound; if any are noted, have the veterinarian assess wound prior to application of dressing. Dispose of soiled dressings appropriately. Always use sterile or aseptic technique when dressing wounds and wash hands immediately at the conclusion of the procedure. *A variety of dressings may be used depending on the type/status of the wound. Follow the veterinarian's orders exactly for type and frequency of dressing change. Irrigation and flushing promote removal of necrotic tissue and exudates. Certain medications such as antibiotics may be applied to the wound to prevent infection. Monitoring the wound status will indicate whether healing is occurring. Proper disposal of soiled dressings prevents spread of infection.*
4. Administer medications as ordered by the veterinarian. *Certain medications such as antibiotics/antimicrobials may be ordered to prevent or treat infection that has spread systemically.*
5. Monitor lab results for findings that indicate presence of infection. *Increased WBCs may indicate infection, depending on cell populations. A positive wound culture aids in identifying and choosing appropriate medications to treat infection.*
6. Monitor the animal's temperature at intervals ordered by the veterinarian. *Fever may indicate presence/intensification of infection or activation of the body's defense system.*
7. Provide devices that prevent further injury to the wound. *Many animals instinctively lick wounds. An E-collar prevents animals from harming themselves.*
8. Educate owner regarding care of the wound, sign of infection, administration of medications, and circumstances that require the animal to be seen by the veterinarian immediately (e.g., dehiscence). *An educated owner is more likely to notice complications at an early stage. Timely intervention increases the likelihood of successful treatment.*

INAPPROPRIATE ELIMINATION

Definition/Characteristics

Inappropriate elimination is urination and/or defecation that occurs inside the home (and for cats, outside the litter box). This is associated primarily with cats and dogs as other species do not have unrestricted access to owners' homes. Inappropriate elimination can be classified as behavioral or territorial.

Inappropriate Elimination Feline Behavioral

Inappropriate elimination can be medically induced or can occur as a result of behavioral abnormalities.

Numerous medical problems can induce inappropriate elimination. Possible causes include urinary tract infections (UTIs), cystitis, idiopathic (interstitial) cystitis, diabetes, renal failure, irritable bowel syndrome, neurologic disease, or parasites. Medications such as steroids and antibiotics also can affect elimination. Cause assessment of inappropriate elimination has identified the following probabilities: 50 percent, cause is idiopathic; 10 percent, known medical cause is identified; 20 percent, medical cause is suspected; 10 percent, cause is due to fear of another cat; 10 percent, cause is related to a specific known event.

Abnormal behavioral elimination can be related to stress or to dislike of the litter box. Stress/anxiety-related problems are numerous and varied. Stress can be related to new people, objects, or situations in the home. Litter box aversion/dislike is typically related to litter type, depth of litter, frequency of cleaning, cleaners used, and box type. These conditions generally fall into the category of idiopathic cystitis, and inappropriate defecation may accompany the inappropriate urination.

Treatment is most likely to be effective when only one cat is in the household, the behavior has been exhibited for less than 1 month's duration, inappropriate elimination is limited to a few areas in the home, and the instigating cause has been identified and eliminated.

Desired Resolution

Cat uses appropriate area for elimination at a frequency that is acceptable to owner.

Interventions with Rationale/Amplification

1. Obtain laboratory samples as directed by the veterinarian. *Common laboratory tests used to rule out medically related causes of inappropriate elimination include urinalysis, fecal examination, CBC, and serum chemistries.*

2. Obtain a complete elimination history/pattern. *Specific questions that should be asked include the following: What is the age and sex of the cat? How many cats are in the household? How was the cat identified as being the inappropriate eliminator? What is the number and location of litter boxes? How frequently is the litter box cleaned? What chemical/protocol is used in cleaning the litter box? What style of litter box is used? What type of litter is used? Are liners used? In what areas does inappropriate elimination occur? How often does it occur? What chemicals/protocols are used in cleaning inappropriate sites? Have any changes taken place in the household (such as a new pet or child, divorce, or new carpet)? What measures have already been undertaken to control the problem? What resolution does the client want? Conclude by asking the client whether there is anything that you haven't specifically addressed that may be relevant to the case.*
3. Provide client education/recommendations based on results of history and medical diagnosis identified by the veterinarian. *The instituted therapy typically involves a combination of aversion and attraction principles. Aversion therapy is designed to decrease a behavior frequency, while attraction therapy is intended to increase the frequency of a behavior. Possible recommendations include the following:*

Attraction Therapies

- Increase number of litter boxes in home to N + 1. (N is the number of cats in the household.)
- Change location of litter box to preferred location or site that cat frequently visits. Placing the box on the site of inappropriate elimination is ideal. Do not place boxes near glass doors or other locations that allow the cat to see outside the home. Try to avoid high-traffic areas of the home.
- Determine the litter that the cat prefers by placing different types of litter in different boxes and seeing which one the cat uses most frequently. If needed, change litter type. Most cats prefer a nonscented clumping litter. Do not use plastic liners.
- Increase depth of litter to 2 inches.
- Use boxes with easy access that have low sides, especially when dealing with debilitated or arthritic cats.
- Use uncovered (nonhooded) boxes.
- Change and/or clean the litter box more frequently.

Decreasing Stress Therapies

- Provide hiding places in the home for the cat, such as cat condos and fabric dens.
- Provide for more predatory behaviors, such as using food release and chasing toys.

- *Play with the cat more frequently.*
- *Provide perches for the cat at different heights and locations in the home.*

Aversion Therapy

- *After cleaning up sites where the cat has urinated or defecated, use an odor neutralizer.*
- *Use aluminum foil to cover areas from which you want to deter the cat. Tape foil to the floor.*
- *Apply lemon or citrus scent to areas from which you want to deter the cat.*

4. Administer medication as directed by the veterinarian. *No drugs have been approved for treatment of inappropriate elimination in cats. Drugs that have been used to treat this problem include antidepressants/anxiolytics (Buspirone), tranquilizers, and hormones (DES and Ovaban).*
5. Assist owner with positive identification of the animal if needed. *Positively identifying the culprit cat in a multicat household is often difficult. The following "tracer" can be used to identify the correct cat. Place 6 ophthalmic fluorescent test strips in a gelatin capsule and administer to cat in the late afternoon. Scan home in the dark under ultraviolet illumination. (A Wood's lamp will work.)*

Inappropriate Elimination Feline Territorial

Territorial marking is most commonly associated with adult sexually intact animals. Cats demonstrate specific postures such as spraying on a vertical surface while standing, moving tail rapidly/shaking, and releasing very small amounts of urine. Marking behaviors are most frequently associated with intact males although dominant females may occasionally engage in the behavior. Territorial marking does not occur secondary to a disease process or medication.

Treatment of territorial elimination is most likely to be effective when only one cat/animal is in the household, the behavior has been exhibited for less than 1 month's duration, inappropriate elimination is limited to a few areas in the home, and the instigating cause has been identified and eliminated.

Desired Resolution

Cat discontinues marking behaviors.

Interventions with Rationale/Amplification

1. Obtain laboratory samples as directed by the veterinarian. *Although territorial marking is not medically induced, clients may initially present the animal as a behavioral eliminator. Common laboratory tests include urinalysis, CBC, and serum chemistries.*

2. Obtain a complete elimination history/pattern. *Refer to "Inappropriate Elimination Feline Behavioral" (page 93) for specific questions.*
3. Provide client education/recommendations based on results of history and medical diagnosis identified by the veterinarian. *Refer to "Inappropriate Elimination Feline Behavioral" (page 93) for additional recommendations. In addition, address the following:*
   - Consider castration or an ovariohysterectomy. Neutering the animal resolves the problem in 50–60 percent of cats. This treatment is most likely to be effective in younger animals.
   - Scold the cat only when "caught in the act." Do not use punishment "after the fact."
   - Place the cat's food and water dishes at the location the cat is marking.
   - Cover the marking site with clear plastic and spray with citrus orange. Apply carpet tape (double-sided tape) around the marking area.
   - Cover windows, especially the lower portion, as cats may be marking in response to seeing other cats outside.
   - Consider pulling carpet back from the subflooring and applying a sealant to the subflooring to fully eliminate odors that may draw the cat back to that location.
4. Administer medication as directed by the veterinarian. *No drugs have been approved for treatment of territorial marking in cats. Drugs that have been used to treat this problem include antidepressants/anxiolytics (Buspirone) and hormones (Ovaban).*
5. Assist owner with positive identification of the animal if needed. *Refer to "Inappropriate Elimination Feline Behavioral" (page 93) for technique protocol.*

Inappropriate Elimination Canine Behavioral

Inappropriate elimination in dogs is most frequently associated with a lack of or partial housetraining skills. Dogs with known housetraining skills that suddenly exhibit incontinence should have a full diagnostic workup as medically induced incontinence is highly probable in these situations.

Desired Resolution

Dog urinates and defecates in appropriate area at a frequency that is acceptable to owner.

Interventions with Rationale/Amplification

1. Obtain laboratory samples as directed by the veterinarian. *Common laboratory tests used to rule out medically related causes of inappropriate elimination include urinalysis, fecal examination, CBC, and serum chemistries.*

2. Obtain a complete elimination history/pattern. *Specific questions should include the following: What is the age of the animal? When did the problem begin? How many dogs are in the home? How many dogs have inappropriate elimination behaviors? What measures have already been undertaken to control the problem? What resolution does the client want?*
3. Provide client education/recommendations based on results of history and medical diagnosis identified by the veterinarian. Refer to "Urinary Incontinence" (page 116) and "Bowel Incontinence" (page 69) for interventions related to medically induced inappropriate elimination.

Housetraining recommendations for animals without medical problems:

- *Take the animal outside to eliminate immediately after meals, after play time, upon waking, after owner returns home, or upon noticing preelimination behaviors such as sniffing and circling.*
- *Automatically take the dog to designated area every 30–120 minutes. A kitchen timer works well as a reminder.*
- *Feed on a regular schedule.*
- *Institute a designated "outdoor potty area."*
- *Provide lavish praise after the dog eliminates outside. Do not play with the dog prior to elimination.*
- *Remain outside with the dog until urination or defecation occurs.*
- *Confine or crate the dog when owner is not home. Confinement time should be limited to 5 hours.*
- *Verbally reprimand the dog when it is "caught in the act." Do not physically punish the animal. Immediately take the animal outside to the designated potty area.*
- *Ensure that the animal is not present when the "accident" is cleaned. Use a commercial odor eliminator to minimize scent.*
- *Do not verbally or physically punish the animal "after the fact."*

Inappropriate Elimination Canine Territorial

Inappropriate urination is associated with a raised leg posture and small urine volume (a few drops to a short stream). Specific marking sites on a vertical surface are usually selected. Ninety percent of territorial marking is associated with males. The majority of males are intact or were intact when the problem began. Ease of identifying canine territorial markers makes testing and diagnostics for medical incitors often unnecessary.

Desired Resolution

Elimination of marking behaviors or reduction of inappropriate sites.

Interventions with Rationale/Amplification

1. Obtain a complete history/pattern of marking. *Most territorial markers initiated the behavior prior to 2 years of age. Specifically address the location of elimination and volume of urine.*
2. Provide client education/recommendations to reduce incidence of marking.
   - *Consider castration. Neutering males is 30–50 percent effective depending on the age and length of time since problem began.*
   - *Thoroughly clean and deodorize the marked areas. Use a commercial deodorizer. Apply a sealant on subflooring where carpet is present or reseal hard surfaces.*
   - *Scold or punish the dog only when "caught in the act." Place/take the animal to appropriate area for elimination.*
   - *Place food and water dishes in areas that are commonly marked.*
   - *Do not permit dog access to previously marked areas in the home unless under direct supervision.*

INEFFECTIVE NURSING

Definition/Characteristics

Ineffective nursing is the inadequate ingestion of nutrients by a nursing infant. Ineffective nursing may originate with maternal behavior or physiology (e.g., rejection of young and reduced milk flow) or with infant behavior or physiology (e.g., "dummy foal" and neonate that is too weak to nurse).

Desired Resolution

Infant obtains adequate nutrition from nursing to meet metabolic and growth needs.

Interventions with Rationale/Amplification

1. Educate owner regarding monitoring and care of infant animals, preferably prior to birth of infants. Most animal births do not occur in a clinical setting; therefore, owners of pregnant females need education to recognize potential problems. *An educated owner is more likely to notice signs of inadequate nutrition at an earlier stage. Timely interventions often can save the lives of fragile infants.*
2. Observe the female animal prior to parturition for signs of adequate milk production (i.e., teats enlarging with fluid). Also look for signs of leaking colostrum. *Most mammals will "bag up" just prior to giving birth. This is a sign that the animal is beginning milk production. Leaking of colostrum is not uncommon in certain species. However, if a substantial*

amount of colostrum is lost, the offspring may receive inadequate immune protection.
3. Observe offspring as soon as possible after birth; note both maternal and infant behavior. Notify the veterinarian immediately if the mother shows any signs of rejecting her young or if any infants do not nurse vigorously within ½ hour of birth. *Some female animals refuse to nurse their young and may even try to harm any offspring that persists in attempts to nurse. The veterinarian will determine what interventions are appropriate. Possible interventions include restraining the mother while the infant(s) nurse and removing infant(s) after each feeding, mildly tranquilizing the mother prior to nursing, or removing the young from the mother entirely and raising them on a bottle. Infants that do not attempt to nurse soon after birth should be thoroughly examined for physical abnormalities. Stimulation such as rubbing the infant with a rough towel may rouse it sufficiently to nurse. Bottle feeding should be a last resort. If bottle feeding is attempted, note carefully whether infant is swallowing formula; forcing fluids into an infant's mouth may cause aspiration.*
4. Observe offspring regularly for signs of adequate nutrition, including full and rounded abdomen, periods of activity/movement, "bright" facial expressions (if eyes are open), and weight gain. *The younger and smaller an animal is, the less it can tolerate inadequate nutrition. Accordingly, it is necessary to observe newborns more frequently than older infants. With multiple offspring, the smallest infants in particular need to be monitored as they may be unable to compete with larger siblings for their share of the mother's milk. If this occurs, larger infants should be removed periodically from the mother so that the smaller offspring can nurse without competition. On occasion, a mother is unable to provide adequate milk for all offspring. In this instance, the young may be supplemented with commercial formula.*

NONCOMPLIANT OWNER

Definition/Characteristics

Noncompliant owner is the failure of an owner to follow established recommendations regarding an animal's health care. Failure may be due to noncompliance (knowing disregard of plan of care) or to learning deficits (not understanding condition of the animal and/or treatment plan).

Desired Resolution

Owner verbalizes understanding of plan of care and implements plan effectively.

Interventions with Rationale/Amplification

1. Assess owner's level of knowledge regarding the animal's condition. Provide education as needed. Thoroughly review plan of care. Ask owner to repeat information in his or her own words. *On occasion, an owner may be given information in an initial visit but does not fully comprehend the information. Repetition increases the likelihood that owner will understand the animal's condition.*
2. Demonstrate any techniques owner may need in order to implement plan of care (e.g., giving oral medications, bandaging, and giving injections). Have owner return-demonstrate. *Many people learn better by seeing and doing rather than by hearing.*
3. In cases of obvious or stated noncompliance, reiterate the importance of following the plan of care. Review health consequences if plan is not followed. Ask owner if the veterinarian, technician, or clinic can do anything to help owner achieve compliance with plan of care. *The technician must be careful to avoid seeming judgmental. This may be difficult since most people who provide health care to animals have a deep-seated concern for the animals' welfare. However, becoming angry or judgmental is counterproductive as the owner may in turn reject all recommendations.*

OBSTRUCTED AIRWAY

Definition/Characteristics

Obstructed airway is the ineffective clearance of secretions and/or obstructions from the trachea/bronchial system, resulting in impeded airflow. Causes may be physiological or mechanical. Obstructed airway is associated with dyspnea, orthopnea, adventitious lung/bronchial sounds, gagging or coughing, increase or decrease in respiratory rate/rhythm, decreased oxygen saturation, and mucous membrane cyanosis. Behaviorally, the animal may exhibit restlessness and anxiety, with lethargy and coma following severe oxygen deprivation.

Desired Resolution

Reestablishment of patent airway, return to normal or baseline oxygen saturation levels and respiratory rate and rhythm, absence of adventitious airway sounds, and absence of mucous membrane cyanosis.

Interventions with Rationale/Amplification

1. Check airway for mechanical obstructions; assist the veterinarian in removal of foreign objects. If possible, position the animal for maximum airway clearance. *Foreign objects that the pet has tried to*

swallow may lodge in the airway and impede airflow. *A head-to-chest position also impedes air flow; extension of the head allows maximum airflow.*
2. Suction secretions that may be obstructing airway. *Thick mucus impedes optimum airflow.*
3. Assemble all necessary emergency equipment, including endotracheal tubes, tracheotomy sets, and suction catheters. *In emergency situations, easy access to emergency equipment improves delivery of care to the animal.*
4. Provide supplemental oxygen as ordered. Provide humidification as ordered. *Supplemental oxygen may increase oxygen saturation; humidification will aid in thinning secretions. Humidification may be increased with a room humidifier or nebulizer. Animals requiring more than ½ hour of oxygen therapy should receive humidified oxygen.*
5. Monitor and record oxygen saturation levels and other signs of decreased air flow, including cyanosis, dyspnea, adventitious lung sounds, and changes in LOC. Notify the veterinarian immediately of any decrease in oxygenation and/or lack of response to treatment. *Persistently low oxygenation indicates treatment has been ineffective and other options need to be considered.*
6. Monitor and record quality and quantity of respiratory discharge. Monitor and record vital signs. *A change in color and/or thickness of respiratory secretions may indicate onset of infection. Likewise, a rise in temperature may indicate infection.*
7. Provide fluids at rate and route ordered by the veterinarian. *Adequate hydration thins secretions from the respiratory tract, allowing the animal to expectorate more easily.*
8. Administer medications and treatments as ordered by the veterinarian. *Various medications aid in thinning secretions and dilating airways. Medications may be administered orally or intravenously or via nebulizer.*

OVERWEIGHT

Definition/Characteristics
Overweight is the intake of nutrients in excess of body requirements. Nutrient intake in excess of body requirements is characterized by increased body weight and fat accumulation, particularly over the rib cage and loin area.

Desired Resolution
Intake that meets but does not exceed nutritional requirements; animal at ideal body weight.

Interventions with Rationale/Amplification

1. Record current body weight and body condition score (BCS). Record the ideal body weight as determined by the veterinarian.
   - BCS 1: Very thin. Ribs clearly visible, severe waist, pelvic bones clearly visible and protruding, severe abdominal tuck.
   - BCS 2: Underweight. Minimal fat covering ribs, waist easily visible, minimal fat around base of tail, prominent abdominal tuck.
   - BCS 3: Ideal weight. Ribs palpable with moderate fat cover, waist visible, tail base smooth but pelvic bones palpable, abdominal tuck.
   - BCS 4: Overweight. Ribs difficult to palpate, very slight waist, no abdominal tuck, back slightly broadened.
   - BCS 5: Obese. Ribs extremely difficult to palpate with extreme amount of fat covering, no visible waist, no abdominal tuck with adipose hanging below abdomen, very broad back.

2. Determine and record the animal's daily caloric requirement. *Maintenance requirements for a mature animal are 60–85 kcal ME/kg of ideal body weight.*

3. Assist the veterinarian in selecting appropriate diet and determining amount/volume of the diet to be fed. *As a general rule, the calories consumed should be reduced to 60 percent (dogs) and 70 percent (cats) of the calories needed to maintain the ideal weight. This can be achieved by reducing the amount of the current diet fed or changing to a calorie-restricted diet.*

   Calculating the pet's caloric requirements should begin the process. This number is based on the target weight and a desirable rate of weight reduction of approximately 1 percent per week. Some pet food manufacturers provide tools that make these calculations quite easy. There are several types of low-calorie diets: those that are higher in protein and other essential nutrients in comparison to energy and those that are higher in fiber.

   Diets with low caloric density and low fiber:

   - *Royal Canin CALORIE CONTROL CC HIGH PROTEIN*
   - *Iams Reduced Calorie*

   Diets with low caloric density and high fiber:

   - *Hill's r/d*
   - *Royal Canin CALORIE CONTROL CC HIGH FIBER*
   - *Purina OM*

   The volume of food that can be consumed is determined by the caloric density of the diet selected. Highest client compliance is achieved when the client understands the exact amount and type of

food that should be given to the animal. Once the type and amount of food is determined, the owner should be instructed to give the food in three smaller meals per day instead of one feeding per day.
4. Recognize the psychological impact/implications of owning an obese animal and be sensitive to owners' feelings. *Owners of obese animals are often overweight themselves. In addition, many owners view their pets as a reflection/extension of themselves. Obesity is a delicate issue and should be addressed with great tact.*
5. Initiate a client education program that addresses health consequences of obesity, selection of appropriate diet and exercise regimes, setting of goals, establishment of a protocol for monitoring progress, anticipated time for weight loss, and helpful suggestions designed to minimize the discomfort/anxiety experienced by owners and their pets.
    - *Consequences of obesity: Obesity can contribute to increased risk during anesthesia, arthritis, intervertebral disc disease, degenerative joint disease, heart failure, exercise intolerance, diabetes, heat intolerance, pancreatitis, hepatic lipidosis, and increased susceptibility to infection. An owner who understands the consequences to his or her pet's health may be more compliant with dietary orders.*
    - *Selection of diet: See Item 3 under the previous heading "Interventions with Rationale/Amplification."*
    - *Exercise: Obese animals are reluctant to exercise. Low-impact exercise such as swimming is ideal but is often difficult to accommodate. Owners should strive to provide two or three short (5–10 minutes) walks per day, then gradually increase the length of each walk. Animals should never exhibit signs of heightened distress such as difficulty breathing or reluctance to move. Exercise is extremely difficult to enforce with cats; consequently, most weight loss programs depend on diet alone. However, you can encourage the client to play chasing games with the cat, recommend toys that allow the cat to use its predatory behaviors, and tell the client to place food in different locations around the house to encourage "hunting" behaviors that passively increase the cat's exercise.*
    - *Anticipated time for weight loss: Obtaining the ideal weight should take a minimum of 8–10 weeks; but depending on the amount of weight to be lost, it may take longer.*
    - *Protocol for monitoring progress: Advise owner to weigh the animal at home weekly and bring the pet to the clinic for a bimonthly "official" weight. Owners who keep a journal of the amount of exercise, volume of food, and weekly weights are highly motivated and most likely to succeed. Owners may keep a graph on each pet that shows the progress his or her pet is making. This graph also helps to identify lack of compliance or the need to readjust the caloric intake early in the program.*

- *Helpful suggestions:*
    - Low-calorie diets are less palatable, and the animal may initially refuse to eat the food. Mixing the new diet with the previous diet for the first week may ease the transition for the animal.
    - Measure the food to be fed one time during the day and feed only from that measured amount, no matter the number of meals fed during the day. Multiple feedings can help with weight loss.
    - Remove the animal from the kitchen when preparing or eating food.
    - Ask all family members to participate in the diet plan. Specifically address the need to stop feeding the animal treats and table scraps.
    - Educate the client that begging behaviors may have nothing to do with the hunger of the pet, but reflects the pets need for attention, especially on the part of dogs. Cats exhibit predatory behaviors regardless of their hunger.
    - For owners who insist on feeding the pet treats, advise using low-calorie treats made by Hill's, Royal Canin, or Iams. Help the client calculate the number of treats allowed during the day and deduct those calories from the food that goes into the bowl.
    - Low-calorie diets that are high in fiber are associated with an increased amount of stool due to the high fiber content.

POSTOPERATIVE COMPLIANCE

Definition/Characteristics

Postoperative compliance is the provision of continued care during the postoperative period. Adequate postoperative care ensures patient safety and minimizes discomfort following surgical intervention.

Desired Resolution

Patient successfully recovers from anesthesia and surgical procedure.

Interventions with Rationale/Amplification

1. Monitor and record vital signs. *Postoperatively, many animals have a low temperature. A normal body temperature helps hasten anesthetic recovery. An increase in heart rate and respiration rate may indicate the presence of pain.*
2. Administer any postoperative prescribed medications and diets. *Examples of medications include analgesics, antibiotics, and antimicrobials. Diets provided specifically for recovery from surgery and trauma include Royal Canin RECOVERY RS and Hill's a/d. Meeting pets' caloric requirements early in the postoperative process has been proven to speed recovery and diminish morbidity. During periods of trauma, including postoperative recovery, the pet cannot utilize carbohydrates well, making*

recovery-type diets extremely useful and effective. (Note: Some surgeries such as spaying and neutering may ultimately necessitate a reduction in caloric intake due to altered/reduced metabolic rates. Owners should be provided with long-term dietary recommendations at the time of discharge.)

3. Remove IV catheter when deemed appropriate by the attending veterinarian. *Wrap the catheter if it is to remain in place for a longer duration of time. If the patient is a persistent chewer, place an E-collar on the animal.*
4. Provide water when the animal no longer shows signs of anesthesia or sedation. *When the animal is judged to be capable of eating, verify with the veterinarian the appropriate amount and type of food.*
5. Monitor surgical site for any signs of self-mutilation. *Place an E-collar or wrap on the animal if warranted. Refer to "Self-Inflicted Injury" (page 112).*
6. Provide owner education. *Provide both oral and written home care instructions concerning postoperative patient care (e.g., no running or jumping). Instruct owner regarding signs of potential complications such as bleeding, swelling, infection, or pain. Schedule a checkup and/or appointment for suture removal. Owners should be given instructions prior to the animal being released.*

PREOPERATIVE COMPLIANCE

Definition/Characteristics

Preoperative compliance encompasses the activities undertaken to ensure appropriate preparation for anesthesia and surgery. Adequate preparation serves to positively identify the patient and correct surgical procedure; minimize anesthetic risks through preoperative tests, medications, or procedures; and ensure that all equipment is in working order.

Desired Resolution

Appropriate preparation for anesthesia and surgery.

Interventions with Rationale/Amplification

1. Educate owner regarding preoperative instructions. *If the animal will be dropped off the morning of surgery, owner must understand the importance of removing food and water prior to the surgical procedure. Owners also should be provided with a written description of the scheduled surgery. Consent and release forms must be obtained prior to owner leaving the animal.*

2. Obtain appropriate consent forms. *Surgery should never be performed without a signed owner consent form.*
3. Collect samples and perform laboratory tests as requested by the attending veterinarian. *Common survey and preanesthetic tests include CBC or PCV, TP, and ECG depending on age of the patient.*
4. Perform a preanesthetic check on all anesthetic machines and monitoring equipment. *Anesthetic machines should be examined daily for leaks and adequate levels of oxygen and anesthetic levels. Scavenging systems and CO_2 absorbers should be changed when indicated.*
5. Remove food and water at appropriate times. *Fast adult cats and dogs 12 hours prior to anesthesia; neonates and animals under 2 kg should not be fasted.*
6. Perform a complete physical exam. *Technicians frequently monitor animals under anesthesia. Baseline values should always be obtained so that values under anesthesia can be better interpreted.*
7. Clean the animal. *Make sure the coat is free of debris (fecal material, mud, dirt, grass, leaves, etc.) and pests (ticks and fleas). These can lead to contamination of the surgical site.*
8. Place an intravenous catheter as directed by the veterinarian. *Catheters provide easy, instant vascular access for anesthetic and emergency drugs. IV catheters may need to be wrapped to prevent damage from chewing. For persistent animals, an E-collar may be warranted.*
9. Administer preoperative medications prescribed by the attending veterinarian. *These can include prophylactic analgesics, antibiotics, antimicrobials, and sedatives.*
10. Provide an opportunity for the animal to relieve itself. *The bladder may also be expressed after anesthesia has been induced and the surgical site has been clipped.*

REDUCED MOBILITY

Definition/Characteristics

Reduced mobility is a decreased ability to ambulate or move extremities. This state is associated with changes in gait (e.g., tripping, shuffling, ambulating using a rolling motion, and abnormally moving extremities during ambulation), decreased range of motion in joints, decreased balance/stability, and/or signs of pain with movement. In extreme cases (e.g., hind-end paralysis), the animal may show signs of skin abnormalities (e.g., pressure ulcers and callus formation) resulting from continued, unrelenting pressure on sensitive areas; joint contractures and loss of muscle mass also may develop following long periods of disuse.

Desired Resolution

Animal ambulates with normal gait or maintains baseline gait; no signs of pain with movement; skin condition normal.

Interventions with Rationale/Amplification

1. Assess and record the animal's functional level. Note any gait abnormalities or balance problems. Obtain history from owner regarding onset, duration, and extent of immobility. *This information provides a baseline to assess the efficacy of treatment. Note that some cases of acute and chronic immobility are caused by veterinary treatment. For example, anesthetized animals are completely immobile during surgery and while recovering from anesthesia. During this period, the animal may develop pressure points on the body that receive reduced circulation. Reduced mobility lasting a substantial period of time may follow surgery or other veterinary interventions such as casting.*
2. Assess the animal for signs of skin breakdown, including non-blanchable reddened areas and open sores. If the animal is unable to change position independently, reposition the animal every 2 hours. If necessary, provide pillows, blankets, or other devices to support the animal's position. *Heavier breeds of dogs (e.g., German shepherds, Great Danes, and Newfoundlands) are particularly prone to skin abnormalities from prolonged immobility.*
3. Assess and record signs of pain with ambulation. *This information will assist the veterinarian in diagnosing the cause of decreased mobility.*
4. Encourage the animal to ambulate following surgery as ordered by the veterinarian. Encourage exercise in nonsurgical animals as tolerated. *Early postsurgical ambulation will assist the animal in maintaining normal body systems. Regular exercise of nonsurgical animals will have the same effect.*
5. Perform passive range-of-motion exercises as ordered by the veterinarian. *Joints that remain immobile for long periods of time may develop contractures or become permanently stiffened. Range-of-motion exercises can prevent contractures and maintain joint mobility.*
6. Administer medications as ordered. *Certain medications such as analgesics, corticosteroids, and NSAIDs may be ordered to treat conditions that contribute to impaired mobility.*
7. Educate owner regarding medication administration; skin care; and exercise, including range-of-motion exercises. Explain the rationale supporting each recommendation. *Many cases of impaired mobility are chronic and must be treated on a daily basis. An informed and educated owner is more likely to follow a plan of care for the animal.*

REPRODUCTIVE DYSFUNCTION

Definition/Characteristics
Reproductive dysfunction is the inability to produce viable offspring. Reproductive dysfunction may result from preconception problems (e.g., decreased libido and altered or absent estrus cycle) or postconception problems (e.g., spontaneous abortion).

Desired Resolution
Animal produces viable offspring.

Interventions with Rationale/Amplification
1. Obtain a complete reproductive history from owner.
    - Females: Include dates of estrus cycles, dates breeding occurred, previous breedings, pregnancies and/or births, and history of problem pregnancies/births (e.g., dystocia).
    - Males: Include previous breedings or attempted breedings, number of offspring, and history of sexual problems (e.g., decreased libido).

    This information will assist the veterinarian in determining the cause of reproductive dysfunction.
2. Assist the veterinarian in obtaining laboratory samples (e.g., sperm, vaginal swabs, and uterine culture/biopsy). Note results. *This information will assist the veterinarian in determining the cause of reproductive dysfunction.*
3. Administer medications and diets as ordered. *Certain medications such as hormones may be ordered to increase likelihood of producing viable offspring. Deficiencies in particular nutrients can decrease fertility and adversely affect litter size. Assess the type of diet the pet is eating prior to conception and owner's plans for diet during gestation and lactation.*
4. Educate owner regarding treatment and medications given to the animal. Explain necessity for keeping a detailed and accurate chronology of entire reproductive cycle. *An educated owner is more likely to follow instructions.*

RISK OF ASPIRATION

Definition/Characteristics
Risk of aspiration is an increased likelihood of food, fluids, or internal secretions entering the trachea. Absence of intact gag reflex and swallowing reflex is a major factor. Animals that have a decreased LOC from any cause may exhibit reduced or absent protective reflexes. In addition, any damage to the cranial nerves controlling gag and swallowing reflexes may increase the animal's risk for aspiration.

Desired Resolution
Intact gag and swallowing reflexes.

Interventions with Rationale/Amplification
1. Do not offer food or fluids to a sedated or anesthetized animal until return of gag reflex has been verified. *Verification is obtained by observing a strong swallow reflex. Animals that need hydration prior to return of gag reflex may be given subcutaneous or intravenous fluids.*
2. Animals with conditions leading to decreased or absent gag/swallowing reflex (e.g., senility and injury) can be given thickened liquids as these are less likely to be aspirated. *Syringeable and tubeable diets are available commercially; they include Royal Canin RECOVERY RS and Hill's a/d.*
3. Animals should be placed or held in natural eating/drinking position if possible prior to ingestion of food/fluids. *Recumbent animals are more likely to aspirate foods/fluids.*

RISK OF INFECTION
Definition/Characteristics
Risk of infection is an increased susceptibility to colonization by microorganisms. It is associated with impaired or altered primary defenses (e.g., broken skin, change in pH of gastric or vaginal secretions, and fluid stasis) and/or impaired altered secondary defenses (e.g., altered immune response).

Desired Resolution
Animal remains free of signs of infection, including fever, discharge, reddened or warm skin, inflammation, productive cough, and/or increased WBC.

Interventions with Rationale/Amplification
1. Note and record presence of risk factors for infection, including but not limited to broken skin, altered secretions, fluid stasis in lungs or bladder, immune suppression (either naturally occurring or medically induced through steroid or immunosuppression therapy), malnutrition, exposure to pathogens, and invasive procedures. *Any condition that impairs primary or secondary defenses increases the risk of infection.*
2. Administer fluids and medications as ordered by the veterinarian. *Antibiotics/antimicrobials may be administered prophylactically (e.g., post-surgically). Proper hydration allows the animal to mobilize secretions in the lungs, increases urine production, and increases perfusion of tissues.*

3. Monitor and record the animal's urinary output. If necessary, catheterize the animal as ordered by the veterinarian to prevent bladder distention. *Fluid stasis in the bladder increases the risk of a UTI.*
4. Provide reverse isolation if indicated by the animal's condition. *A severely weakened or immunosuppressed animal needs to be kept separate from other animals that may harbor pathogens.*
5. Use effective hand washing techniques prior to and after handling all animals. *Proper hand washing is a simple yet extremely effective method of preventing nosocomial infections.*
6. Assess the animal daily for signs of localized or systemic infection, including but not limited to fever, purulent discharge, productive cough, inflammation, pain, altered LOC, altered lab values (e.g., increased WBC count and changes in urine (cloudy or foul-smelling). Notify the veterinarian immediately if signs of infection are noted. *Timely intervention increases the likelihood of successful treatment of infection.*
7. Educate owner regarding risk factors for infection and signs of infection. *Owners who actively monitor the animal's status are more likely to identify problems at an early stage. Timely intervention increases the likelihood of successful treatment.*

RISK OF INFECTION TRANSMISSION

Definition/Characteristics

Risk of infection transmission is an increased likelihood of spreading microorganisms to another animal or a human. It occurs when a suspected or proven carrier of microorganisms is or may be exposed to uninfected and susceptible animals or humans.

Desired Resolution

Animal is free of contagious disease or will be isolated from other animals/human contact if disease cannot be effectively treated.

Interventions with Rationale/Amplification

1. Assess the animal for signs of infection, including but not limited to fever, anorexia, weakness, purulent discharge, diarrhea, cough, inflammation, altered consciousness, changes in behavior, increased WBC count, and positive blood tests for specific diseases. Notify the veterinarian immediately if signs of active infection are present. *It is important to identify carriers of microorganisms immediately to prevent spread of infection to other animals in the facility and to protect human caregivers from possible zoonotic infections.*
2. Institute proper level of precautions immediately. *Standard precautions include proper hand washing before and after handling the animal*

and use of gloves; contact precautions include hand washing and use of gloves and gowns; droplet precautions include hand washing and use of surgical mask and gloves; airborne precautions include hand washing and use of gloves and an N-95 mask. When contact or droplet precautions are instituted, the animal should be isolated from other animals. If airborne precautions are instituted, the animal should be isolated in a negative pressure room if available or housed in a separate building from other animals. The use of chlorine and water foot baths also helps to stop transmission.

3. Thoroughly sanitize and disinfect all surfaces with which the animal had contact, such as examining table, cage, and run. *Thorough disinfection reduces transmission of microorganisms. If cleaning is normally done by other staff, the technician should oversee the process of disinfection.*
4. Administer medications as prescribed by the veterinarian. *Antibiotics/antimicrobials may be ordered by the veterinarian. Certain diseases are not transmissible after a prescribed course of antibiotics/antimicrobials.*
5. Educate owner regarding diagnosis and possible outcomes for the animal (i.e., resolution of infection and carrier status) and risks to other animals and humans if the animal is not isolated. *Certain diseases (e.g., feline leukemia) have no effective treatment, and an infected animal remains a threat to other animals if strict isolation is not maintained.*

SELF-CARE DEFICIT

Definition/Characteristics

Self-care deficit is the lack of ability to maintain nutrient intake and hygiene through feeding, grooming, and/or toileting. Common causes include pain, impaired mobility, obesity, and depression. Common signs associated with self-care deficit include anorexia, greasy/matted or soiled hair coat, urine scald, and distended abdomen due to enlarged bladder or fecal impaction.

Desired Resolution

Maintain a clean, well-fed patient until it regains self-care ability.

Interventions with Rationale/Amplification

Feeding

1. Present food in an accessible manner.
   - *If the patient cannot ambulate, place water/food bowls within reach. For severely impaired animals, monitor during feeding for signs of aspiration. Promptly remove any bowls once the patient is finished.*
   - Elevate the bowls to assist animals that are unable to lower their heads to the floor due to pain, stiffness, or bandaging.

- Hand-feed depressed animals.
- **Provide** privacy for nervous animals to eat. *Placing a towel over the front of a cage can provide a degree of privacy for the animal.*
2. Weigh the animal daily. *These animals are at a high risk for weight loss and malnutrition.*

Grooming

1. Maintain a clean, unmatted, soil-free coat through brushing, bathing, and occasional clipping.

- Brush the coat as often as needed. *Frequent brushing helps distribute body oils throughout the coat and decreases the formation of mats.*
- Bathe and or spot-clean the animal as needed. *Initially, the animal may need a full bath with periodic touch-ups. Be sure to use the appropriate shampoo. The attending veterinarian may prescribe a medicated shampoo: colloidal oatmeal (Epi-Soothe) as a soothing and antipuritic shampoo, tar-sulfur (Lytar) for oily seborrhea, sulfur and salicylic acid (SebaLyt) for dry seborrhea, and benzoyl peroxide (OxyDex) for pyoderma.*
- Clip hair to a shorter length to help maintain a debris- and soil-free coat. *However, do not clip the hair without veterinary and owner approval.*

Toileting

1. Provide ambulatory support as needed for patients to eliminate outside. *A towel or a sling around the abdomen may be used to assist dogs. Caution should be taken not to cause further harm or injury to the animal.*
2. Express bladder as needed.
3. Place the animal on a grate and/or absorbent pads to minimize exposure to urine and feces. *Remove any collar to prevent injury to the animal.*

Owner Education

1. Discuss and establish a routine for long-term care. *An owner who is educated about the daily care required is more likely to be compliant.*

SELF-INFLICTED INJURY

Definition/Characteristics

Self inflicted injury is self-mutilation in response to stress, pain, or injury. Animals may inflict injury on themselves through biting, compulsive licking, excessive scratching, rubbing, or kicking. This state can also be associated with isolated or confined animals that lack an adequate outlet for energy and stress.

Desired Resolution

Animal remains free of any self-inflicted injury.

Interventions with Rationale/Amplification

1. Examine the animal for physical injury due to self-mutilation. Document location, severity, and stage of healing for all injuries. *Refer to various safety technician assessments (e.g., impaired tissue integrity) for treatment of injury.*
2. Obtain a complete history from owner, including date of onset of self-injury, duration of episodes, and severity and triggers of self-mutilation. *This information will assist the veterinarian in determining the cause of self-mutilation.*
3. Administer medications as ordered by the veterinarian. *Most cases of self-mutilation result from the animal's response to painful or irritating events such as surgery, foreign bodies, or allergies. Medications such as analgesics (opiod, NSAID) antihistamines, or tranquilizers may be ordered to reduce or eliminate episodes of self-mutilation.*
4. When appropriate, employ physical measures to prevent self-injury (e.g., placing an E-collar and lightly bandaging area of mutilation). *Letting the lesion heal often limits further mutilation.*
5. Educate owner regarding treatment plan, including recommended exercise, socialization, diet, medication, and management techniques. Explain the rationale for each recommendation. *An educated owner is more likely to adhere to the treatment plan.*

SLEEP DISTURBANCE

Definition/Characteristics

Sleep disturbance is the lack of adequate and/or high-quality sleep due to internal stimuli (e.g., pain) or external stimuli (e.g., intensive care unit [ICU] environment). The animal may exhibit behavioral changes, altered sensory responses (including increased pain response), lethargy, and delayed wound healing.

Desired Resolution

Animal receives adequate high-quality sleep and exhibits no signs of sleep deprivation/disturbance.

Interventions with Rationale/Amplification

1. During admission examination, note internal stimuli that may lead to disturbed sleep, such as trauma or injury causing pain and shortness of breath in recumbent position. Refer to various technician assessments for appropriate interventions. *Physical causes that lead to sleep disturbance should be the first priority in treatment.*

2. Reduce external stimuli that prevent the animal from sleeping. *ICUs in clinics/hospitals and loud kennels are especially disturbing to animals. Actions to reduce the impact on the animal include dimming lights during periods of natural darkness to simulate night (except for periods when care is provided to the animal); reducing noise; clustering care activities so that the staff disturbs the animal less frequently; allowing the animal to have articles of bedding, toys, and/or pieces of owner's clothing from home (reduces psychological stress); and/or providing adequate exercise if tolerated by the animal. Cover the cage door to provide privacy for nervous cats.*

STATUS WITHIN ACCEPTABLE PARAMETERS

Definition/Characteristics

Status within acceptable parameters acknowledges that the animal's state or its reaction to the prescribed treatment, procedure, surgery, or general status meets the required expectations for the individual case. Example scenarios include a postspay incision that heals without any incidences, a diabetic animal that is well regulated, a boarding animal that does not exhibit any signs of disease, and an infection that appears to be resolving with appropriate treatment.

Desired Resolution

Maintain the animal's status within acceptable parameters.

Interventions with Rational/Amplification

1. Continue to implement initially established interventions until further notice from the attending veterinarian. *It is imperative to complete the prescribed treatment and medications. Careful monitoring of animals remains vital to their continued successful recovery.*
2. Monitor identified parameters for any changes in status.

UNDERWEIGHT

Definition/Characteristics

Underweight is the intake of nutrients below body requirements and can be characterized by decreased body weight, slow healing, visible protrusion of spinal processes and ribs, altered patterns of bowel elimination, and weakness.

Desired Resolution

Intake that meets nutritional requirements; animal at ideal body weight.

Interventions with Rationale/Amplification

1. Record current body weight and BCS. Record the ideal body weight as determined by the veterinarian. *Thorough documentation of malnutrition cases associated with impoundment or prosecution is advised. Photographs are warranted in cases involving animals impounded by the state and brought to a veterinary facility.*
   - *BCS 1: Very thin. Ribs clearly visible, severe waist, pelvic bones clearly visible and protruding, severe abdominal tuck.*
   - *BCS 2: Underweight. Minimal fat covering ribs, waist easily visible, minimal fat around base of tail, prominent abdominal tuck.*
   - *BCS 3: Ideal weight. Ribs palpable with moderate fat cover, waist visible, tail base smooth but pelvic bones palpable, abdominal tuck.*
   - *BCS 4: Overweight. Ribs difficult to palpate, very slight waist, no abdominal tuck, back slightly broadened.*
   - *BCS 5: Obese. Ribs extremely difficult to palpate with extreme amount of fat covering, no visible waist, no abdominal tuck with adipose hanging below abdomen, very broad back.*
2. Determine and record the animal's daily caloric requirement. *Maintenance requirements for a mature animal are 60–85 kcal ME/kg of ideal body weight.*
3. Select appropriate diet and determine amount/volume of the diet to be fed. *Malnourished animals should be placed on calorically dense diets. Suitable diets are high in fat and protein, easily digestible, and very palatable. Examples include diets such as Hill's a/d and p/d and Royal Canin RECOVERY RS, INTESTINAL HE, EARLY CARE Puppy, and HYPOALLERGENIC HP. Although it is not recommended, a second alternative is to feed more of the current diet.*

   Malnourished animals benefit from several smaller meals given throughout the day. Palatability should be a major concern as high palatability induces the animal to ingest more nutrients. Hospitalized animals can be enticed to eat via heating of canned food or adding low-sodium broth to dry foods.

   Supplements such as vitamins and minerals or probiotics also may be ordered. Once the optimal body weight has been achieved, a new diet may need to be selected to prevent excessive weight gain.
4. Note and record quality and quantity of the animal's stool output. *Extremely malnourished animals may initially produce a less-than-normal amount of stool. Introduction of increased nutrients may result in diarrhea until the animal's digestive system acclimates to the new diet.*
5. Weigh the animal daily if hospitalized. *Trending of the animal's weight over time will enable the veterinarian to determine whether treatment is effective. Animals should be weighed on a weekly basis when sent home. Owners should record weight changes to assess success of increased nutrition.*

6. Educate owner about the diet necessary for the animal to reach and maintain ideal body weight. Also provide education about health consequences of malnutrition. *The vast majority of malnutrition results from illness-related anorexia/cachexia or unintentional neglect secondary to lack of understanding on owner's part.*

   Malnutrition can contribute to increased risk during anesthesia, heart failure, exercise intolerance, heat intolerance, ketosis, and increased susceptibility to infection and increases the healing time for any disease. Remember, an owner who understands the consequences to his or her pet's health may be more compliant with dietary orders.

   Lactating females are especially at risk for this condition due to the high caloric requirements of milk production. If possible, educate owners during the animal's pregnancy regarding the nutritional demands of lactation.

URINARY INCONTINENCE

Definition/Characteristics

Urinary incontinence is the inability to voluntarily control urinary elimination and is often noted as dribbling/dripping urine or "urge" urination (the constant need to urinate). Incontinence can stem from neurogenic or nonneurogenic causes. Neurogenic causes involve the nervous system, whether trauma-related or genetic. Examples include trauma to the spinal cord, disc disease, pelvic or pudendal (external genital) nerve lesions, atonic bladder, and dementia. Nonneurogenic can include UTI, ectopic ureter, urethral sphincter incompetence, or neoplasia.

Incontinence may be total (complete loss of control) or partial, such as when the animal is subject to emotional stress. Total incontinence is associated with unpredictable, sometimes constant dribbling of urine. Partial incontinence is often characterized by loss of urine in somewhat predictable patterns (e.g., when the animal is disciplined or excited and when the animal is not toileted for several hours during overnight period).

Desired Resolution

Animal achieves voluntary control of urinary elimination or achieves a pattern of elimination that is acceptable to owner.

Interventions with Rationale/Amplification

1. Obtain a complete history from owner, including times and amounts of urinary output, length of time problem has been occurring, and circumstances surrounding incontinence. *A complete history will assist the veterinarian in determining the type of incontinence (total versus partial) and*

possible management techniques. *The most common forms of incontinence are urge/inflammatory and hormone-responsive.*

Partial incontinence associated with behavioral or submissive urination typically involves incontinence when owners are greeting the animal after an absence of several hours. The animal's body language is usually submissive with the head lowered, ears flat, and the abdomen close to the ground. Alternatively, the animal may roll over onto its dorsum.

2. Monitor for urine scald and implement environmental modifications as needed to minimize contact with urine. *Animals that are totally incontinent are particularly susceptible to irritation in the perineal area. A mild soap and water mixture can be used to clean away urine and urine stains on the animal. Once the animal has been cleaned and dried, a protective layer such as petroleum jelly can be applied to prevent urine scald. To prevent contact with urine, animals can be elevated off the cage floor using a PVC-coated grate. When using a grate, remove the collar to prevent injury to the animal. All bedding should be frequently monitored for wetness and replaced as needed.*

3. Monitor the urine for blood, discoloration, or other signs of UTI. *Urine should be a straw yellow color with no clouding (sediment) and a slight odor. Concentrated urine is often dark yellow with a strong odor. Underconcentrated urine resembles water, with no color or smell. Visible blood causes the urine to appear red or brownish in color. Small traces of blood and biochemicals can be found by using reagent strips/dipsticks such as Multistix.*

4. Obtain a urine sample for analysis and culture as directed by the veterinarian. *UTIs may cause a previously continent animal to develop incontinence. Physical causes for incontinence should always be ruled out or treated if found. Urine samples may be collected by observing the animal and catching a sample or by needle aspiration of the bladder. Sterile samples must be collected via cystocentesis or catheterization. Voided samples obtained via free catch or manual expression have a very high incidence of contamination. Urine samples should be processed within 30 minutes of collection. Samples that cannot be processed immediately can be refrigerated for up to 4 hours. Urine evaluation typically involves evaluation of physical and chemical properties, sediment analysis, and culture. Quick evaluations can be performed using dipsticks, also known as reagent strips. These test strips provide a quick urine evaluation for the presence of biochemicals such as ketones and glucose, along with the presence of protein, blood, and leukocytes.*

5. Estimate and record urine output. *In addition to volume, note the timing and frequency of urination. The normal daily urine output for dogs and cats ranges from 20–45 mL/kg/day.*

6. Palpate and express bladder as ordered by the veterinarian (often ordered t.i.d.). *Whenever incontinence is noted, the technician should*

palpate the bladder and determine whether the incontinence is accompanied by a large, distended bladder or a nonpalpable bladder. A history of bladder size relative to periods of incontinence can assist the veterinarian in diagnosis. Animals suffering from spinal cord injuries should be expressed three times per day or catheterized. Strict aseptic techniques should be followed.

7. Place and maintain urinary catheter as directed by the veterinarian. *An animal unable to adequately express the bladder may require catheterization. A urinary catheter greatly increases the chance of a UTI. Animals with urinary catheters should wear an E-collar and be closely monitored to ensure patency of the catheter. Catheterized animals should be palpated q.i.d. to ensure adequate drainage of the bladder. Alternatively, a closed system may be used to monitor output.*
8. Administer/dispense medications as ordered by the veterinarian. *Certain medications such as antibiotics/antimicrobials may be ordered by the veterinarian to treat UTIs. Salves or ointments may be ordered to treat urine scald. Steroids may be indicated in certain neurologic conditions, and various hormones may be used to treat hormone-responsive incontinence.*
9. Provide educational materials and discuss possible management techniques with owner. *Partial incontinence may be managed by toileting the animal more often and more regularly, by limiting fluids at certain times of day (e.g., after evening meals), and by eliminating sources of stress/excitement that provoke incontinence (e.g., excessive discipline and exciting play).*

    Recommendations for eliminating behavioral or submissive incontinence include:
    - *Keeping the greeting brief and calm.*
    - *Arranging for greetings to take place outside the home.*
    - *Not greeting the dog immediately upon return. Ignore the animal until the dog has acclimated and is calm; then acknowledge the animal.*
    - *Not picking up the dog or attempting to soothe/stop submissive behaviors. Picking up the pet will only serve as a reinforcement.*
    - *Not punishing the dog for submissive behavior or urination, Punishing the dog will serve to aggravate the problem.*
    - *Remaining neutral and "unaware" of the behavior.*
    - *Cleaning up urine when the animal is out of the area.*

Total incontinence presents more of a management problem since release of urine occurs throughout the day rather than at somewhat predictable times. It is possible to teach some owners to help express urine by application of abdominal pressure. Absorbent potty pads and disposable and washable pet diapers also are available to aid owner with daily cleanup and care. In addition, animals can be crated during periods when owner is absent and unable to immediately *clean up after the animal. However, many owners elect to euthanize animals that have become totally incontinent.*

VOMITING/NAUSEA

Definition/Characteristics
Vomiting is the forceful propulsion of food from the stomach often accompanied by increased salivation, tachycardia, gastric stasis, and diarrhea. Inciting causes are numerous and can include infectious diseases, toxins, anesthetics, metabolic disturbances, pharmaceuticals, and GI irritation.

Desired Resolution
Relief from vomiting.

Interventions with Rationale/Amplification
1. Provide an environment that is well ventilated and free from odors. Avoid sudden movement of the animal. *Strong or offensive odors and movement can trigger vomiting.*
2. Clean environment and affected hair coat if emesis occurs. *Very debilitated animals often lie in vomitus that remains in the kennel. Skin that is in prolonged contact with vomitus can become irritated and scalded. Placing sick animals on a cage rack can help prevent contact with the fluids. Always remove collars from animals placed on a rack.*
3. Assess frequency and approximate fluid loss from each vomiting episode. *Vomiting contributes to ongoing fluid loss, thus contributing to deficits in fluid volume.*
4. Assess hydration status q.i.d. for all vomiting animals. *See "Hypovolemia" (page 90). Weight, CRT, skin turgor, and mucous membrane color and moisture are most frequently assessed.*
5. If vomiting is occurring during postoperative period, assess for pain and note whether opioids were used in the anesthetic protocol. *See "Acute Pain" (page 58). Opioids can induce nausea and vomiting. If vomiting occurs in the postoperative period, notify the veterinarian immediately.*
6. Initiate n.p.o. status and immediately notify the veterinarian when identifying emesis in an animal that was not previously vomiting. *The length of time the animal remains n.p.o. will be determined by the veterinarian. n.p.o. status provides a "rest period" for the GI system. Animals on n.p.o. status may need to receive IV or subcutaneous fluids to prevent dehydration.*

7. Once n.p.o. status is removed, reintroduce clear fluids followed by soft, moist foods. *Small volumes of water or sugar solutions are least likely to trigger vomiting episodes. The amount of fluid given to the animal should be noted. Foods specifically formulated for recovery include Hill's a/d and Royal Canin RECOVERY RS.*

BIBLIOGRAPHY

Aiello, S. E. (Ed.). (1998). *The Merck veterinary manual* (8th ed.). Whitehouse Station, NJ: Merck & Co.

Askew, H. R. (1996). *Treatment of behavior problems in dogs and cats. A guide for the small animal veterinarian.* Cambridge, MA: Blackwell Science.

Berg, M. (2005). Educating clients about preventative dentistry. *Veterinary Technician, 26*(2), 102–112.

Bonagura, J. D. (2000). *Kirk's current veterinary therapy XIII: Small animal practice.* Philadelphia: W. B. Saunders.

Carpenito, L. J. (1995). *Nursing care plans & documentation* (2nd ed.). Philadelphia: J. B. Lippincott.

Dodman, N. (2006). Inappropriate elimination. Retrieved November 2006, from http://www.petplace.com

Doenges, M. E., Moorhouse, M. F., & Murr, A. C. (2006). *Nurse's pocket guide: Diagnoses, prioritized interventions, and rationales* (10th ed.). Philadelphia: F. A. Davis.

Hand, M. S., & Novotony, B. J. (2002). *Pocket companion to small animal clinical nutrition* (4th ed.). Topeka, KA: Mark Morris Institute.

Holloway, N. M. (1999). *Medical-surgical care planning* (3rd ed.). Springhouse, PA: Springhouse Corporation.

Jack, C. M, Watson, P. M., & Donovan, M. S. (2003). *Veterinary technician's daily reference guide: Canine and feline.* Baltimore: Lippincott Williams & Wilkins.

Lagoni, L., Butler, C., & Hetts, S. (1994). *The human-animal bond and grief.* Philadelphia: W. B. Saunders.

McCurnin, D. M. (1998). *Clinical textbook for veterinary technicians* (4th ed.). Philadelphia: W. B. Saunders.

Nelson, R. W., & Couto, C. G. (1999). *Manual of small animal internal medicine.* St. Louis, MO: Mosby.

Phipps, W. J., Monahan, F. D., Sands, J. K., Marek, J. F., & Neighbors, M. (2003). *Medical-surgical nursing: Health and illness perspective* (7th ed.). St. Louis, MO: Mosby.

Potter, P. A., & Perry, A. G. (2001). *Fundamentals of nursing* (5th ed.). St. Louis, MO: Mosby.

Royal Canin USA, Inc. *Focus Special Edition: Royal Canin's guide to feline obesity, Focus Special Edition: Pitfalls in GI disorders in the dog,* and *Focus Special Edition: A behavioral approach to canine obesity.* Periodontal Disease in Dogs

Royal Canin USA, Inc. *Nutrition in practice: The ROYAL CANIN veterinary diet guide to gastrointestinal disease* and *The ROYAL CANIN veterinary diet guide to HYPOALLERGENIC HP®.*

Royal Canin USA, Inc. *Veterinary diet product guide.*

Sink, C. A., & Feldman, B. F. (2004). *Laboratory urinalysis & hematology for the small animal practitioner.* Jackson, WY: Teton NewMedia.

Sirois, M. (2004). *Principles and practice of veterinary technology* (2nd ed.). St. Louis, MO: Mosby.

Tranquilli, W. J., Grimm, K. A., & Lamont, L. A. (2000). *Pain management for the small animal practitioner* (2nd ed.). Jackson, WY: Teton NewMedia.

Wingfield, W. E. (1997). *Veterinary emergency medicine secrets.* Philadelphia: Hanley & Belfus.

Chapter 4
Medical Conditions and Associated Technician Evaluations

INTRODUCTION

This chapter describes selected medical conditions affecting dogs and cats. A brief description of each condition is followed by technician evaluations that are commonly associated with the condition. Additional comments are included with certain technician evaluations; this information typically relates to the specific medical condition under discussion. However, the student technician still must refer to Chapter 3 for general information regarding each technician evaluation. This chapter is for reference purposes only; students wanting a broader, more detailed description of any medical condition are referred to standard veterinary technician or veterinary medical textbooks.

CONDITIONS INVOLVING THE GASTROINTESTINAL (GI) SYSTEM

Colitis

Colitis is an inflammation of the large intestine or colon; causes of colitis are numerous and can include parasitic, bacterial, nutritional, neoplastic, foreign body, or unknown etiologies. Clinical signs associated with colitis include diarrhea, tenesmus, hematochezia, and occasionally vomiting. Animals often defecate small volumes with very high frequency.

- Diagnostic Tests: CBC, serum chemistry, radiographs, colonoscopy, biopsy, dietary trials, fecal float, and culture.
- Potential Treatments: Antibiotics, IV fluids, diet modification, and antiparasitics.

Technician evaluations commonly associated with this medical condition:

Diarrhea
- *Diarrhea associated with colitis frequently contains mucus and frank blood.*

Hypovolemia
- *The majority of adult animals compensate for colitis-associated fluid losses through the consumption of additional fluids. Animals experiencing vomiting and very young animals with limited fluid reserves are particularly at risk for dehydration.*

Bowel Incontinence
- *Previously housetrained animals may experience bouts of apparent incontinence during a colitis episode. Colitis-associated incontinence typically resolves when the colitis etiology is identified and appropriately treated.*

Acute Pain
- *Animals suffering from colitis may experience abdominal pain or discomfort. Signs of pain are often exhibited just before or during defecation.*

Underweight
- *Animals suffering from chronic or idiopathic cases of colitis that are unresponsive to treatment may eventually suffer from weight loss.*

Dental Disease

Dental disease is a very broad condition encompassing a large number of disorders. Abnormalities can affect the supporting structures or the tooth. Various branches of dentistry that deal with these issues include endodontics (tooth pulp, root, and periapical tissues), periodontics (supporting structures), and orthodontics (bite evaluation/malocclusion). Examples of potential dental abnormalities include gingivitis, periodontitis, gingival hyperplasia, neck lesions, tumors, foreign bodies, stomatitis complex, fractures, supernumerary teeth, retained deciduous teeth, and malocclusions. Clinical signs are variable and reflect the underlying disease process. Generalized clinical signs of dental disease can include anorexia, halitosis, ptyalism, facial swelling, oral ulceration, and pain.

- Diagnostic Tests: Physical examination, radiographs, and CBC.
- Potential Treatments: Extremely variable depending on dental condition. Surgery, dental prophylaxis, antibiotics, analgesics, anti-inflammatories, diet modification, and home-care programs.

Technician evaluations commonly associated with this medical condition:

Altered Oral Heath
Underweight
- *Oral pain or inability to prehend/masticate food can result in weight loss.*

Client Knowledge Deficit
- *Appropriate client education is fundamental to dental health. Technicians should spend adequate time educating owners regarding home-care/prophylactic measures.*

Acute Pain or Chronic Pain
- *Owners often mistakenly believe that animals are not experiencing oral pain if the animals still eat hard food.*

Cardiac Insufficiency
- *Chronic dental disease and associated bacteremia are inciting causes of valvular insufficiency.*

Aggression
- Animals experiencing oral pain may become aggressive secondary to pain. Many cat owners report a noticeable improvement in personality once dental issues are resolved.

Preoperative and Postoperative Compliance

Dysphagia

Common causes of dysphagia include oral pain, oral trauma, neoplasia, cleft palate, foreign bodies, and neuromuscular disease. Oral pain can occur secondary to any condition that affects the health of the oral cavity or its underlying support structures. Examples of conditions that result in oral pain include fractures (bone and tooth), periodontitis, osteomyelitis, or infectious diseases that trigger stomatitis/glossitis. Neuromuscular diseases associated with dysphagia can include localized myasthenia, temporal masseter myositis, cranial nerve dysfunction, and rabies.

Dysphagia, or difficulty swallowing, can be accompanied by a variety of clinical signs, such as swollen and painful or atrophied muscles of mastication, regurgitation, anorexia, weight loss, and inadvertent aspiration during swallowing. Prognosis and treatment of this condition depend on the inciting cause.

- Diagnostic Tests: CBC, serum chemistry, radiographs, radiographic contrast studies, muscle biopsy, and edrophonium challenge test.
- Potential Treatments: Extremely variable depending on inciting cause. Surgery, analgesics, anti-inflammatories, steroids, and diet modification.

Technician evaluations commonly associated with this medical condition:

Hypovolemia
Underweight
Risk of Aspiration
 - An impaired ability to swallow frequently precedes aspiration.
Altered Oral Health
Acute Pain or Chronic Pain
Client Knowledge Deficit

Foreign Body Obstruction

Consumption of foreign objects occurs with relative frequency in the small animal patient. Young animals typically consume foreign materials at a higher frequency than aged animals. String-type foreign bodies are more common in the cat. Objects can become lodged in any portion of the GI system with clinical signs reflecting the affected portion. Vomiting, abdominal discomfort, tenesmus, and dehydration may all be noted. Objects occluding a more proximal (oral) location in the gastrointestinal tract tend to cause a higher frequency of vomiting than objects more aborally located.

Prognosis for this condition reflects the degree of intestinal or gastric damage caused by the foreign body, necessity of surgical intervention, and the animal's general systemic condition. Animals in which perforation or necrosis has occurred are less likely to have favorable recovery than those experiencing the uncomplicated enterotomy or unassisted passage of the foreign body through the GI system.

- Diagnostic Tests: CBC, serum chemistry, radiographs, radiographic contrast studies, and endoscopy.
- Potential Treatments: Surgery, antibiotics, IV fluids, and dietary management if a large section of bowel is removed.

Technician evaluations commonly associated with this medical condition:

Abnormal Eating Behavior
- *Young animals often engage in pica.*

Vomiting

Hypovolemia
- *Dehydration occurs secondary to vomiting and decreased water intake.*

Acute Pain
- *Abdominal pain is common.*

Constipation
- *Foreign bodies located in the distal colon can result in clinical signs consistent with constipation.*

Gastric Dilatation Volvulus (GDV)

GDV is a medical emergency involving malpositioning and rapid accumulation of gas in the stomach. Rotation of the pyloric portion of the stomach from a right ventral position to a more left dorsal location results in outflow obstruction and ischemia. Concurrent shock and torsion of the spleen is common. GDV tends to affect large, deep-chested breeds and is rarely found in small breeds of dogs and cats. Clinical signs include nonproductive retching, abdominal distension, abdominal pain, dehydration, abnormal mucous membranes (pale, injected, or dark), and tachycardia.

- Diagnostic Tests: Physical examination, radiographs, CBC, serum chemistry, ECG, and ultrasound.
- Potential Treatments: Gastric decompression (using an orogastric tube or trocar), IV fluids, shock therapy, surgery (repositioning

of stomach, gastropexy, or possible splenic removal), antibiotics, and analgesics.

Technician evaluations commonly associated with this medical condition:

Hypovolemia

Electrolyte Imbalance
- *Acid/base status and serum electrolytes should be closely monitored. Hypokalemia is often noted and may exacerbate cardiac arrhythmias.*

Vomiting
- *Retching or dry heaving is typical.*

Acute Pain
- *Analgesics, typically opioids, are required for pain control. Flunixin meglumine (NSAID) has anti-endotoxic properties.*

Decreased Tissue Perfusion
- *Excessive distension/pressure and torsion of the stomach decrease tissue perfusion. Exercise extreme caution when passing an orogastric tube to relieve gastric pressure. Forcing the tube can rupture the stomach or esophagus.*

Cardiac Insufficiency
- *Associated cardiac arrhythmias are common, and monitoring is recommended. Premature ventricular contractions and ventricular tachycardia are typically seen.*

Risk of Infection
- *Mesenteric congestion predisposes animals to infection and endotoxemia.*

Abnormal Eating Behavior
- *Affected breeds often eat large meals rapidly (wolf food).*

Client Knowledge Deficit
- *To reduce further incidences of bloat (gastropexy prevents future volvulus), the owner should be instructed to:*
  - *Feed several small meals instead of one large meal each day.*
  - *Use a highly digestible diet.*
  - *Minimize exercise immediately before and after eating.*
  - *Keep emergency contact numbers available and be cognizant of early signs of bloat.*

Gastritis

Inflammation of the stomach lining is a common malady of small animals. Gastritis can occur in acute, chronic, or toxic forms. Acute gastritis is often associated with dietary indiscretion; foreign objects; or bacterial, parasitic, or viral infections. Chronic gastritis is often linked to allergies, immune-mediated conditions, or any disease capable of inciting inflammation. Clinical signs of gastritis include anorexia, altered appetite, nausea, vomiting, and hematemesis. Fever and abdominal pain occur infrequently.

- Diagnostic Tests: CBC, serum chemistry, radiographs, endoscopy, allergy testing, and biopsy.
- Potential Treatments: Supportive therapies such as antibiotics, IV fluids, antiemetics, and steroids.

Technician evaluations commonly associated with this medical condition:

Vomiting
- *Vomiting associated with acute gastritis frequently resolves within 24 hours of administering fluids parenterally and initiating an n.p.o. status. The presence of "coffee ground" vomitus is indicative of gastric bleeding.*

Underweight
- *Weight loss (excluding dehydration water loss) is not associated with acute conditions. Weight loss can occur secondary to chronic, poorly managed gastritis conditions.*

Hypovolemia

Abnormal Eating Behavior
- *Animals with acute gastritis may attempt to eat grass or exhibit pica.*

Noncompliant Owner
- *Dogs suffering from food allergies must remain on low-allergen diets. Feeding of table scraps or other dietary indiscretions can trigger a flare-up in the patient with chronic gastritis.*

Intussusception

Intussusception is the prolapsing of a segment of intestine into the lumen of an immediately adjacent portion of intestine. This condition is frequently referred to as a "telescoping" of the intestines and is likely to result in intestinal obstruction. Although intussusception can occur anywhere in the intestines, the ileocolic location is most common. Inciting causes of intussusception are numerous and can include enteritis (bacterial or viral), parasites, neoplasia, or a postoperative complication of surgery. Clinical signs of intussusception include vomiting, abdominal pain, diarrhea, and hematochezia. Prognosis is dependent upon the underlying inciting cause, amount of intestine resected, and presence of peritonitis.

- Diagnostic Tests: Abdominal palpation, radiographs, contrast studies, ultrasound, CBC, serum chemistry, and fecal evaluation.
- Potential Treatments: Surgical intervention, supporting antibiotics, IV fluids, anti-inflammatories, analgesics, and dietary management in cases of short bowel syndrome.

Technician evaluations commonly associated with this medical condition:

Acute Pain

Hypovolemia
- Dehydration can result from reduced water consumption, vomiting, or excessive diarrhea.

Vomiting
- Vomiting frequently resolves soon after surgical intervention.

Diarrhea

Risk of Infection
- Animals undergoing abdominal surgery and intestinal resection are at a high risk for peritonitis.

Ileus

Ileus is an intestinal obstruction due to a lack of peristalsis or functional obstruction. Causes of ileus secondary to mechanical obstruction include neoplasia (intra- or extraluminal), a foreign body, intussusception, or incarceration. Paralytic ileus can be associated with torsion, peritonitis, inflammation, postoperative states, intoxications, or any type of abdominal trauma.

- Diagnostic Tests: Extremely variable depending on inciting cause. Tests can include ultrasound, radiographs, CBC, and serum chemistries.
- Potential Treatments: Treatment is highly variable depending on cause. It can include symptomatic medical treatment or surgical intervention.

Technician evaluations commonly associated with this medical condition:

Hypovolemia
Vomiting
Risk of Infection
Self-Care Deficit
Decreased Tissue Perfusion—Gastrointestinal
Acute Pain

Malabsorption

Malabsorption is the inability to absorb nutrients across the intestinal wall. Causes are numerous and can include food allergies, inflammatory bowel disease (IBD), parasitism, neoplasia, or bacterial overgrowth. Clinical signs can include weight loss, vomiting, diarrhea, and lethargy. Prognosis is dependent upon the inciting cause.

- Diagnostic Tests: CBC, serum chemistry, trypsin-like immunoreactivity (TLI), folate, cobalamin, fecal flotation, allergy testing, and intestinal biopsy.
- Potential Treatments: Extremely variable depending on inciting cause. Most can benefit from dietary management.

Technician evaluations commonly associated with this medical condition:

Underweight
- *Failure to absorb nutrients results in caloric reduction and weight loss. Diet selection is dependent upon the cause of malabsorption. Animals suffering from food allergies are often placed on hypoallergenic diets such as Hill's d/d or z/d, Royal Canin DIGESTIVE LS, or HYPOALLERGENIC HP. Animals suffering from neoplastic-induced malabsorption may be placed on high-calorie diets such as Hill's Growth or Royal Canin INTESTINAL HE.*

Diarrhea
- *Malabsorption induces an osmotic diarrhea. The unabsorbed food particles remaining in the intestinal lumen act to pull water into the intestines, thus increasing the fluidity of the feces. It is important to provide adequate oral fluids; otherwise, these animals can rapidly dehydrate.*

Vomiting
- *Vomiting secondary to food allergy diminishes once the appropriate diet is selected.*

Maldigestion

Maldigestion is the inability to break down nutrients into an absorbable form. Exocrine pancreatic insufficiency (EPI) is the most common type of maldigestion seen in animals. Clinical signs can include chronic diarrhea, increased appetite, polydipsia, coprophagia, pica, flatulence, and weight loss. Fecal material produced by these animals is often foul-smelling with a high fat content. Lactate deficiencies have been reported occasionally. Older animals may gradually lose their ability to digest milk sugars, thereby becoming lactose intolerant as they age.

- Diagnostic Tests: CBC, serum chemistry, TL1, PL1, cobalamin, serum B12, triglyceride challenge test, and qualitative fecal analysis (trypsin, amylase, or lipase activity).
- Potential Treatments: Pancreatic enzyme replacement, supportive treatments such as antibiotics, H_2 receptor antagonists, and vitamins.

Technician evaluations commonly associated with this medical condition:

Underweight
- *Animals lacking the appropriate pancreatic enzymes lack the ability to digest fats and carbohydrates. Pancreatic enzyme supplements are often sprinkled directly onto the animal's food. Diets should be very low in fiber and highly digestible, such as Hill's i/d, Royal Canin HYPOALLERGENIC HP, or Royal Canin INTESTINAL HE. The total daily ration should be divided into 2 or 3 smaller portions.*
- Dogs should receive a daily multivitamin to ensure adequate levels of fat-soluble vitamins.
- Dogs should never receive table scraps or high-fat drippings to increase palatability of therapeutic diets.

Diarrhea
- *Owners should be instructed to monitor quantity and consistency of fecal output once dietary changes are made and supplements have been initiated.*

Megaesophagus

Defined as a dilatation and hypomotility of the esophagus, megaesophagus can be a congenital or acquired condition. Regurgitation, which is often confused with vomiting, is the most common clinical sign. Additional signs can include salivation, halitosis, dyspnea, and generalized muscular weakness. Megaesophagus is rarely seen in cats. Aspiration pneumonia remains a constant concern for animals suffering from this condition.

- Diagnostic Tests: CBC, serum chemistry, radiographs, acetylcholine receptor antibody titer, and endoscopy.
- Potential Treatments: Supportive treatments such as diet modification, antibiotics, H_2 blockers, and promotility agents.

Technician evaluations commonly associated with this medical condition:

Underweight
- *Emaciation occurs secondary to chronic regurgitation of ingesta. Dietary modification remains the mainstay of successful technician intervention. The following interventions are recommended:*
    1. *Initiate calorically dense diet.*
    2. *Elevate food dish using step stool or adapted baby walker so that animal's front end is elevated during meal consumption. This facilitates gravitational flow through the esophagus.*

3. Provide food in small, frequent meals throughout the day (4 or 5 meals per day). Canned food formed into small "meatballs" often facilitates passage.
4. Elevate dog's front end for 10–15 minutes after each meal.

Risk of Aspiration
- Animals must be closely monitored for signs of aspiration pneumonia. Instruct the owner to observe daily for signs of nasal discharge, coughing, or dyspnea.

Abnormal Eating Behavior
- Animals may attempt to "wolf down" food.

Vomiting
- Animals suffering from this condition are most likely regurgitating as opposed to vomiting. Regurgitation is not associated with signs of abdominal thrusting. Feeding modifications remain the most effective means of preventing regurgitation.

Oral Trauma

Trauma to the oral cavity can include maxillary or mandibular fractures, temporomandibular dislocation, laceration of tongue or other soft tissues, tooth fractures, or penetration of a foreign body (a fish hook or stick). Trauma often occurs secondary to automobile accidents or to animal bites/kicks. Clinical signs can include blood in the oral cavity, dysphasia, anorexia, drooling, and lack of facial symmetry.

- Diagnostic Tests: Physical examination and radiographs.
- Potential Treatments: Variable depending on injury. Surgery, antibiotics, anti-inflammatories, analgesics, and nutritional support.

Technician evaluations commonly associated with this medical condition:

Altered Oral Health
Risk of Infection
- The bacterial level in the oral cavity is very high.

Underweight
- Oral trauma can result in anorexia. Appropriate nutritional support is critical to recovery.

Acute Pain
Hypovolemia
Self-Inflicted Injury
- Animals often use their paws in an attempt to remove a foreign body from the oral cavity. This activity typically exacerbates the injury.

Ineffective Nursing
- Neonates will experience difficulty nursing if oral trauma occurs.

Salivary Mucocele

A salivary mucocele is formed by the accumulation of saliva and mucus in the subcutaneous tissue surrounding a salivary gland. This soft, nonpainful swelling, which usually occurs secondary to damage of the salivary ducts, is the most common malady associated with salivary glands. The sublingual gland is most frequently affected and results in a large swelling at the base of the tongue. Clinical signs include drooling, blood-tinged saliva, anorexia, open-mouth appearance, and abnormal licking or tongue motions.

- Diagnostic Tests: Physical examination, fine needle aspiration, and sialography (rarely used).
- Potential Treatments: Monitoring of condition and surgery: marsupialization (create a fistula or drain in the gland) and salivary gland removal.

Technician evaluations commonly associated with this medical condition:

Anorexia
- Mucoceles can form as a large mass in the mouth, making eating difficult. Providing nutritional support or altering diet consistency may be of value.

Altered Oral Health
- Large masses can push the tongue to the side of the oral cavity. Damage to the tongue and mass by the teeth is common.

Obstructed Airway
- Although it is rare, mucoceles associated with the pharyngeal wall can impair swallowing and respiration. Animals suffering from this complication are likely to require surgical intervention.

CONDITIONS INVOLVING THE LIVER, GALLBLADDER, AND PANCREAS

Cholangiohepatitis

Cholangiohepatitis is an inflammation of the biliary system and, by extension, of the local hepatic parenchyma. This condition is a relatively common liver disease of cats and is seen in both acute (suppurative) and chronic (nonsuppurative) forms. Clinical signs include fever, lethargy, anorexia, weight loss, diarrhea, abdominal, pain, icterus, dehydration, and vomiting. Cats suffering from the chronic condition also may develop cirrhosis with loss of normal architecture and deposition of fibrous tissue in the liver.

- Diagnostic Tests: CBC, serum chemistries, ultrasound, radiographs, bile culture, postprandial bile acid measurement, abdominocentesis, and biopsy.

- Potential Treatments: Antibiotics, antioxidants, IV fluids, antiemetics, and analgesics.

Technician evaluations commonly associated with this medical condition:

Vomiting
Hyperthermia
Self-Care Deficit
Hypovolemia
Risk of Infection
- Bile is often infected with Escherichia coli or Pseudomonas due to reflux of digestive enzymes associated with vomiting and retrograde migration of bacteria.

Underweight
- In cats, cholangiohepatitis often occurs in association with pancreatitis or inflammatory bowel disease. Chronically affected animals often lose weight. Adequate nutritional support will assist in preventing hepatic lipidosis. High-protein diets are recommended. Examples of suitable diets include Hill's i/d, Royal Canin HYPOALLERGENIC HP or INTESTINAL HE, Iams Intestinal Low-Residue, and Purina EN GastroENteric. Vitamin supplementation, specifically vitamin K, is usually warranted in chronic cases.

Acute Pain
Client Knowledge Deficit
- Educate owners regarding clinical signs and need of the chronic patient.

Hepatitis

Hepatitis is an inflammation of the liver resulting from infectious or toxic insult to hepatic cells. Toxic insult can encompass exposure to drugs, both therapeutic drugs and drugs accidentally administered (anticonvulsants, glucocorticoids, antifungals, and acetaminophen), or environmental toxins (pesticides, plant toxins, herbicides, and cleaning agents). Infectious agents associated with hepatitis include feline infectious peritonitis (FIP), canine infectious hepatitis, histoplasmosis, leptospirosis, or any agent capable of inducing septicemia.

Clinical signs may include jaundice, fever, vomiting, diarrhea, pale mucous membranes, disorientation, ataxia, and seizures.

- Diagnostic Tests: Serum chemistry, serum bile acids, CBC, coagulation tests, radiographs, ultrasound, abdominocentesis, and biopsy.
- Potential Treatments: Supportive and symptomatic treatments such as fluids, antibiotics, antiemetics, antivirals, or steroids and removal of exposure to toxic agent.

Technician evaluations commonly associated with this medical condition:

Anorexia
Acute Pain
Vomiting
Diarrhea
Hyperthermia
Risk of Infection Transmission
- *Some causes of hepatitis can be zoonotic.*

Hepatic Encephalopathy

Hepatic encephalopathy, a life-threatening complication of liver failure, results from the liver's inability to properly metabolize proteins and detoxify. Ammonia is a normal by-product of protein catabolism. The liver of a healthy animal converts ammonia to urea, which is then excreted by the kidneys in urine. Failure to remove ammonia results in toxic accumulation, which is damaging to neurons and causes various neurologic deficits. Clinical signs associated with this condition include seizures, depression, ataxia, and anorexia.

- Diagnostic Tests: Tests reflect diagnosis of portal systemic shunts (PSSs). CBC, serum chemistry, postprandial ammonia, ammonia tolerance test, serum bile acids, coagulation tests (PT, PPT, and ACT), plain film radiographs, positive contrast portography, ultrasonography, and nuclear scintigraphy.
- Potential Treatments: n.p.o. status, enemas, lactulose, IV fluids (no lactate), and potassium supplementation.

Technician evaluations commonly associated with this medical condition:

Hypovolemia
- *Any fluids containing lactate should be avoided. Lactate requires metabolization by the liver. Saline at 0.45 percent (half strength) in 2.5 percent dextrose is commonly selected. Potassium at 20 mEq/L may be added to fluids depending on electrolyte status.*
- *Dehydration and hypokalemia exacerbate hepatic encephalopathy.*

Underweight
- *Protein-restricted diets are imperative. Carbohydrates should constitute the primary energy source. All protein should be of high biologic value to minimize ammonia production. Examples of suitable diets include Hill's k/d, Royal Canin HEPATIC LS or HYPOALLERGENIC HP, and Pro Plan CNM NF.*

Vomiting
- *Vomiting occurs secondary to accumulation of ammonia. In addition, lactulose, which is used during treatment, can contribute to emesis.*

Diarrhea
- Lactulose is converted to lactic acid by colonic bacteria. The resultant acidic environment draws ammonia from the blood and assists with excretion. Side effects of treatment are common and include diarrhea, vomiting, anorexia, and hypokalemia.
- Animals on lactulose treatment should have two or three soft stools per day.

Altered Mentation
- Neurologic signs frequently occur after meals, especially those high in protein, when larger amounts of ammonia are produced.

Client Knowledge Deficit
- Education of the owner regarding appropriate dietary restrictions is critical.

Hepatic Lipidosis

Hepatic lipidosis is almost exclusively a malady of cats and remains the most common liver disease in this species. Also referred to as "fatty liver disease," this condition results in accumulation of excessive triglycerides in the liver. Any cat experiencing protein catabolism or inadequate protein and carbohydrate intake is at risk for developing hepatic lipidosis. Obese cats in particular are susceptible to this condition. Typically, most cases are precipitated by a prolonged period (1 week) of anorexia. Clinical signs include anorexia, lethargy, vomiting, jaundice, hepatomegaly, and generalized weakness.

- Diagnostic Tests: CBC, serum chemistry, radiographs, coagulation tests, liver biopsy, and ultrasound.
- Potential Treatments: Supportive therapies such as diet modification, antibiotics, antiemetics, IV fluids, total parenteral nutrition (TPN), partial parenteral nutrition (PPN), placement of feeding tube, and vitamin supplementation.

Technician evaluations commonly associated with this medical condition:

Vomiting
- Nutritional support is fundamental to recovery. Cats typically become anorexic after episodes of nausea or vomiting and refuse all food. Use of nasogastric intubation or forced oral feeding is often required. Placement of gastrostomy or esophagotomy tubes may be warranted if nutrient requirements remain unmet using less invasive measures.
- Animals should initially receive 1/3 to 1/2 of the required daily nutrient requirements. Cats suffering from lipidosis have gastric stasis and are often unable to adequately digest protein. Administering the daily ration over 3 or 4 small feedings per day helps minimize discomfort and increase digestibility. Fully calculated rations should be administered by Day 3.

- Cats with hepatic lipidosis need immediate dietary intervention with a diet high in protein and low in carbohydrates. For tube feeding, choices include Hill's a/d and Royal Canin RECOVERY RS. Once the cat is able to take in food normally Hill's l/d or k/d or Royal Canin feline GROWTH can supply the level of protein that the cat needs.
- Nasogastric or esophagotomy tube recommendations:
  - Warm food to body temperature. If a microwave is utilized, ensure that there are no hot spots.
  - Remove wrap to expose tube if the animal is bandaged.
  - Inspect entrance sight for any signs of infection or discharge.
  - Remove tube cap, attach syringe, and aspirate to ensure that stomach is empty. (Note: If more than 3–5 cc of fluids/ingesta are aspirated, inform the veterinarian prior to administering food.)
  - Ensure tube patency via infusing 5 cc of warm water.
  - Infuse food slowly over 2–5 minutes.
  - Clean and flush tube by infusing 5 cc of warm water. Allow a column of water to remain in the tube after flushing. (This will minimize air intake and refluxing of gastric contents.) Recap tube.
  - Wipe external tube surface with alcohol.
  - Clean skin with chlorhexidine solution and allow to dry, doing so 3 times per week or as needed.
  - Reapply wrap if warranted.

Client Knowledge Deficit
- Owner education regarding proper care of esophagostomy or gastrostomy tubes is imperative. Compliance is greatly improved if owners are provided demonstration and then given the opportunity to administer the food under supervision. In addition, owners should be provided with written instructions that address food preparation, volume, method and frequency of administration, proper techniques for tube cleaning, and signs indicating that the veterinarian should be contacted.

Hypovolemia
- Avoid using fluids high in dextrose as cats are metabolically challenged. Avoid use of LRS.

Diarrhea
- Caloric intake is of utmost importance. Less palatable bland diets used for treatment of diarrheas may not be warranted.

Constipation
- Cats that have not consumed adequate oral fluids may become constipated.

Overweight
- Many animals suffering from lipidosis are overweight or diabetic or have concurrent diseases such as IBD or pancreatitis. Animals should be brought to a normal weight gradually to minimize the probability of reoccurrence. It is imperative that weight loss be a

gradual process; otherwise, the lack of caloric intake can trigger a reoccurrence of lipidosis.
- Instruct owner to closely monitor cat for anorexia during times of stress. Periods of anorexia require immediate intervention.

Jaundice

Also termed *icterus* and *hyperbilirubinemia*, jaundice is a yellowing of the skin, sclera, and/or muocus membranes. This distinct yellow coloring results from an accumulation of bilirubin. Bilirubin, which is the result of RBC hemolysis, is normally removed from circulation by the liver and excreted into bile. Accumulation of bilirubin can result from obstruction of bile flow (gallstone or pancreatitis), liver disease (cholangiohepatitis, neoplasia, or hepatic lipidosis), or sudden exaggerated RBC hemolysis (immune-mediated or infectious diseases). Jaundice is a clinical sign that can be associated with many disease conditions.

- Diagnostic Tests: Varied depending on inciting cause. CBC, serum chemistries, urinalysis, coagulation tests, bile acids, ultrasound, liver biopsy, and serology.
- Potential Treatments: Varied depending on inciting cause. UV light therapy, blood transfusions, surgery, treatment for liver failure, and supportive therapy (fluids).

Technician evaluations commonly associated with this medical condition:

Vomiting
Hypovolemia
Altered Mentation
Acute Pain
Hyperthermia
Risk of Infection
Risk of Infection Transmission

Pancreatitis

Defined as an inflammation of the pancreas, pancreatitis is a painful condition that affects both dogs and cats. Dogs are more likely to be affected, and the two species often exhibit marked differences in their presenting clinical appearance. Clinical signs commonly seen in dogs include depression, vomiting, anorexia, abdominal pain, dehydration, and diarrhea. In comparison, cats are less likely to experience vomiting or demonstrable abdominal pain and more typically present with a history of anorexia, dehydration, and lethargy. The condition is most likely multifactorial with nutrition (obesity and high-lipid diets seen in dogs), ischemia (dehydration and disseminated intravascular coagulation, or DIC), inflammatory bowel disease (cats), abdominal trauma, infection, hypercalcemia, or drugs potentially contributing to the onset of the disease.

- Diagnostic Tests: CBC, serum chemistry (amylase/lipase evaluation), diagnostic imaging (ultrasound and radiographs), urinalysis, fPLI (pancreatic function test), cobalamin, and folate (GI function tests).
- Potential Treatments: Supportive treatments such as IV fluids, TPN, PPN, analgesics, antiemetics, antibiotics, and steroids.

Technician evaluations commonly associated with this medical condition:

Acute Pain
- *Extreme abdominal pain can be associated with this condition. Dogs often assume a "praying position." Analgesics frequently prescribed include opioids such as butorphanol and oxymorphone. It is often difficult to assess abdominal pain in cats. Many cats present without apparent signs of abdominal pain, although this may be due to insufficient detection. Morphine is avoided as it can incite pancreatic duct spasm.*

Hypovolemia
- *Restoration of normal intravascular volume and pancreatic perfusion is fundamental to successful treatment. Fluids should be administered at such a rate as to restore the dehydration deficit volume, meet daily maintenance requirements, and replace ongoing losses. When IV support is no longer necessary, fluids should be gradually tapered off over a 24-hour period.*

Vomiting
- *Animals suffering from acute pancreatitis are typically held n.p.o. for 48–72 hours. This provides the gland adequate time to "rest" and reduces pancreatic secretion. (The pancreas produces many digestive enzymes and hormones.) Animals should not be permitted to see or smell food during the n.p.o. status as this will trigger secretion of the digestive enzymes produced by the pancreas.*
- *Reintroduction of food should not occur until at least 24 hours after the last vomiting episode. Initial food offering should include a very small amount of a bland, low-fat diet such as Hill's w/d or r/d, Royal Canin DIGESTIVE LOW FAT LF, or Iams Intestinal Low-Residue. Animals should remain on a q.i.d. feeding schedule for several days.*
- *If the n.p.o. status must be extended past 5 days (dogs), TPN must be instituted.*
  Note: Due to the high incidence of hepatic lipidosis in cats, alternative nutrition (TPN or PPN) must be initiated within 2–3 days. Animals placed on TPN or PPN should be "tapered off" gradually to reduce incidences of metabolic crashes.

Diarrhea
- Diarrheas associated with pancreatitis can be of large or small bowel origin. Cats are especially likely to present with concurrent inflammatory bowel disease or cholangiohepatitis.

Electrolyte Imbalance
- Intravenous administration of balanced electrolyte solutions is typically initiated.

Overweight
- In dogs, obesity frequently contributes to pancreatitis. Dogs suffering from recurrent episodes of pancreatitis should be fed calorie-restricted diets until the ideal body weight is obtained. Long-term dietary management should include a fat-restricted diet with no access to table scraps. Dogs suffering from recurrent episodes of pancreatitis may be placed on pancreatic enzyme supplements.
- Cats may be overweight or underweight and typically do not present with a history of recent diet change or high-fat intake.

Self-Care Deficit

Pancreatic Insufficiency

The pancreas has both endocrine (hormones) and exocrine (enzymes) functions. EPI refers to inadequate enzyme production and is associated with inadequate levels of amylase, lipase, or protease. These enzymes are responsible for digestion of carbohydrates, lipids, and proteins, respectively. Lack of adequate enzymes results in maldigestion and malabsorption of dietary nutrients. Clinical signs of this condition can include diarrhea, weight loss, decreased muscle mass, ravenous appetite, vomiting, coprophagia, pica, flatulence, and a rough hair coat.

- Diagnostic Tests: CBC, serum biochemistry, TLI assay, and fecal proteolytic activity.
- Potential Treatments: Pancreatic enzyme replacements, H_2 blockers, and antibiotics.

Technician evaluations commonly associated with this medical condition:

Vomiting

Underweight
- Inadequate nutrient absorption results in weight loss. Diets should be low in fiber and highly digestible. Examples of appropriate diets include Hill's i/d; Iams Intestinal Low-Residue; Purina EN GastroENteric; and Royal Canin DIGESTIVE LOW FAT LF, HYPOALLERGENIC HP, and INTESTINAL HE. Moist diets should be used when applying pancreatic powders to facilitate mixing. Divide daily ration into three small meals. Mix pancreatic enzyme replacer with food and allow to stand

20 minutes prior to offering. Administration of a daily multivitamin is highly recommended. Animals are often deficient in fat-soluble vitamins, cobalamin, and tocopherol.

Client Knowledge Deficit
- Consistency of dietary management is important. Educate owners regarding importance of selected protocol.

Diarrhea
- Undigested food particles act as an osmotic laxative, pulling fluid into the intestinal lumen.

Bowel Incontinence
- Animals may have difficulty if unable to defecate frequently. Correction of the diarrhea typically resolves the issue.

Abnormal Eating Behavior
- Pica and coprophagy are frequently noted.

Self-Care Deficit
- Continual diarrhea that is high in fat predisposes the animal to a greasy anal area.

Portosystemic Shunts (PSSs)

Defined as a vascular abnormality, PSSs permit blood from the intestinal vessels to bypass the liver and proceed directly into systemic circulation. Further classification of PSS is dependent upon the location of the shunting vessel. Shunts located in the liver parenchyma are identified as intrahepatic, while those external to the liver are termed extrahepatic.

Lack of vascular support induces reduced hepatic mass and essentially limits the liver's metabolic and detoxification capabilities. Ultimately, blood that bypasses the liver fails to undergo hepatic metabolism, thus resulting in accumulation of excessive amounts of ammonia.

Clinical signs that arise from secondary damage to the gastrointestinal, neurologic, and urinary systems can include vomiting, diarrhea, anorexia, ataxia, altered mentation, head pressing, circling, seizures, polyuria, hematuria, and stranguria.

- Diagnostic Tests: Postprandial ammonia, ammonia tolerance test, serum bile acids, CBC, serum chemistry, coagulation tests (PT, PPT, and ACT), plain film radiographs, positive contrast portography, ultrasonography, and nuclear scintigraphy.
- Potential Treatments: Surgery (cellophane band or ameroid ring): Cellophane bands induce a perivascular fibrosis. Ameroid rings contain ameroid, which is compressed casein (protein) that expands with moisture. Both methods gradually occlude the

extrahepatic vessel, thus avoiding a life-threatening portal hypertension as would be caused by rapid ligation.
- Medical treatment for inoperable cases or those requiring stabilization prior to surgery include IV fluids, oral lactulose (or enema), stomach protectants, and antibiotics.

Technician evaluations commonly associated with this medical condition:

Hypovolemia
- Dehydration and hypokalemia exacerbate hepatic encephalopathy.

Vomiting
- Vomiting occurs secondary to accumulation of ammonia. In addition, lactulose, which is used during treatment, can contribute to emesis.

Diarrhea
- Diarrhea is a common manifestation of maldigestion caused by inadequate bile production.
- Lactulose, which is used to treat PSS, is converted to lactic acid by colonic bacteria. The resultant acidic environment draws ammonia from the blood and assists with excretion. Side effects of treatment are common and include diarrhea, vomiting, anorexia, and hypokalemia.

Underweight
- Loss of hepatic function can reduce bile production. Inadequate bile hinders digestion, effectively causing maldigestion. Failure to gain weight is often noted in affected puppies.
- Diets should contain high-quality protein from non-meat sources to minimize ammonia production, highly digestible complex carbohydrates, elevated levels of dietary fiber from soluble and insoluble fibers, with enriched amounts of antioxidants, B vitamins, and omega-3 fatty acids.
- Animals undergoing surgery can be placed on Hill's k/d or l/d or Royal Canin HEPATIC LS during the immediate postoperative period (up to 6 months). After 6 months post-op, dogs are usually fed a senior diet that contains 8–10 percent protein.

Altered Mentation
- Hepatic encephalopathy is a condition that occurs secondary to the liver's inability to detoxify. The debilitated liver fails to convert ammonia to urea. (Ammonia is a normal by-product of protein catabolism. The kidneys then excrete the urea in urine.) Excessive ammonia levels damage neurons, causing various neurologic deficits.
- Neurologic signs frequently occur after meals, especially those high in protein, when larger amounts of ammonia are produced.

Altered Urinary Production
- Failure to metabolize uric acid and high ammonia concentrations predisposes dogs to ammonium urate uroliths.

Risk of Infection
— *Debilitated animals are at risk for secondary infections. Reverse isolation techniques should be utilized during the postoperative period.*

Self-Care Deficit

CONDITIONS INVOLVING THE CARDIOVASCULAR AND RESPIRATORY SYSTEMS

Cardiac Arrhythmia

Cardiac arrhythmias, variations from normal rhythm, are frequently encountered in veterinary medicine. Abnormal rate and rhythm patterns most commonly encountered include tachycardia, bradycardia, atrial fibrillation, respiratory sinus arrhythmia, ventricular fibrillation, ventricular premature contractions, and ventricular tachycardia. Conditions that manifest this clinical sign are numerous and can include congenital heart defects, dilated or hypertrophic cardiomyopathy, drug reactions, anesthesia, electrolyte disturbances, hyperthyroidism, shock, myocarditis, trauma, valvular disease, neoplasia, pericarditis, and myocardial infarction. Arrhythmias can be subclinical or can appear as syncope, weakness, lethargy, tachypnea, dyspnea, increased CRT, or cyanosis.

- Diagnostic Tests: Variable depending on underlying etiology. ECG, radiographs, ultrasound (echocardiogram or Doppler), CBC, serology, and halter monitor.
- Potential Treatments: Highly variable depending on inciting cause.

Technician evaluations commonly associated with this medical condition:

Cardiovascular Insufficiency
— *Cardiac output can decrease secondary to arrhythmia.*

Exercise Intolerance
— *Reduced cardiac output incites exercise intolerance.*

Acute Pain

Decreased Tissue Perfusion
— *Infarcted tissues become ischemic.*

Sleep Pattern Disturbance
— *Pulmonary edema, or fluid accumulation, may preclude normal recumbent resting positions. Animals may experience coughing during the night or change sleeping positions frequently to facilitate oxygenation.*

Client Knowledge Deficit
— *Many cardiac conditions resulting in arrhythmia require long-term management. Educating owners regarding medications, exercise regimes, weight control, and clinical signs of disease progression is fundamental to successful management.*

Cardiomyopathy

Cardiomyopathy is a general diagnostic term used to describe a primary myocardial disease of unknown cause. In the small animal patient, cardiomyopathy is often classified as hypertrophic (primarily cats), dilated (primarily dogs), or restrictive. The overriding effect of this condition is a net decrease in cardiac output. Clinical signs of cardiomyopathy parallel those of congestive heart failure (CHF) with anorexia, muscle wasting, ascites, abdominal distension, pulmonary edema, lethargy, pallor, syncope, increased CRT, weak pulse, and cyanosis being noted.

- Diagnostic Tests: Radiographs, ECG, ultrasound, echocardiography, Doppler, CBC, and serum biochemistry.
- Potential Treatments: Symptomatic treatment with many drugs designed to minimize signs of CHF, including diuretics, positive inotropes, bronchodilators, vasodilators, ACE inhibitors, beta blockers, calcium channel blockers, anticoagulants, taurine supplementation, oxygen therapy, and IV fluids. Dietary management should be compatible with drug therapies and have appropriate sodium restriction.

Technician evaluations commonly associated with this medical condition:

Cardiovascular Insufficiency
- *Use extreme caution when administering IV fluids. Acute decompensation secondary to cardiac overload can occur.*

Hypervolemia
- *Decreased glomerular filtration rate secondary to vasoconstriction, increased sodium retention, and ADH production can result in fluid volume excess.*

Decreased Tissue Perfusion
- *Hypertrophic and restrictive cardiomyopathies are associated with a higher incidence of thromboembolism.*

Altered Gas Diffusion
- *Fluid collection or shifts into the interstitial space/alveoli impair gas exchange.*

Exercise Intolerance
- *Reduced oxygenation and cardiac output precludes activity.*

Risk for Overweight

Client Knowledge Deficit
- *Educate owners regarding the following:*
  - *Use of appropriate diets with restricted salt is indicated. Examples include Hill's h/d, Purina CV Cardiovascular, and Royal Canin EARLY CARDIAC EC.*
  - *Exercise restriction may be required.*
  - *A low-stress environment should be provided.*

Congestive Heart Failure (CHF)

CHF is reduced cardiac output due to impaired pumping activity by the heart. This condition can be right- or left-sided. Congestion, or accumulation of fluid, due to right ventricular failure is associated with systemic venous congestion evidenced by ascites, hepatic congestion, and peripheral edema. Left ventricular failure is typically associated with pulmonary edema. Causes of CHF are numerous and can include valvular insufficiency, congenital disease, hypertension, myocarditis, dilated cardiomyopathy, neoplasia, arrhythmias, and hypertrophic cardiomyopathy. Clinical signs associated with CHF reflect the underlying disease process and can include anorexia, muscle wasting, ascites, pulmonary edema, lethargy, pallor, syncope, murmurs, increased CRT, weak pulse, and cyanosis.

Clinical classification of CHF using the New York Heart Association standard is helpful when describing the disease.

- Class I: No obvious exercise limitations.
- Class II: Slight exercise limitations with routine physical activity.
- Class III: Comfortable at rest. Clinical signs develop with minimal physical activity.
- Class IV: Clinical signs apparent at rest, all activity limited.
  - Diagnostic Tests: Radiographs, ECG, CBC, serum chemistry, echocardiograms, and heartworm test.
  - Potential Treatments: Variable depending on underlying etiology. Beta blockers, ACE inhibitors, diuretics, positive inotropes, arterial dilators or venodilators, oxygen therapy, and IV fluids.

Technician evaluations commonly associated with this medical condition:

Cardiovascular Insufficiency
- *Loss of cardiac output can be related to decreased contractility, arrhythmias, abnormal conduction, or structural changes.*
- *Use extreme caution when administering IV fluids. Acute decompensation secondary to cardiac overload can occur.*

Hypervolemia
- *Decreased glomerular filtration rate secondary to vasoconstriction, increased sodium retention, and ADH production can result in fluid volume excess.*

Altered Gas Diffusion
- *Fluid collection or shifts into the interstitial space/alveoli impair gas exchange.*

Exercise Intolerance
- *Reduced oxygenation and cardiac output precludes activity.*

Client Knowledge Deficit
- Educate owners regarding the following:
  - Use of appropriate diets with restricted salt is indicated. Examples include Hill's h/d and Purina CV Cardiovascular.
  - Exercise restriction may be required.

Heartworm Disease

Caused by the parasite *Dirofilaria immitis*, heartworm disease is a well-known infectious cardiovascular disease affecting dogs and cats in the United States. Adult parasites reside in the pulmonary artery and right ventricle of infected animals. Female worms release microfilaria, which are ingested when mosquitoes feed. Larvae that develop in the mosquito subsequently infect the next host (dog, cat, etc.) bitten by the infected arthropod. Clinical signs reflect the stage of infection and can include coughing, dyspnea, fever, weight loss, muscle wasting, ascites, tachycardia, and syncope.

- Diagnostic Tests: ELISA for adult antigen, CBC, Knott's test, radiographs, ECG, and immunofluorescent antibody.
- Potential Treatments: Antiparasitic drugs (adulticides or microfilaricides), anti-inflammatories (NSAID and steroid), supportive symptomatic therapy such as oxygen, and IV fluids. Cats often receive only symptomatic therapy as adulticide treatment is associated with a high death loss.

Technician evaluations commonly associated with this medical condition:

Noncompliant Owner
- Lack of appropriate prophylactic therapy often precedes infection.

Exercise Intolerance
- Extensive lung injury is associated with this condition. Strict exercise restriction 4–6 weeks posttreatment is vital to recovery.

Hyperthermia
- Fever, jaundice, vomiting, icterus, anorexia, and swelling at injection sites can be associated with toxicity to adulticides.

Cardiovascular Insufficiency
- Adult worms (both alive and dead) impede normal blood flow through the heart and lungs, impair pulmonary function, and damage vascular lining. Cardiac output is reduced. Platelet consumption and other coagulopathies such as DIC often develop.

Altered Gas Diffusion
- Supplemental oxygen is often required. Animals undergoing treatment should be closely monitored for dyspnea or other signs of thromboembolism.

Risk of Infection
- Debilitated patients are at high risk for secondary infections.

Risk of Infection Transmission
- *Prior to treatment, animals can infect mosquitoes, which can subsequently infect other animals.*

Acute Pain

Hypertension

Hypertension is persistently high blood pressure. This condition can be primary (rare in animals) or secondary to another disease process such as chronic renal or hepatic disease, hyperthyroidism, diabetes mellitus, pain/fear, head trauma, or Cushing's disease. Clinically normal, unanaesthetized animals exhibit pressures of 160/100 mm Hg (systolic/diastolic). Indirect blood pressures exceeding 200/110 mm Hg (dog) and 190/140 mm Hg (cat) are considered elevated. Organs most sensitive to adverse effects of hypertension include the eye, heart, brain, and kidneys. Clinical signs associated with this condition can include blindness, retinal hemorrhage, retinal detachment, polyuria/polydipsia (PU/PD), vomiting, diarrhea, ascites, syncope, ataxia, disorientation, tachycardia, dyspnea, tachypnea, cyanosis, and epistaxis.

- Diagnostic Tests: Blood pressure measurement (indirect: Doppler and Dinamap; direct: arterial catheter with pressure transducer), CBC, serum biochemistry, urinalysis, radiographs, and ECG.
- Potential Treatments: Treat underlying disease. Antihypertensive drugs such as beta blockers, ACE inhibitors or calcium channel blockers, and diuretics.

Technician evaluations commonly associated with this medical condition:

Overweight
- *Many hypertensive animals are clinically overweight.*

Exercise Intolerance

Cardiac Insufficiency

Inappropriate Elimination
- *Hypertension secondary to diabetes mellitus, Cushing's disease, and hyperthyroidism is associated with PU/PD. Polyuric animals may eliminate inappropriately. Animals placed on diuretics also may urinate inappropriately. Inappropriate urination typically resolves with treatment of the underlying disease or discontinuance of diuretics.*

Altered Sensory Perception
- *Retinal detachment or hemorrhage can result in loss of sight.*

Altered Mentation
- *Hypertension can induce cerebrovascular incidents (strokes), resulting in disorientation and altered states of consciousness.*

Client Knowledge Deficit
- Long-term control of the hypertensive patient may require client education regarding:
  - *Reducing dietary sodium.*
  - *Addressing weight issues.*
  - *Monitoring blood pressure every 2 weeks. Once acceptable pressures are obtained, monitor BP every 1–3 months.*
  - *Watching for signs of hypotension if the animal is receiving drug therapy.*

Pericarditis

The pericardium is a sac or tunic surrounding the heart that functions to prevent cardiac overdilation, provide lubrication, protect from infection or adhesions, and maintain the heart in a fixed position in the thoracic cavity. Inflammation of this tunic is termed *pericarditis*. Pericarditis is often classified as restrictive, in which fibrous adhesion develops (secondary to neoplasia, idiopathic causes, or bacterial or fungal infections) or traumatic (secondary to penetration of a foreign body). Clinical signs are often vague, with exercise intolerance, anorexia, muffled heart sounds, dyspnea, and a prominent jugular pulse being noted. Pericarditis is often accompanied by pericardial effusion.

- Diagnostic Tests: ECG, ultrasound, centesis, radiographs, CBC, and serum biochemistry.
- Potential Treatments: Surgery, antibiotics, diuretics, and vasodilators.

Technician evaluations commonly associated with this medical condition:

Acute Pain
- *Inflammation and effusion can induce the pain response.*

Risk for Cardiac Insufficiency
- *Decreased cardiac output can result from pericardial constriction. Occurring in conjunction with pericardial effusion, pericarditis may initially auscultate as a friction rub, then progress to a muffled sound as fluid accumulates around the heart.*

Decreased Perfusion
- *Hypertrophic and restrictive cardiomyopathies are associated with a higher incidence of thromboembolism.*
- *Animals experiencing pericarditis often exhibit signs similar to those found in right-sided heart failure.*

Exercise Intolerance
- *Reduced oxygenation and cardiac output precludes activity.*

Pneumothorax

This life-threatening condition is caused by accumulation of free air in the chest cavity. Loss of negative pressure in the thoracic cavity results in lung collapse, thereby prohibiting ventilation. A pneumothorax is classified as open, in which there is a tear or an opening of the chest wall, or closed, in which the chest wall is intact. A closed pneumothorax can result from a rupture of a pulmonary bulla, the esophagus, the trachea, or the bronchus. This condition can be further classified as spontaneous or traumatic. A spontaneous pneumothorax is closed and occurs without an apparent inciting cause. Traumatic pneumothorax results from an identified trauma to the chest cavity or lungs. Clinical signs of pneumothorax include dyspnea, panting, an increased respiratory rate, abdominal effort with respiration, decreased lung sounds, tachycardia, and cyanosis.

- Diagnostic Tests: Radiographs.
- Potential Treatments: Thoracocentesis, chest tube, surgery, and antibiotics. Supportive treatments include IV fluids, analgesics, antiemetics, antibiotics, and steroids.

Technician evaluations commonly associated with this medical condition:

Altered Ventilation
- *Lung collapse prohibits ventilation. An untreated tension pneumothorax is rapidly lethal as air continues to enter the chest cavity via a one-way valve/lesion after the lung has collapsed. Ventilation can resume only after normal intrathoracic pressure is reestablished.*
- *Appropriate positioning of the animal to facilitate ease of ventilation is of utmost importance. Avoid placing the animal in dorsal or lateral recumbency. Many animals will naturally assume a sitting position.*
- *Cage rest is important for recovery. Animals should be prevented from engaging in exercise until the condition is resolved.*

Acute Pain
- *Animals experiencing trauma will have associated pain. Animals receiving opioid analgesics should be monitored very carefully due to potential respiratory compromise. Animals developing spontaneous pneumothorax may or may not exhibit signs of pain.*

Risk of Infection
- *Penetration of the thoracic cavity or collapse of the lung tissue makes the animal extremely susceptible to infection. Careful monitoring for nasal discharge, coughing, hyperthermia, and other signs of infection is imperative.*

Impaired Tissue Integrity
- Animals experiencing traumatic pneumothorax may have concurrent dermal wounds. These wounds should be cleaned and cared for as determined by the veterinarian.
- A chest tube may be placed to reestablish negative pressure in the thoracic cavity. A one-way Heimlich valve assists in evacuating air. This valve allows air to leave the thoracic cavity during the expiratory phase of the respiratory cycle. Animals in which chest tubes have been placed should receive continuous monitoring. Tube malfunction results in immediate and severe respiratory compromise.
- Examine the tube for presence of fluid and appropriate functioning of valve. Wipe the external tube surface daily with alcohol. Examine surrounding dermis for signs of redness, creptius, discharge, or swelling. Clean area using diluted chlorhexidine as needed. Apply chest wrap as directed by the veterinarian. Chest wraps should be applied such that expansion of the chest cavity is not reduced. Application of an E-collar helps minimize disturbance of the tube.
Note: Properly secured tubes do not need to be clamped when the animal is moved.

Pneumonia

Pneumonia, which is an inflammation of the lungs, is a common respiratory condition. Causes of pneumonia are numerous and can include viral (canine distemper virus, adenovirus, parainfluenza, and calicivirus), bacterial (*Bordetella*), parasitological (filaroides and aelurostrongylus) or fungal infections, aspiration of ingesta/foreign bodies, neoplasia, contusion, and chemical insult (inhaled). Clinical signs may include lethargy, anorexia, exercise intolerance, cyanosis, dyspnea, tachypnea, coughing, crackles, wheezing, nasal discharge, and fever.

- Diagnostic Tests: Physical examination, radiographs, ultrasound, thoracocentesis, bronchoalveolar lavage, CBC, bronchoscopy, and culture.
- Potential Treatments: Antibiotics, oxygen therapy, pulmonary physiotherapy (nebulization or cuppage), bronchodilators, antitussives, and IV fluids.

Technician evaluations commonly associated with this medical condition:

Hyperthermia
- Viral and bacterial pneumonias can be associated with high fevers.

Altered Gas Diffusion
- Cellular necrosis and inflammatory changes impair gas exchange.
- Oxygen therapy may be necessary to improve gas exchange at the alveolar level. Oxygen levels of 30–50 percent are appropriate for most O_2 cages.

- *Airway humidification via nebulization facilitates removal of respiratory secretions. Nebulization also can be used to administer antibiotics or bronchodilators.*
- *Alter position every 2–4 hours to decrease dependent fluid accumulation in the lungs. Adequate postural drainage facilitates recovery.*
- *Restrict activity. Mild exercise such as slow walking can improve respiratory function but should be discontinued immediately if respiratory distress is noted.*

Acute Pain

Risk of Infection Transmission
- *Many causes of pneumonia are infectious. Utilize appropriate precautions depending on etiology.*

Underweight
- *Animals suffering from extreme respiratory difficulty have a hard time eating. Aspiration of foodstuff can occur. Provide food that is easy to swallow, nutrient-dense, and easily digested.*

Hypovolemia
- *IV therapy may be required to maintain adequate fluid balance; however, caution must be exercised to avoid overhydration and worsening of pulmonary condition.*

Altered Sensory Perception
- *Cats suffering from reduced olfactory senses secondary to nasal discharge may refuse to eat.*

Septal Defects

Also termed intracardiac shunts, atrial septal defects (ASDs) and ventricular septal defects (VSDs) are effectively "a hole" in the wall separating the cardiac chambers. Defects, which are congenital, result in abnormal blood flow patterns through the heart. Clinical manifestations of septal defects reflect the size, location, and pressure differential at the site of the defect. Exercise intolerance and other clinical signs of CHF are typically noted. Animals afflicted by mild defects may remain asymptomatic for years.

- Diagnostic Tests: Physical examination (murmur), radiographs, echocardiography, and Doppler.
- Potential Treatments: Surgery, symptomatic therapy: beta blockers, ACE inhibitors, diuretics, positive inotropes, arterial dilators or venodilators, bronchodilators, oxygen therapy, and IV fluids.

Technician evaluations commonly associated with this medical condition:

Cardiac Insufficiency
- *Overall cardiac output is reduced as blood is shunted from one side of the heart to the other (high to low pressure).*

Hypervolemia
- Fluid retention associated with CHF induces hypervolemia.

Altered Gas Diffusion
- Pulmonary edema interferes with gas exchange.

Exercise Intolerance
- Reduced oxygenation and cardiac output precludes activity.

Risk of Infection
- Certain procedures such as prophylactic dentals are associated with a large bacterial release. Extreme caution should be exercised when treating or anesthetizing animals with a known septal defect.

Client Coping Deficit
- Clients are often distressed to learn the pet is affected by a congenital abnormality. Many owners may consider returning newly acquired pets to the seller/breeder.

Shock

Shock, an inadequate delivery of oxygen and nutrients to the cells, is caused by acute peripheral circulatory failure secondary to loss of circulatory control or intravascular fluids. Shock is not a disease in and of itself; it is more accurately viewed as a clinical manifestation of any disease, condition, or injury that results in a critical decrease in blood flow. The following four categories of shock have been identified:

- Hypovolemic: Is most common. Caused by a decrease in intravascular fluid volume. Secondary to hemorrhage or other fluid volume loss (third spacing and diuresis).
- Cardiogenic: Is associated with decreased cardiac output and high venous pressures. Secondary to heart failure, cardiomyopathy, arrhythmias, and valvular insufficiencies.
- Distributive: Is vasogenic in origin. Secondary to anaphylaxis, sepsis, trauma, and drug reactions.
- Obstructive: Is the physical blocking of cardiac output secondary to conditions such as heartworms, neoplasia, and embolism.

In addition to the classification being identified, shock is further categorized with regard to stage or progression. Stages reflect the clinical appearance and response to treatment as opposed to the underlying etiology.

- Stage One/Compensated shock: Pale mucous membranes, increased CRT, tachycardia, decreased pulse strength, and cool extremities. (Note: In septic shock, hyperemic mucous membranes, fever, and pounding pulses can be noted.)
- Stage Two/Decompensated shock: Tachycardia, decreased pulse strength, cool extremities, muddy mucous membranes, decreased

blood pressure, depressed mental status, hypothermia, and oliguria.
- Stage Three/Irreversible shock: Severe systemic hypoperfusion, petechiation of skin or mucous membranes, hematochezia, and coma.
  - Diagnostic Tests: Variable depending on underlying etiology. Physical examination, blood pressure, ECG, pulse oximetry, blood gas analysis, and CBC.
  - Potential Treatments: Ensure ventilation, oxygen therapy, IV fluids (crystalloid/colloid), positive inotropes, sympathomimetics (dopamine), and steroids.

Technician evaluations commonly associated with this medical condition:

Hyperthermia
- *Hyperthermia is often associated with septic shock.*

Hypothermia
- *The presence of hypothermia implies a poor prognosis.*

Hypovolemia
- *Fluids are critical to recovery. Venous access via catheterization should be established immediately. Crystalloid fluids are most commonly used for initial treatment. Shock fluid delivery rates/types include the following:*
  - *Crystalloids*
    Dog: 80–90 mL/kg/h
    Cat: 50–55 mL/kg/h
  - *Synthetic Colloids*
    Dog: 10–40 mL/kg
    Cat: 5 mL/kg
  - *Plasma: 10–40 mL/kg/day*
  - *Whole Blood: 10–30 mL/kg*
- *Use of synthetic colloids during fluid therapy can reduce the amount of crystalloid fluids required by 40–60 percent.*

Cardiac Insufficiency
- *A reduction in cardiac output accompanies most cases of shock.*

Altered Urinary Output
- *Decreased glomerular filtration and acute renal failure (ARF) result in oliguria and anuria.*

Decreased Perfusion
- *Vascular collapse results in loss of adequate tissue perfusion. Increasing oxygen saturation of blood via supplemental oxygen and maintenance of adequate pressures via use of fluid administration facilitate perfusion. Monitoring central venous pressure (CVP) is useful for assessing efficacy of fluid administration. Normal CVP is 7.5 mm Hg. (Note: To convert mm Hg to cm H_2O, multiply the value by 1.36.) CRT, mucous membrane color, limb temperature, and urine output are useful indicators of perfusion.*

Altered Gas Diffusion
- *Supplemental oxygen administration is useful in maintaining blood oxygen saturation. Oxygenation can be accomplished via use of a mask, a nasal catheter, or an O_2 cage. Monitoring percent saturation via pulse oximetry is very helpful. Normal SpO_2 is > 95 percent. Values < 90 percent imply severe hypoxemia. ABG analysis also can be utilized. A PaO_2 value of < 80 mm Hg is considered hypoxic.*

Altered Mentation
- *LOC can be classified as alert, obtund (dull), stupor, or coma.*

Valvular Insufficiency

Valvular abnormalities are typically an acquired condition, although congenital malformations such as mitral/tricuspid valve dysplasia have been documented. Regardless of the inciting cause, thickened, shrunken, or eroded valve leaflets are unable to prevent backwashing of blood, which results in abnormal blood flow patterns. Any valve can be affected, although chronic degeneration of the mitral valve (right atrioventricular) remains the most common cause of heart failure in dogs. Valvular insufficiency secondary to a bacteremia typically afflicts the aortic or mitral valves. Clinical manifestations of this condition can range from subclinical to CHF. Signs consistent with CHF include anorexia, muscle wasting, ascites, pulmonary edema, lethargy, pallor, syncope, murmurs, increased CRT, weak pulse, coughing, tachypnea, dyspnea, exercise intolerance, and cyanosis.

- Diagnostic Tests: Radiographs, echocardiograms, Doppler, ECG, blood culture, and CBC.
- Potential Treatments: Diuretics, positive inotropes, vasodilators, beta blockers, ACE inhibitors, oxygen therapy, IV fluids, dietary modification, and exercise restriction.

Technician evaluations commonly associated with this medical condition:

Cardiac Insufficiency
- *Use extreme caution when administering IV fluids. Acute decompensation secondary to cardiac overload can occur.*

Hypervolemia
- *Decreased glomerular filtration rate secondary to vasoconstriction, increased sodium retention, and ADH production can result in fluid volume excess.*

Decreased Perfusion
- *Thromboembolism can be associated with endocarditis.*

Altered Gas Diffusion
- *Fluid collection or shifts into the interstitial space/alveoli impair gas exchange.*

Exercise Intolerance
- Reduced oxygenation and cardiac output precludes activity.

Client Knowledge Deficit
- Procedures that induce a bacteremia (prophylactic dental, endoscope exam, and urethral catheterization) and conditions associated with high levels of bacterial shedding (dental disease and recurring infections of skin, prostate, and urinary tract) predispose animals to infectious endocarditis and valvular insufficiency.
- Provide owner education regarding:
  - Use of appropriate diets with restricted salt such Hill's h/d, Purina CV Cardiovascular, or Royal Canin EARLY CARDIAC EC.
  - Appropriate exercise regimes.
  - Clinical signs of prescribed medication toxicities or further cardiac decompensation.

CONDITIONS INVOLVING THE BLOOD AND SPLEEN AND THE LYMPHATIC AND IMMUNE SYSTEMS

Allergic Reactions

Allergies are an abnormal overreactive response by the immune system. The level of the immune response—and, hence, the severity—can vary from mild chronic conditions as is seen with some skin allergies to the extreme life-threatening state of anaphylaxis. Allergic reactions often occur secondary to isolated events such as blood transfusion, venomous insect stings, and drugs (vaccines, hormones, and anesthetics). Alternatively, many allergic conditions are chronic and ongoing in nature. Food hypersensitivity, allergic bronchitis, flea allergy, and atopy are representative of the conditions that require long-term management.

Clinical signs are dependent upon the inciting cause:

- Anaphylaxis: life-threatening; characterized by respiratory difficulty, tachypnea, urticaria (hives), angioedema, pruritis, erythema, excitement, and ataxia. Potential gastrointestinal signs include vomiting, diarrhea, and tenesmus.
- Isolated allergic reaction: pruritis, edema, urticaria, and piloerection.
- Food hypersensitivity: pruritis often involving the head, ears, axillae, and feet; diarrhea; weight loss; vomiting; and military dermatitis (cats).
- Atopy (allergy to inhaled substances such as pollen, grass, and mold): pruritis, face rubbing, alopecia, foot chewing, edema, dermatitis, and otitis externa. Often seasonal initially.
  - Diagnostic Tests: Clinical signs, serology, CBC, and intradermal skin tests.

- Potential Treatments: Extremely variable depending on etiology. Remove inciting cause if possible. Epinephrine, steroids, antihistamines, fluids, hypoallergenic diets, immunotherapy (desensitization), medicated shampoos, and flea control (animal and environment).

Technician evaluations commonly associated with this medical condition:

Noncompliant Owner
- *Food allergies require that all family members ensure that the animal receives only the approved diet. Feeding of treats or table scraps will diminish or eliminate the effectiveness of therapy.*
- *Routine bathing in hypoallergenic or soothing oatmeal shampoos can help remove pollens and other allergens from the animal's coat.*

Altered Gas Diffusion
- *Respiratory distress associated with anaphylaxis typically occurs within minutes of exposure to the inciting cause. Oxygenation through intubation, tracheotomy, or application of a face mask may be warranted.*

Decreased Perfusion
- *The increased vascular permeability and fluid leakage associated with anaphylaxis results in hypotension and tissue hypoxia. Placement of an intravenous catheter and rapid administration of fluids are fundamental to recovery.*

Vomiting
- *Cats in particular are likely to demonstrate gastrointestinal symptoms during acute allergic reactions. Animals suffering from food hypersensitivity also may experience vomiting and diarrhea.*
- *Numerous hypoallergenic diets are available through Hill's, Iams, Royal Canin, and Purina.*

Altered Mentation
- *Apparent excitement or bizarre behavior can accompany anaphylaxis.*

Obstructed Airway
- *Laryngeal edema and bronchial spasm can obstruct the airway.*

Cardiac Insufficiency
- *Decreased cardiac preload and increased capillary permeability (third spacing) can be associated with acute anaphylaxis.*

Anemia

Anemia, which is a decrease in red blood cells, can be caused by any disease capable of decreasing red blood cell production, increasing RBC destruction, or frank blood loss. Anemias are classified morphologically with regard to size (normocytic, microcytic, or macrocytic) and hemoglobin concentration (normochromic or hypochromic). In addition, anemias are classified as regenerative/nonregenerative depending on the presence or absence of reticulocytes (immature RBCs). Clinical signs of anemia can include weakness, exercise intolerance, dyspnea, tachycardia, pale mucous membranes, anorexia, melena, splenomegaly, icterus, pica, and petechial hemorrhages.

- Diagnostic Tests: CBC, serum chemistry, urinalysis, fecal, Coomb's test, thyroid function tests, and bone marrow biopsy.
- Potential Treatments: Extremely variable depending on inciting cause. Blood products, antibiotics, steroids, chemotherapy, iron, erythropoietin, and IV fluids.

Technician evaluations commonly associated with this medical condition:

Hypothermia
 - Anemic animals often experience difficulty maintaining body temperature.

Exercise Intolerance
 - Exercise intolerance results from hypoxia secondary to diminished blood oxygen carrying capacity.

Risk for Infection

Risk for Underweight
 - Chronic hypoxia can result in reduced digestion and decrease appetite.

Altered Sensory Perception
 - Hypoxia secondary to severe anemia can affect the occipital cortex, thereby resulting in cortical blindness.

Self-Care Deficit
 - Depression and lethargy caused by the anemia can lead to lack of self-grooming behaviors.

Abnormal Eating Behavior
 - Anemic animals often exhibit pica.

Atopy

Also termed *allergic inhalant dermatitis*, atopy remains a common allergy of dogs and cats. Susceptible animals inhale allergens (pollen, grass, mold, etc.) and subsequently mount an inflammatory allergic response. Pruritis, which is initially seasonal, then progresses to a year-round affliction, is the primary clinical sign. Foot chewing, edema, alopecia, otitis externa, hyperpigmentation, and eczema also may be noted. A small percentage of animals exhibit rhinitis and asthma.

- Diagnostic Tests: CBC, serology, and intradermal testing (ELISA and RAST) to identify offending allergen.
- Potential Treatments: Hyposensitization therapy, symptomatic control (antihistamines and steroids), dietary management, and avoidance of offending allergen.

Technician evaluations commonly associated with this medical condition:

Self-Inflicted Injury
- *Intense pruritis, especially of the face, axillae, and abdomen, can induce repetitive scratching and self-mutilation.*

Risk of Infection
- *Self-traumatized areas are susceptible to secondary bacterial* (Staphylococcus) *and yeast* (Malassezia) *infections.*

Impaired Tissue Integrity
- *Self-trauma leads to dermal lesions. Use of soothing oatmeal shampoos can provide some relief from discomfort and removes allergens from the hair coat. Weekly bathing may be warranted. Dietary supplementation of fatty acids or use of diets such as Royal Canin SKIN SUPPORT SS is often indicated.*

Altered Oral Health
- *Continual use of the incisors to scratch pruritic areas leads to excessive wear and loss of enamel surface.*

Client Knowledge Deficit
- *Owners often become frustrated due to the chronic nature of the problem. The majority of dogs (75 percent) eventually develop year-round pruritis. Owners' reports of sleep loss due to continual scratching noises are common. Removal of pets' collars prior to bedtime will decrease the amount of noise during sleeping hours.*

Coagulopathies

For normal hemostasis (blood clotting) to occur, an animal must have adequate platelets, adequate coagulation proteins, and a normal vascular response. Abnormalities in any of these confines results in coagulopathy. Coagulopathies can be congenital or acquired. Examples of congenital disorders include Factor VII, Factor VIII (hemophilia A), and Factor IX (hemophilia B) deficiencies; cyclic hematopoiesis of gray collies; Chediak-Higashi syndrome; and von Willebrand's disease. Acquired coagulopathies are known to result from DIC, ingestion of rodenticides, immune- and drug-related thrombocytopenias, and vasculitis.

Clinical signs of both congenital and acquired coagulopathies include epistaxis, melena, hematuria, prolonged bleeding times (incisions or injection sites), petechial and ecchymotic hemorrhages, lethargy, pallor, and depression.

- Diagnostic Tests: Bleeding times, activated clotting times (ACTs), prothrombin time (PT), activated partial thromboplastin time (APTT), fibrinogen, von Willebrand factor (vWF) assay, platelet count, CBC, and serum chemistry.

- Potential Treatments: Variable depending on cause. Treat underlying condition and toxicity. Supportive blood/component transfusion, vitamin K, IV fluids, and steroids.

Technician evaluations commonly associated with this medical condition:

Exercise Intolerance
- *Coagulopathies can induce concurrent anemia. Lethargy and exercise intolerance occur secondary to hypoxia. Exercise appropriate caution when selecting a physical activity suitable for an animal with known coagulopathies.*

Impaired Tissue Integrity
- *Exercise extreme caution during parenteral drug administration and placement of catheters. Apply pressure wraps when appropriate. Monitor surgical incision sites for prolonged bleeding times.*

Risk for Infection

Risk for Hypovolemia

Altered Mentation
- *Animals may exhibit lethargy and depression.*

Disseminated Intravascular Coagulation (DIC)

DIC is a medical condition that occurs secondary to another disease process. Neoplasia, renal failure, Cushing's disease, peritonitis, hepatitis, FIP, heartworm disease, blood transfusions, pancreatitis, heatstroke, and snake bites have all been associated with this life-threatening condition. DIC is considered a consumption coagulopathy in that inadvertent activation of the clotting mechanisms initiates rapid consumption of platelets and clotting factors. This leaves the animal susceptible to severe bleeding. Clinical signs associated with DIC include epistaxis, oral bleeding, melena, hematuria, ecchymoses, and petechial hemorrhage.

- Diagnostic Tests: CBC, PT, APTT, ACT, FDP assay, latex agglutination test, and urinalysis.
- Potential Treatments: Treat inciting cause; fluid therapy; blood, plasma, or blood component transfusion; heparin; and aspirin.

Technician evaluations commonly associated with this medical condition:

Impaired Tissue Integrity
- *Inadequate platelet numbers leave animals susceptible to profuse bleeding from injection and catheter sites. Intramuscular injections should be avoided, and only peripheral blood vessels should be used for catheters and blood draws (avoid jugular).*
- *Apply adequate compression postinjection to minimize hemorrhage or hematoma formation. Application of a compression bandage postinjection*

is advisable. Provide direct animal monitoring for several minutes postinjection.

Decreased Perfusion
- DIC is associated with reduced perfusion of many organ systems. Fluid replacement therapy is critical to maintaining adequate perfusion and promoting recovery. Care should be taken, however, to monitor for volume overload. Volume overload is a critical concern for animals experiencing an associated anuria.

Risk of Infection
- DIC renders animals especially vulnerable to secondary bacterial infection. Reverse isolation techniques should be implemented to minimize secondary infections.

Decreased Perfusion
- Severe anemia inhibits oxygen carrying capacity of the blood, resulting in poor tissue oxygenation. Animals may require supplemental oxygen via mask, cage, or nasopharyngeal catheter.

Altered Oral Health
- Oral bleeding is a common sequela of this condition. Feed only soft foods. It also is advisable to discontinue oral health care regimes such as teeth brushing until the condition has resolved.

Exercise Intolerance
- DIC renders animals extremely susceptible to hemorrhaging as a result of normal physical activities. Concurrent anemia makes exercise very stressful. Strict confinement should be instituted during episodes of DIC.

Hemorrhage

Hemorrhage is the escape of blood from a ruptured vessel, and its causes are as numerous as they are varied. Hemorrhage, which can occur from arteries and veins, can be classified as internal (into body cavities) or external or directly into tissues. Events/conditions commonly associated with hemorrhage include blunt trauma, laceration, exercise-induced pulmonary hemorrhage, and coagulopathies. Hemorrhage can be manifested as the presence of frank blood, melena, bruising, tachycardia, or lethargy.

- Diagnostic Tests: Varied depending on inciting cause. CBC, coagulation tests, radiographs, ultrasound, and chemistry profile.
- Potential Treatments: Determine cause and treat appropriately.

Technician evaluations commonly associated with this medical condition:

Client Coping Deficit
- Blunt trauma such as being hit by a car often causes extreme injury to the animal. Providing a calm, secure environment for owners is

critical after they have experienced a sudden, emotionally stressful event.

Hypothermia
- Animals experiencing excessive blood loss have a difficult time maintaining body temperature.

Risk of Infection
- Blood loss debilitates animals, making them susceptible to other diseases.

Tissue Integrity Impaired

Cardiac Insufficiency
- Excessive blood loss diminishes cardiac output. IV fluids are required to maintain intravascular pressures. Selection of crystalloid versus colloid fluids is subject to the inciting cause.

Exercise Intolerance
- Loss of blood impairs oxygen carrying capacity.

Altered Ventilation
- Animals can experience tachypnea.

Decreased Perfusion
- Decreased blood volume can result in decreased blood pressure. Low pressures are associated with loss of tissue perfusion and a decreased glomerular filtration rate.

Acute Pain

Immunodeficiencies

Immunodeficiency is the lack of adequate response by the humoral- or cell-mediated components of the immune system. Causes of this condition are numerous and can be congenital or acquired. Examples of congenital causes include selective immunoglobulin A (IgA) deficiency in beagles, cell-mediated deficiency in Birman cats, and granulocytopathy syndrome in Irish setters. Acquired causes of immunodeficiency are more prevalent and can include FIV, FeLV, canine distemper virus, parvovirus, and numerous antineoplastic drugs. Recurrent or persistent infection is the hallmark of immunodeficiency.

- Diagnostic Tests: Highly variable depending on inciting cause. CBC and serology.
- Potential Treatments: Antimicrobials, antifungals, and immunomodulators.

Technician evaluations commonly associated with this medical condition:

Risk of Infection
- *Immunodeficient animals should remain current on vaccinations. Modified live vaccinations are avoided.*

Vomiting
Diarrhea
Dehydration
Underweight
Risk of Infection Transmission
Acute Pain
Reproductive Dysfunction

Immune-Mediated Hemolytic Anemia (IMHA)

Defined as an autoimmune disease in which autoantibodies are directed against the red blood cells, IMHA is a relatively common hemolytic disorder of dogs. Clinical signs can include weakness, pale mucous membranes, icterus, vomiting, anorexia, tachycardia, dyspnea, pounding pulses, melena, and discolored urine. This condition can be classified as primary, in which antibodies target the RBC cellular membrane directly, or secondary, in which antibodies mistakenly destroy the RBC.

- Diagnostic Tests: CBC with RBC morphology, reticulocyte count, serum chemistry profile, urinalysis, coagulation panel, direct antibody test, and serology (secondary IMHA).
- Potential Treatments: Blood transfusion, immunosuppressive therapy (prednisone), anticoagulant therapy, gastric protectants, antibiotics, and chemotherapy (secondary IMHA).

Technician evaluations commonly associated with this medical condition:

Hypothermia
- *Anemic animals often experience difficulty maintaining body temperature.*

Exercise Intolerance
- *Exercise intolerance results from hypoxia secondary to diminished blood oxygen carrying capacity.*

Altered Gas Diffusion
- *Anemia reduces the oxygen carrying capacity of the blood and results in inadequate saturation at the alveolar level despite adequate availability of oxygen.*

Decreased Perfusion
- *Thromboembolism remains one of the leading causes of death secondary to IMHA.*

Risk for Infection
- *Secondary IMHA results from antibodies that mistakenly confuse the membrane antigens of infectious agents or neoplasia with the surface antigens on the RBC. Infectious agents known to incite this condition include* Dirofilaria, Ehrlichia, Anaplasma, Rickettsia, *Babesia, FeLV, and FIV.*

Self-Care Deficit

Pemphigus

Classified as an autoimmune response to the basal layer of the dermis (skin), pemphigus is an uncommon disease complex causing dermal ulceration. Autoantibodies present in the skin induce a separation of the cells within the skin layers. The resultant separation is manifested as ulceration and erosion of the dermis. Several distinct forms of pemphigus have been identified and include foliaceus, vulgaris, erythematosus, vegetans, and bullous. Although the location, size, and severity of lesions may vary, all forms are characterized by erosions, ulcerations, and encrustation of the skin along mucocutaneous junctions. Additional clinical signs can include alopecia, erythema, and pustules.

- Diagnostic Tests: Cytology, biopsy with histopathology, and antinuclear antibody (ANA) testing.
- Potential Treatments: Steroids, immunosuppressive (cyclophosphamide), and vitamin E.

Technician evaluations commonly associated with this medical condition:

Risk of Infection
- *High doses of immunosuppressives required to control the disease can predispose animals to secondary infections. Instruct clients to monitor for pyodermas, candidiasis, UTI, and respiratory infections.*

Impaired Tissue Integrity
- *Ulcerations typically occur on face (lip, ears, and eyes) and perineal mucocutaneous junctions. Mild antiseptics such as diluted chlorhexidine can be used to gently remove encrustations.*
- *Animals should avoid direct UV light (between 10 AM and 2 PM). Application of topical sunscreens to depigmented areas may be of value in treatment of erythematosus.*

Acute Pain
Self-Inflicted Injury
- *Animals with severe ulcers may lick or rub affected areas, worsening the condition.*

Rheumatoid Arthritis

Rheumatoid arthritis is a chronic inflammatory autoimmune joint disease of unknown etiology. It is characterized by progressive erosion and destruction of affected joints. Clinical signs can include lameness, joint pain, joint crepitus, joint swelling, fever, depression, and anorexia.

- Diagnostic Tests: Radiographs, synovial fluid evaluation, rheumatoid factor (RF) test, and synovial biopsy.
- Potential Treatments: Symptomatic treatment. Medications can include steroids, immunosuppressive agents, gold salts, and analgesics.

Technician evaluations commonly associated with this medical condition:

Acute Pain or Chronic Pain
Reduced Mobility
Client Coping Deficit
- *Rheumatoid arthritis is a progressive, debilitating disease for which there is no cure.*

Septicemia

Septicemia is a systemic disease associated with the presence of pathogenic microorganisms or their toxins in the vascular system. Clinical signs associated with septicemia include fever, petechial hemorrhages, tachycardia, hyperemic mucous membranes, and depression. Additional clinical signs can reflect the pathogen's primary site of infection. A percentage of septic animals may progress to septic shock, which results in circulatory collapse and a mortality approaching 90 percent. Bacteria implicated in septicemias include *Escherichia coli, Klebsiella, Pseudomonas, Salmonella, Staphylococcus,* and *Streptococcus.*

- Diagnostic Tests: CBC, blood culture (aerobic and anaerobic), radiographs, and cerebrospinal fluid (CSF) tap and culture.
- Potential Treatments: IV fluids and antibiotics.

Technician evaluations commonly associated with this medical condition:

Hyperthermia
Hypovolemia
Risk of Infection Transmission
Vomiting
Diarrhea
- *Diarrhea is often bloody when animals are in decompensatory septic shock.*

Cardiac Insufficiency
- *Initially, severely septic animals may exhibit tachycardia, pounding pulse, rapid CRT, and hyperemic mucous membranes. Animals that have decompensated progress to bradycardia, thready pulses, pale mucous membranes, prolonged CRT, peripheral edema, and cool extremities.*

Splenomegaly

Splenomegaly, an enlargement of the spleen, can result from a myriad of causes. Neoplasia, inflammation (secondary to bacterial, virusal, or fungal infections), torsion, abscessation, accumulation of hemolyzed red blood cells, and congestive splenomegaly secondary to drugs can all incite splenic enlargement. Clinical signs reflect the underlying cause

and vary from subclinical to acute abdominal pain with vomiting and depression.
- Diagnostic Tests: CBC, biopsy, ultrasound, radiographs, and exploratory surgery.
- Potential Treatments: Splenectomy, antibiotics, antifungals, IV fluids, antiemetics, and analgesics.

Technician evaluations commonly associated with this medical condition:

Vomiting
Hypovolemia
- *Reduced fluid intake and vomiting lead to dehydration.*

Acute Pain
- *Splenic torsion is particularly noted for inducing a pain response. Torsion often occurs concurrent to GDV.*

Risk for Infection
- *Splenectomized animals are highly susceptible to sepsis. The immune response of splenectomized animals is depressed; therefore, caution should be exercised and reverse isolation techniques considered.*

Cardiac Insufficiency
- *Splenic torsion impairs blood supply.*

Underweight
- *Catabolic states associated with neoplasia result in weight loss.*

Systemic Lupus Erythematosus (SLE)

SLE is an immune-mediated disease; a definitive cause has not been identified. High levels of circulating antibodies attack erythrocytes, thrombocytes, and leukocytes. Inflammation associated with these antibodies affects multiple body systems and leads to vasculitis, glomerulonephritis, polyarthritis, and dermatitis. Clinical signs reflect the organ system affected by the inflammation. Shifting leg lameness, painful joints, muscle wasting, skin and oral ulcerations, petechiae, icterus, ascites, edema, lethargy, anorexia, fever, and weakness are potential signs of this condition.

- Diagnostic Tests: CBC, serum chemistry, urinalysis, antinuclear antibody (ANA) test, LE cell preparation, Coomb's test, arthrocentesis, and muscle biopsy.
- Potential Treatments: Symptomatic treatment based on reducing inflammation. Steroids, immunosuppressive agents, and analgesics.

Technician evaluations commonly associated with this medical condition:

Acute Pain or Chronic Pain
- *Arthritis associated with SLE is very painful.*

Altered Oral Health
- Animals with SLE are likely to experience oral ulceration. Rinsing the animal's mouth with a dilute chlorhexidine solution can help minimize ulcer contamination. Soften diet as necessary.

Client Coping Deficit
- SLE is a severe, incurable disease. Forty percent of dogs diagnosed with this condition are deceased within the year. Educating owners regarding treatment protocols and the necessity of minimizing secondary infections is imperative.

Hyperthermia
Exercise Intolerance
Risk of Infection
- Animals suffering from an autoimmunity have an increased susceptibility to infectious agents. Bronchopneumonia and septicemia are the most frequent causes of death associated with SLE.

Thrombosis

Thrombosis is the formation of a blood clot, or thrombus, in the vascular system. This condition can occur secondary to blood stasis, abnormal endothelium (vessel lining), or decreased fibrinolysis. Although it is seen in both arteries and veins, thrombosis occurs with greater frequency in veins. Thrombi leaving the site of formation and traveling through the vascular are termed *emboli*. Cardiac and pulmonary tissues are frequently affected by the emboli. Clinical signs, which can include edema, cool extremities, paralysis, coughing, dyspnea, or sudden death, reflect the location, degree of vascular occlusion, and level of associated tissue ischemia.

- Diagnostic Tests: Physical examination, CBC, serum biochemistry, ultrasound, and radiographs.
- Potential Treatments: Supportive: anticoagulants, surgery, analgesics, and antibiotics.

Technician evaluations commonly associated with this medical condition:

Exercise Intolerance
- Physical activity can encourage detachment of a thrombus. Exercise restriction is critical.

Decreased Perfusion
- Inadequate perfusion is a hallmark of thrombus formation.

Cardiac Insufficiency
- Many animals suffer from underlying cardiovascular disease.

Hypovolemia
Acute Pain
- Ischemia induces a pain response.

Altered Gas Diffusion
- Pulmonary emboli prevent oxygen exchange.

Bowel Incontinence
- A distal aorta thrombus (saddle thrombus) results in ischemic neuromyopathy.

Reduced Mobility
- A distal aorta thrombus (saddle thrombus) results in ischemic neuromyopathy and paralysis. Thrombosis of the distal limb results in localized ischemia, acute pain, and loss of use.

CONDITIONS INVOLVING THE REPRODUCTIVE AND ENDOCRINE SYSTEMS

Abortion

Abortions can occur secondary to fetal abnormalities (chromosomal defects and abnormal development), maternal problems, or infectious disease. Maternal reasons for abortion are numerous and include any generalized systemic disease, nutritional deficiency, decreased progesterone production, hypothyroidism, or toxic insults. Infectious agents commonly cited include *Brucella canis, Escherichia coli,* herpesviruses, *streptococci, Toxoplasma gondii,* FeLV, *Campylobacter,* and *Parvovirus.* Clinical signs associated with spontaneous abortion can include vaginal discharge, fever, listlessness, and abdominal pain.

Abortifacients (mismate injection) or agents intentionally administered to induce abortion include estrogens and prostaglandins. Ovariohysterectomy (OHE) also is an option.

- Diagnostic Tests: CBC, serum chemistry, radiographs, and ultrasound. Tests used to determine cause of abortions include vaginal culture, serum antibody titers, fetal and placental tissue cultures, and histology.
- Potential Treatments: Antibiotics, IV fluids, and hormones (prostaglandins).

Technician evaluations commonly associated with this medical condition:

Reproductive Dysfunction
- Obtain a complete medical history relating to previous and current pregnancy. Specifically question owner regarding any medications, supplements, or substances administered during the previous week.

Hyperthermia
Risk of Infection

Addison's Disease

Also referred to as hypoadrenocorticism, Addison's disease results from reduced hormone production by the adrenal glands. Hypoadrenocorticism is classified as primary or secondary. Problems arising directly from within the adrenal gland are considered primary, while those resulting from pituitary or hypothalamus abnormalities are termed secondary. The normal hormonal axis for these glands involves release of corticotropin-releasing hormone (CRH) from the hypothalamus, which stimulates pituitary release of adrenocorticotropic hormone (ACTH), stimulating the release of mineral and glucocorticoid hormones from the adrenal gland. Clinical signs can include lethargy, anorexia, vomiting, PU/PD, muscle weakness, weight loss, melena, and hypothermia.

Uncontrolled cases may fulminate in a life-threatening state termed *addisonian crisis*. Shock results from the severe hyperkalemia and hyponatremia, which induce hypotension, hypovolemia, azotemia, and cardiac arrhythmias.

- Diagnostic Tests: Serum chemistries, serum cortisol, and ACTH stimulation test.
- Potential Treatments: Supportive treatment such as IV fluids (imperative) and steroids. Long-term control: corticosteroids (prednisone) and mineralocorticoids (fludrocortisone).

Technician evaluations commonly associated with this medical condition:

Vomiting
Diarrhea
Hypovolemia
  - *Normal saline (0.9 percent NaCl) is often the fluid of choice.*
Electrolyte Imbalance
  - *Hyperkalemia and hyponatremia are present in 80–90 percent of animals. Monitoring of these electrolytes during fluid administration is advised.*
Hypothermia
  - *Addisonian animals frequently experience difficulty maintaining body temperature and will exhibit heat-seeking behaviors. Providing a heat source such as warmed blankets will comfort animals.*
Underweight
  - *Use of high-calorie diets is indicated until condition is stabilized and appropriate weight has been attained. A high-energy adult diet such as Royal Canin INTESTINAL HE or Hill's p/d may be used. If a growth diet is used to allow the animal to regain weight, most animals can return to age-appropriate diets once they are stabilized.*

Reduced Mobility
- Profound muscle weakness can result in ataxia.

Client Knowledge Deficit
- Client education is imperative as this disease requires lifelong treatment of the animal. In addition to understanding the necessity of prescribed medications and routine blood work, owners should learn how to monitor animals carefully for vomiting, diarrhea, PU/PD, anorexia, or heat-seeking behavior. Blood work is indicated every 6–12 months once the animal is stabilized.

Agalactia

Agalactia is a lack of milk production in the absence of mammary disease. This condition can occur secondary to stress, disease conditions excluding the mammary tissue, premature parturition, or hormonal abnormalities. Agalactia can be temporary, as is seen with most stress-induced cases of the condition. True agalactia is rarely seen in dogs or cats. However, these species can experience a stress-induced decrease in milk production or be hereditarily prone to poor milk production.

- Diagnostic Tests: CBC and serum chemistry.
- Potential Treatments: Tranquilizers and hormones (oxytocin).

Technician evaluations commonly associated with this medical condition:

Ineffective Nursing
- Neonates must be hand-raised using commercial milk replacer but should remain with dam.

Cushing's Disease

Also referred to as hyperadrenocorticism, Cushing's disease results from increased hormone production (cortisol) by the adrenal glands. Hyperadrenocorticism is classified as pituitary-dependent (tumor of the pituitary gland), adrenocortical-dependent (tumor of the adrenal gland), or iatrogenic. Pituitary-dependent is more common than adrenocortical forms. Iatrogenic Cushing's is associated with inappropriate (increased) administration of steroids. Steroids used for treatment of chronic skin and ear infections are frequently implicated. Clinical signs can include PU/PD, polyphagia, panting, a pendulous abdomen, alopecia, muscle wasting, lethargy, calcinosis cutis, hyperpigmentation, and bruising.

- Diagnostic Tests: CBC, serum chemistry, urinalysis, endogenous ACTH, ACTH stimulation, dexamethasone suppression test, ultrasound, radiographs, and MRI.
- Potential Treatments: Mitotane, ketoconazole, adrenalectomy, Anipryl, and radiation therapy of pituitary.

Technician evaluations commonly associated with this medical condition:

Urinary Incontinence
- *Previously housetrained animals may urinate inappropriately as a result of polyuria caused by increased cortisol levels. These animals are likely to resume normal behaviors once the disease is controlled. During the interim, owners should be advised to modify the animal's environment to accommodate the increased frequency and urgency of urination.*

Risk of Infection
- *Animals suffering from Cushing's frequently experience concurrent urinary tract infections. Owners should be instructed to monitor for signs of cystitis. Routine monitoring of urine for infection may be advised.*

Impaired Tissue Integrity
- *Cutaneous bruising is a common complication of this disease. Caution should be exercised when obtaining blood samples or placing intravenous catheters.*

Underweight
- *Muscle wasting is associated with elevated steroid (cortisol) levels. The pendulous abdomen, or potbelly, of many animals results from loss of normal abdominal musculature. Animals should receive diets containing a high-quality protein.*

Vomiting
- *Animals receiving treatment frequently experience vomiting.*

Client Knowledge Deficit
- *Client education is imperative as this disease requires lifelong treatment of the animal. In addition, the owner should be advised that complications associated with treatment and medications are frequent.*
- *Prognosis is dependent upon the inciting cause of hyperadrenocorticism. Adrenalectomized animals require daily hormone replacement therapy (prednisone), and owners should be instructed to monitor for signs of hypoadrenocorticism. Animals receiving mitotane should be monitored for signs of continued Cushing's, Addison's, vomiting, neurologic abnormalities, and gastric irritation. Frequent blood work to monitor the animal should be anticipated.*

Diabetes Insipidus

Diabetes insipidus is a complex disease resulting from inadequate production of antidiuretic hormone (ADH) or renal insensitivity to ADH. In the normal animal, ADH functions to promote water reabsorption from the renal tubules. Therefore, inadequate ADH results in excessive water

excretion and dilute (low specific gravity) urine. Clinical signs include PU/PD, nocturia, weight loss, and incontinence.
- Diagnostic Tests: CBC, serum chemistry, urinalysis, water deprivation test, ADH response test, and CT of pituitary.
- Potential Treatments: ADH (vasopressin) supplementation and fluids.

Technician evaluations commonly associated with this medical condition:

Hypovolemia
- *Animals should be provided unlimited access to water. Restricted access to water, even for a short period of time, can lead to catastrophic results in these animals. Instruct owners to monitor closely for signs of dehydration during strenuous exercise, hot weather, or any circumstance that increases water demands.*

Electrolyte Imbalance
Urinary Incontinence
- *Inability to reabsorb water results in large volumes of dilute urine. Providing environmental modifications such as dog doors minimizes destruction of property while simultaneously meeting the animals increased urine volume and frequency needs. Most animals can be brought to an acceptable level of continence with proper treatment.*

Underweight
Sleep Pattern Disturbance
- *In many cases, animals must urinate several times during the night. Providing environmental modifications such as dog doors or absorbent pads decreases the frequency of disruptions.*

Diabetes Mellitus

Diabetes mellitus is a complex disease involving abnormal glucose uptake and utilization. A decrease in cellular uptake of glucose results in hyperglycemia and glycosuria. This condition can result from a lack of insulin production by the pancreas (insulin-dependent diabetes mellitus [IDDM], type 1) or cellular insensitivity to the effects of insulin (non-insulin-dependent diabetes mellitus [NIDDM], type 2). IDDM is seen with greater frequency than NIDDM. Clinical signs associated with diabetes include PU/PD, polyphagia, cataracts, and weight loss.

Diabetic ketoacidosis is a severe, life-threatening condition resulting from untreated, uncontrolled diabetes. Animals failing to metabolize carbohydrates properly begin to mobilize fat and protein stores, thus inducing the formation of ketones. Excessive ketone production induces a state of metabolic acidosis. Acidotic animals can exhibit a variety of clinical signs, including dehydration, vomiting, lethargy, anorexia, tachypnea, and weakness. The severity of clinical signs typically reflects the degree of acidosis.

- Diagnostic Tests: CBC, serum chemistry, urinalysis, and blood glucose curves (treatment evaluation).
- Potential Treatments: Insulin therapy, glipizide, and dietary modification.

Technician evaluations commonly associated with this medical condition:

Client Knowledge Deficit/Client Coping Deficit
- *Treating and managing a diabetic animal is a frightening concept for many people. The prospect of performing injections and checking blood glucose can be daunting. Providing emotional and technical support is fundamental to successful management of this disease. Diet, injection techniques, insulin care, and exercise should be addressed.*
- *The following are common concepts that must be addressed with owners regarding dietary therapy, injection technique, insulin care, and exercise:*
    - Insulin Care
        - Mix insulin prior to administration by gently rolling in hands. Do not shake bottle.
        - Replace insulin every 60 days. Do not use expired products.
    - Injection Information
        - Demonstrate injection technique to client.
        - Provide written instructions for injection techniques to inexperienced owners. (Injections can be administered s.q. anywhere from midneck to caudal aspect of rib cage.)
        - Observe owner administering injection to verify technique and ability to draw up correct dose (units of insulin).
        - Inform owner that:
            - He or she should not "double up the dose" if an injection was missed.
            - He or she should not administer a second injection if unsure whether "first one went in."
        - Provide instruction on proper disposal of used needles and syringes.
        - Instruct owner to inform veterinary staff of diabetic condition if the animal is seen at a clinic other than primary care facility.
    - Diet
        - Animals must be within normal weight. (Obesity is common.) If weight loss is recommended, it should be slow and gradual. Underweight animals should be fed high-quality, calorie-dense diets.
        - High-fiber diets help minimize postprandial fluctuations in glucose (Hill's w/d or Royal Canin DIABETIC HF or CALORIE CONTROL CC HIGH FIBER).
        - High-protein diets seem to work well in cats (Hill's m/d, Royal Canin DIABETIC DS, and Purina DM).
        - Calorie content of each meal must be consistent.

- The animal should not be fed table scraps or a nonspecified diet.
- Timing of meals and insulin injection must be consistent.
- Water should be available at all times. Monitor the animal's volume of water consumption.
- The animal should be fed at the time of insulin injection if insulin is administered b.i.d. If insulin is administered s.i.d., the animal should be fed at time of injection and 8–10 hours postinjection.
  - Exercise
    - Moderate daily exercise assists in maintaining glucose levels within a stable range. Strenuous, sporadic exercise that stresses the dog should be avoided. Exercise can be encouraged in cats through daily play, but this is often difficult to achieve.
  - Monitor for Inadvertent Hypoglycemia
    - Keep corn syrup (any sugar solution can be used) available in case of a hypoglycemic emergency.
    - Hypoglycemia can occur if insulin is administered to an anorectic animal or to an animal after strenuous exercise.
    - Clinical signs of hypoglycemia include lethargy, weakness, ataxia, head tilt, and seizures. Instruct owners to administer corn syrup, keep the animal warm, and notify the veterinarian if signs of hypoglycemia are observed.
  - Monitor for Signs of Uncontrolled or Continued Hyperglycemia, PU/PD, and Cataracts
    - Blood glucose level may require monitoring.

Risk for Infection
  - Glycosuria predisposes animals to urinary tract infections. Owners should be instructed to monitor for signs of cystitis. Routine monitoring of urine for infection is advised.

Overweight
  - See previous dietary recommendations.

Inappropriate Elimination
  - Previously house/litter box-trained animals may urinate inappropriately as a result of PU/PD. These animals are likely to resume normal behavior once the disease is controlled. During the interim, owners should be advised to modify the animal's environment to accommodate the increased frequency and urgency of urination.

Altered Mentation
  - Ketoacidosis can result in "drunken" behaviors prior to more severe signs.

Impaired Mobility
  - Cats with diabetes mellitus polyneuropathy may demonstrate hind limb (plantar grade) ataxia "walking on hocks."

Altered Sensory Perception
  - Blindness secondary to retinitis or decreased visual acuity secondary to cataracts may lead to fearful or aggressive behaviors.

Dystocia

Dystocia is difficult labor. Causes are numerous and can result from maternal or fetal conditions. Decreased pelvic diameter secondary to obesity, prior fracture, or genetics is frequently sited in dams. Fetal abnormalities can include malformed fetus, increased size, or abnormal presentation. Uterine inertia (failure of uterus to contract) secondary to prolonged labor or electrolyte abnormalities also is a commonly sited cause of dystocia. Dystocia is considered when active straining occurs for more than 45 minutes or the resting phase (time between puppies) of delivery lasts longer than 4–6 hours. Clinical signs of dystocia include active straining, licking of vulva, or presence of immobile fetus in vulva.

- Diagnostic Tests: Palpation and radiographs.
- Potential Treatments: Oxytocin, manual extraction, IV fluids, calcium supplementation, and c-section.

Technician evaluations commonly associated with this medical condition:

Reproductive Dysfunction
Acute Pain
- *Labor and delivery is a painful experience for the animal; however, analgesics are rarely administered for routine deliveries in veterinary medicine. Analgesics may be delivered during an episode of dystocia. Uncontrolled pain may be a precipitating cause in the dam's failure to develop offspring attachment.*

Ineffective Nursing
- *Dams experiencing difficult delivery may have reduced milk production or fail to bond with offspring. Monitor very closely for the first 24 hours. First-time mothers have a greater likelihood of experiencing difficulties.*

Constipation
- *Defecation is often painful postparturition. Vulvar tears or swelling, abdominal tenderness, dry stool secondary to decreased fluid intake, or fear of leaving offspring can all contribute to postparturition constipation.*

Risk of Infection
- *Dams should be monitored closely for abnormal vaginal discharge or other signs of infection. Lochia (normal discharge seen after delivery) will initially be red (bloody color). The discharge should have minimal odor. The color will become browner with time, although some dogs may demonstrate a dark green coloration. A septic smell should never be associated with the discharge. The lochia typically become serious by Day 3. Animals may experience a small volume of serous discharge for 2 weeks postdelivery.*

Self-Care Deficit
- *Stressed or painful females may require assistance in removing lochia from perineal hair. Females requiring cesarean sections should have all*

antiseptic washed from teat area prior to puppies or kittens being permitted to nurse.

Anxiety
- *Animals should be discreetly and closely monitored postpartum to ensure appropriate maternal behavior. Extremely anxious females may traumatize offspring.*

Eclampsia

Also termed *periparturient hypocalcemia, lactation tetany,* and *puerperal tetany,* eclampsia results from decreased serum calcium levels. Clinical signs of this life-threatening condition typically develop 1–3 weeks postpartum. Small breed dogs nursing large litters are most commonly affected. Clinical signs include ataxia, muscle weakness, shivering, anxiety, panting, drooling, muscle fasciculations, tachycardia, convulsions, and seizures. Owners should be informed that eclampsia is likely to reoccur with future litters.

- Diagnostic Tests: History, clinical signs, and serum electrolyte (calcium) levels.
- Potential Treatments: IV calcium, IV fluids, and oral calcium supplementation.

Technician evaluations commonly associated with this medical condition:

Electrolyte Imbalance
- *Administration of IV calcium and fluids is the treatment of choice.*

Hyperthermia
- *Muscle fasciculations secondary to low calcium levels cause a rapid and potentially lethal increase in core body temperature.*

Inadequate Nursing
- *Puppies should not be permitted to nurse until the condition has been completely resolved. This may necessitate removing the puppies from the dam for a period of time. Puppies older than 3 weeks should be removed from the dam. If puppies remain on the dam, nutritional supplementation of a commercial milk replacer must be provided to decrease the lactational stress on the mother.*
- *Dams that are able to resume nursing should be placed on a quality diet such as Hill's Growth and provided daily oral calcium supplementation. If a relapse occurs after treatment, puppies must be weaned from the mother immediately and be hand-fed for the remainder of the neonatal period.*

Client Knowledge Deficit
- *Animals experiencing eclampsia are likely to suffer repeat episodes when nursing future litters. Many owners wrongly attempt to avert this condition by supplementing calcium during the pregnancy. This practice is contraindicated and will increase the likelihood of occurrence. Owners should be*

instructed to feed a quality growth diet during the last third of pregnancy (Hill's Growth or Iams for puppies) and throughout lactation. Oral calcium supplementation should begin after parturition.

Hyperthyroidism

Hyperthyroidism results from excessive thyroid hormone production and occurs most frequently in cats. Given that thyroid hormone affects the cellular metabolic rate of many cell types, clinical signs of this disease can be diverse. Clinical signs associated with hyperthyroidism include weight loss, voracious appetite (10 percent may be anorectic), tachycardia, PU/PD, hyperactivity, vomiting, diarrhea, poor hair coat, and vocalizations.

- Diagnostic Tests: Physical examination, CBC, serum chemistry, and serology: thyroid hormone (T4), thyroid-stimulating hormone (TSH), thyroid autoantibody, T3 suppression test, and thyrotropin releasing hormone (TRH) stimulation test.
- Potential Treatments: Thyroidectomy, radioiodine therapy, methimazole, antiemetics, and fluids.

Technician evaluations commonly associated with this medical condition:

Underweight
- *Hyperthyroid animals often eat large amounts of food yet fail to gain or maintain weight. This phenomenon results from the animal's accelerated metabolic rate (caused by increased thyroid hormone levels). Animals should be placed on a high-quality, high-calorie diet such as Hill's p/d or kitten.*

Vomiting
Altered Mentation
Diarrhea
- *Cats may produce large volumes of feces due to increased food consumption and metabolic rate. Until the condition is resolved, the litter box may need to be cleaned more frequently.*

Cardiac Insufficiency
- *Extreme tachycardia and hypertrophic cardiomyopathy can develop secondary to hyperthyroidism. Decreased cardiac output and pulmonary edema is evidenced by dyspnea, tachypnea, exercise intolerance, increased CRT, and pale mucous membranes. Heart rate should be monitored and recorded several times per day in debilitated hospitalized animals.*

Self-Care Deficit
- *Hyperthyroid cats typically have a greasy, matted, unkempt appearance. Areas of alopecia also may develop. Animals should receive grooming assistance to maintain the hair coat properly. More frequent bathing may be warranted to relieve excessive coat greasiness caused by increased sebum production.*

Sleep Disturbance
- Hyperthyroid animals frequently experience difficulty sleeping for normal periods of time. Cats often "night walk" and vocalize. Providing a quiet, secluded area may assist in minimizing sleep awakenings in these animals.

Hypothyroidism

Hypothyroidism results from inadequate release of thyroid hormone. This disease is classified as primary, wherein the abnormality arises directly from the thyroid gland, or secondary, which is associated with disorders of the pituitary gland. (In the normal animal, thyroid-stimulating hormone [TSH] is released from the pituitary; it then stimulates the release of thyroid hormone from the thyroid gland.) This condition is the most common endocrinopathy affecting dogs, yet it is rarely seen in cats. Clinical signs associated with hypothyroidism include lethargy, weight gain, cold intolerance, dermatitis, otitis, alopecia, muscle weakness, hyperpigmentation, bradycardia, anestrus, and infertility.

- Diagnostic Tests: Serology: thyroid (T4) level, TSH level, and TSH stimulation test.
- Potential Treatments: Thyroid hormone supplementation (thyroxine).

Technician evaluations commonly associated with this medical condition:

Reproductive Dysfunction
- Hypothyroid animals frequently experience reproductive problems. Lack of estrus or inability to maintain a pregnancy is common in the female. Males may experience a decrease in conception rates and libido.

Client Knowledge Deficit/Noncompliant Owner
- Client education is imperative as this disease requires lifelong treatment of the animal. The thyroxine level used for most dogs is toxic to humans; therefore, owners should be instructed to comply with use of childproof containers and take appropriate precautions to prevent inadvertent ingestion by children. Owners purchasing larger volumes of medication should be instructed to carefully monitor drug expiration dates.
- Serology should be conducted every 6–12 months (after the animal has been stabilized) to ensure appropriate dose of medication. Blood collection must be within 4–6 hours of hormone administration to ensure accurate results.

Overweight
- Inadequate thyroid hormone levels result in a lowered metabolic rate, making these animals prone to weight gain. Animals can be placed on calorie-restricted diets such as Hill's r/d, Royal Canin CALORIE CONTROL CC

HIGH FIBER, CALORIE CONTROL CC HIGH PROTEIN, or Purina OM until weight is within acceptable parameters. Long-term use of a modified-calorie diet such as Hill's w/d or Royal Canin NEUTERED CAT YOUNG ADULT may be required to maintain the animal's weight.
- Animals placed on appropriate T4 dosage should see improvements with weight within 6–10 weeks.

Mastitis

Mastitis is an inflammation of the mammary gland. Although physical and chemical agents can induce mastitis, bacteria account for the majority of cases. This condition is most prevalent in the postpartum period. Bacteria commonly associated with mastitis include *Escherichia coli*, *Staphylococcus*, and *Streptococcus*. Clinical signs include pain, heat and swelling in the affected gland, reluctance to permit nursing, fever, lethargy, anorexia, and changes in milk appearance (clumping, watery, or bloody).

- Diagnostic Tests: CBC, serum chemistry, and milk culture.
- Potential Treatments: Antibiotics, IV fluids, and anti-inflammatory analgesics.

Technician evaluations commonly associated with this medical condition:

Ineffective Nursing
- *Nursing can be extremely painful for the dam, causing previously tolerant females to exhibit aggressive behaviors toward offspring. Removal of milk from the affected glands is necessary for recovery. Thus, neonates should be permitted to nurse unless Escherichia coli is suspected or the dam becomes aggressive. If neonates must be removed, alternative methods should be used to remove milk manually from the affected glands.*
- *Neonates should be hand-raised or supplemented when inadequate nutrient intake is suspected. Neonates should be weighed daily and examined for signs of inadequate nutrient intake such as listlessness, crying, gaunt abdomen, or skin tenting.*

Acute Pain
- *Nonpharmacological approaches to pain control should be utilized. Warm compresses or a heating pad applied several times a day can provide comfort.*

Hypovolemia
- *Milk production is directly relation to hydration status. Dehydrated females are unable to produce an adequate volume of milk. Fluid consumption should be closely monitored. Neonates also should be monitored for signs of inadequate fluid intake.*

Hyperthermia
- Mastitis can remain a localized condition or progress to a systemic illness. Escherichia coli *mastitis patients are most likely to become septic.*

Underweight
- Ensuring adequate nutrient intake is critical to milk production. Use of high-calorie diets such as Hill's p/d or Royal Canin INTESTINAL HE is advised.

Paraphimosis

Paraphimosis is the inability of the male animal to withdraw the penis into the prepuce. Clinical signs include an engorged protruding penis, excessive licking, stranguria, and hematuria.

- Diagnostic Tests: Physical examination and CBC.
- Potential Treatments: Manual replacement, antibiotics, and IV fluids.

Technician evaluations commonly associated with this medical condition:

Reproductive Dysfunction
- The ability of males to inseminate females after an episode of paraphimosis is dependent upon the inciting cause. Males are unable to perform in a reproductive capacity during an acute episode.

Acute Pain
- Application of cold packs once the penis is returned to the prepuce will help decrease discomfort. Pain will diminish significantly once the penis is returned to the sheath and swelling subsides.

Risk for Self-Inflicted Injury
- Prevention of further trauma from excessive licking or chewing is imperative. E-collars should be utilized until the condition is completely resolved. Access to females and sexual excitement of male should be restricted.

Impaired Tissue Integrity
- The mucosal surface of the penis is very delicate and is frequently traumatized by excessive licking and swelling.

Prostatitis

The prostate is an accessory sex gland that contributes to the fluid component of the semen/ejaculate. Located at the neck of the bladder, the prostate empties into the urethra. Prostatitis, or inflammation of the prostate gland, is usually secondary to infection. This condition can be acute or chronic; and in severe cases, it can lead to prostatic abscessation. Bacteria commonly associated with prostatitis include *Escherichia coli, Pseudomonas, Staphylococcus,* and *Streptococcus.* Clinical signs can include dysuria, hematuria, constipation, gait abnormalities such as moving with apparent stiffness in hind limbs, abdominal pain, fever, preputial discharge, abdominal pain, and vomiting.

- Diagnostic Tests: Physical examination, CBC, urinalysis, urine culture, ultrasound, prostate wash culture, or fine needle aspirate.
- Potential Treatments: Antibiotics, IV fluids, and castration (after episode is resolved).

Technician evaluations commonly associated with this medical condition:

Reproductive Dysfunction
- *Prostatitis is painful, and the majority of males will not copulate during acute episodes. Sperm viability is dramatically reduced, and animals should not be used for reproductive purposes for some time following episodes of prostatitis. Castration is often recommended as it quickens recovery time.*

Constipation
- *Constipation occurs secondary to colon compression caused by the enlarged prostate.*

Acute Pain
Hyperthermia
Hypovolemia
Inappropriate Elimination
- *As urination is extremely painful, previously housetrained animals may begin to urinate inappropriately. In dogs, this behavior will likely discontinue when the prostatitis has resolved.*

Pseudopregnancy

Also termed *false pregnancy* and *phantom pregnancy*, this condition is seen in females 40–60 days postestrus. Females can exhibit all signs of impending parturition, such as mammary development, milk production, depression, personality changes, and nesting behaviors. This condition tends to reoccur within the same females.

- Diagnostic Tests: Radiographs and history.
- Potential Treatments: None required. Condition will spontaneously resolve. OHE will prevent reoccurrence.

Technician evaluations commonly associated with this medical condition:

Client Coping Deficit
- *Clients believing that their animal is pregnant are often disappointed upon learning the actual pregnancy status. Many clients need reassurance/confirmation that the female is not gravid.*

Pyometra

Pyometra, the accumulation of pus within the uterus, is most common in middle-aged females that experienced a heat cycle within the previous 2 months. Pyometras are classified as open or closed. Although both are potentially life-threatening conditions, the closed pyometra is particularly dangerous. Closing of the cervix causes retention of pus within the uterus and dramatically increases the potential for uterine rupture. Clinical signs of pyometra can include vulvar discharge (open), lethargy, vomiting, PU/PD, dehydration, and abdominal distension.

- Diagnostic Tests: CBC, serum chemistry, radiographs, and ultrasound.
- Potential Treatments: Antibiotics, IV fluids, surgery, and hormones (prostaglandin).

Technician evaluations commonly associated with this medical condition:

Reproductive Dysfunction
- *Seventy percent of animals treated medically will experience a second episode of pyometritis within two years. Owners are advised to breed females on the first estrus following treatment. Females undergoing an OHE will be unable to reproduce.*

Hypovolemia
- *Rehydration prior to surgery is imperative.*

Electrolyte Imbalance
- *Correction of electrolyte imbalance prior to surgery is imperative.*

Hyperthermia
- *Fever is present in approximately 30 percent of presenting females.*

Vomiting
Self-Care Deficit

Vaginitis

An inflammation of the vagina, vaginitis is a clinical sign of disease, not a disease in and of itself. Manifestations of vaginitis can include erythema, discharge, odor, and edema of the vaginal walls and vulva. Diseases associated with vaginitis can be bacterial, viral, yeast, or congenital.

- Diagnostic Tests: CBC, vaginal culture, vaginal cytology, and serum antibody titers.
- Potential Treatments: Variable dependent upon causative disease.

Technician evaluations commonly associated with this medical condition:

Self-Inflicted Injury
- Vaginitis associated with excessive discharge or moisture can result in perivulvar dermatitis. Females often demonstrate excessive grooming and licking of the vulvar area.

Reproductive Dysfunction
Risk of Infection Transmission
- Diseases such as herpesviruses and Brucella are sexually transmitted diseases. Due to the risk of disease transmission, breeding of affected females should be avoided until the problem has been resolved.

CONDITIONS INVOLVING THE RENAL AND URINARY SYSTEMS

Antifreeze Intoxication (Ethylene Glycol [EG] Intoxication)

EG is the toxic component of antifreeze solutions used in automobiles. The sweet taste of antifreeze entices animals to consume this highly toxic compound. Exposure typically occurs when animals happen upon fluid inadvertently spilled during filling of radiators or dripped from leaky radiators. Antifreeze is extremely toxic, and consumption of even small amounts can result in renal failure. A variety of clinical signs are seen depending on the stage of intoxication. During the acute period of stage 1, ataxia, vomiting, depression, dehydration, PU/PD, and vomiting are evidenced. Owners often state that the animal "appeared drunk." Stage 2 is associated with a resolution of previous clinical signs and the appearance of tachycardia. Owners often mistakenly confuse stage 2 with a resolution of the problem. Stage 3 commences within 24–72 hours of ingestion and is associated with signs of renal failure including depression, vomiting, and oliguria.

- Diagnostic Tests: History of exposure, PRN rapid blood test (detects EG only, not the toxic metabolites during stage 1), physical examination, CBC, and chemistry profile (stage 3).
- Potential Treatments: Induce vomiting for recent ingestion, alcohol dehydrogenase inhibitors (4-methylpyrazole) in dogs only, IV ethanol (ethyl alcohol), IV fluids, sodium bicarbonate, antiemetics, and nutritional support.

Technician evaluations commonly associated with this medical condition:

Altered Urinary Production
- Oliguria is present during stage 3. Monitoring of urine production is critical due to renal tubular damage associated with this intoxication.

Hypovolemia
- *A metabolic acidosis occurs concurrent to dehydration. Sodium bicarbonate can be administered in fluids to correct the acidosis.*

Electrolyte Imbalance

Vomiting
- *Vomiting is associated with stage 1 (acute intoxication phase) and stage 3 (renal failure phase).*
- *Vomiting can be intentionally induced in animals that consumed antifreeze within the previous 60 minutes. This assists in eliminating the toxin prior to the development of clinical signs.*

Client Coping Deficit
- *Clients may require assistance dealing with feelings of guilt if they inadvertently provided the source of exposure.*

Client Knowledge Deficit
- *Educate clients regarding the correct procedure for cleaning up chemical spills. Prevention is ideal using funnels or other devices. Paper or cardboard can be placed under the vehicle during fluid changes (antifreeze, brake fluid, and oil) and then properly disposed of. An absorbent material such as cat litter or paper towels should be used to remove spills, followed by water flush. Owners also should be instructed to monitor the area under vehicles for leaks.*

Altered Mentation
- *During the acute phase, intoxicated animals may appear "drunk." Severe depression typically accompanies stage 3.*

Self-Care Deficit
- *Intoxicated animals require assistance with self-care.*

Cystitis

Cystitis, an inflammation of the urinary bladder, is a clinical sign commonly associated with urinary tract infections. Ascending bacterial infections of the lower urinary tract account for the majority infectious causes of cystitis. Urease-forming bacteria can alter the urine pH, making it more likely for $MgNH_4PO_4$ (struvite) crystals or uroliths to form. Noninfectious causes of cystitis can include urolithiasis, neoplasia, trauma, chemical insult, or idiopathic disease. Cystitis is frequently associated with pollakiuria, dysuria, stranguria, hematuria, and inappropriate urination.

- Diagnostic Tests: Urinalysis, urine culture, and radiographs (contrast studies).
- Potential Treatments: Varied depending on inciting cause. Can include antibiotics, IV fluids, dietary management, urolith removal, and chemotherapy.

Technician evaluations commonly associated with this medical condition:

Noncompliant Owner
- Educate owners regarding the importance of administering the full course of antibiotics. Inappropriate administration of antibiotics can precipitate recurrent urinary tract infections.

Urinary Incontinence
- Incontinence is usually temporary and resolves with appropriate treatment.
- Cystocentesis is the method of choice for collection of urine samples. Urinalysis should be performed immediately upon collection to ensure accurate results. Postponing examination can result in lysis of RBC, WBC, and crystals. Samples that cannot be examined immediately should be refrigerated. The sample should be discarded after 12 hours.
- Diet modification is often recommended. Potential diets are often low in magnesium, with added antioxidants and urinary acidifiers. Examples include Hill's s/d and c/d; Eukanuba Low pH/S and Moderate pH/O; Purina UR Urinary and Urinary St/Ox; and Royal Canin URINARY SO (struvite and oxalate), CONTROL FORMULA, and DISSOLUTION FORMULA.
- Consumption of adequate water is vital to ensuring urine production and natural flushing of bacteria. Providing animals, especially cats, access to distilled moving water (bubble fountains) can assist in promoting water intake. The goal of water intake is to decrease urine specific gravity below 1.030. Use of moist foods (> 60 percent moisture) has been shown to increase water intake in cats. Increasing feeding frequency, placing a bowl of moist food alongside the previous diet, or using broths as a top dressing also facilitates increased fluid intake.
- Reduction of stress, environmental enrichment, and adequate litter box maintenance are known to reduce signs of cystitis in cats. Environmental enrichment can include providing climbing poles, hiding boxes, and scratching posts. Use of synthetic facial pheromone therapy (Feliway) can facilitate stress reduction. Appropriate litter box cleaning and placement is imperative.(Refer to the inappropriate elimination interventions.)

Acute Pain
- Cystitis is often accompanied by painful urination. Animals often initiate micturition, then stop midstream secondary to pain.

Renal Failure (Acute and Chronic)

Renal failure can be acute or chronic with classification reflecting the rate of renal function decline. ARF is evidenced by a rapid loss of function with accumulation of uremic toxins and disregulation of fluid and electrolytes. In comparison, chronic renal failure (CRF) is gradual, progressive decline with renal loss taking months to years. Clinical signs for both forms can include anorexia, depression, vomiting, diarrhea, dehydration, oral ulceration and halitosis, seizures, and ataxia. Additional clinical signs consistent with ARF include oliguria, polyuria, tachypnea, and bradycardia. PU/PD, nocturia, ascites, subcutaneous edema, and blindness are further indications of CRF.

- Diagnostic Tests: CBC, serum biochemistry, urinalysis, radiographs, ultrasound, biopsy, serology (ARF), and EG levels (ARF).
- Potential Treatments:
  - Both conditions: Sodium bicarbonate, potassium chloride, phosphate binders, and antiulcer medications.
  - ARF: Antiemetics, diuretics, and antibiotics.
  - CRF: ACE inhibitors, Calcitriol, and erythropoietin.

Technician evaluations commonly associated with this medical condition:

Vomiting
Diarrhea
Self-Care Deficit
  - *Animals in ARF may fail to maintain adequate hygiene.*
Hypovolemia
Electrolyte Imbalance
  - *Common electrolyte abnormalities include elevated BUN, creatinine, phosphate, and TP with a metabolic acidosis.*
Altered Oral Health
  - *Uremic ulcers present in the oral cavity can impair food consumption.*
Altered Urinary Production
  - *ARF is often associated with reduction in urine output. Normal output is 1 mL/kg/h. Urine production with oliguria is less than 0.25 mL/kg/h, and anuria is classified as 0.1 mL/kg/h.*
Altered Mentation
  - *Clinical appearance is varied, ranging from depression to seizures. Excessive accumulation of uremic toxins and electrolyte disturbances precipitate altered mentation.*
Risk of Infection
  - *Debilitated animals are more susceptible to infectious disease.*
Underweight
  - *Dietary management of CRF patients is vital to slowing disease progression. Use of diets such as Hill's k/d, Purina NF Kidney Function, Eukanuba Multi-Stage Renal, or Royal Canin Renal LP are recommended.*

Urinary Bladder Perforation

Bladder perforation occurs secondary to trauma (hit by car [HBC], kicks, or falls), unrelieved obstruction (urolithiasis and neoplasia), or iatrogenic causes (urinary catheters and cystocentesis). Although bladder perforation is a severe condition with a mortality rate approaching 40 percent, clinical signs associated with it are often nebulous. Signs, which may be delayed 24–48 hours posttrauma, can include anuria, abnormal micturition, hematuria, painful abdomen, ascites, vomiting, anorexia, and depression.

- Diagnostic Tests: Physical examination, radiographs (plain film and cystography), and urinalysis.
- Potential Treatments: Surgery, IV fluids, antibiotics, and anti-inflammatories.

Technician evaluations commonly associated with this medical condition:

Altered Urinary Production
- *Application of pressure to the abdominal cavity should be strictly avoided. Do not express the bladder. Extreme caution must be exercised during any postoperative bladder palpation. Monitoring of urine content and output is critical.*

Electrolyte Imbalance

Vomiting
- *Uremia and electrolyte imbalances can induce vomiting.*

Hypovolemia
- *Decreased fluid intake, vomiting, and extravascular fluid movement contribute to dehydration. Ensure urinary bladder integrity prior to administering large fluid volumes.*

Risk of Infection
- *Animals with bladder perforation are highly susceptible to peritonitis. Extreme caution should be exercised when placing or caring for an indwelling urinary catheter.*

Acute Pain
- *Exhibition of pain is highly variable. Animals suffering from additional trauma, as with HBC, may exhibit a great deal of pain. Animals perforating secondary to urethral obstruction may temporarily appear to be in less pain immediately following the rupture.*

Uroliths

Uroliths, which are calculi, or "stones," in the urinary tract, are a medical condition common to both dogs and cats. The chemical composition of uroliths varies and can include struvite, calcium oxalate, urate, and cystine. Clinical signs typically reflect the size, chemical content, and location of the urolith. Uroliths located in the bladder manifest as hematuria, pollakiuria, and stranguria, while those lodged in the urethra present with abdominal pain, anuria, stranguria, depression, anorexia, and vomiting. The primary danger of urolithiasis remains obstruction of urine outflow, bladder rupture, and severe electrolyte disturbances. Castrated males are at highest risk for urethral obstruction.

- Diagnostic Tests: Physical examination, radiographs, urinalysis, urolith chemical analysis, serum chemistry, ECG, bile acids, and ultrasound. The Minnesota Urolith Center currently provides free stone analysis.
- Potential Treatments: Relief of urethral obstruction if present, surgery, IV fluids, correct electrolyte imbalances, analgesics, and antiemetics.

Technician evaluations commonly associated with this medical condition:

Acute Pain
- *Irritation to the bladder epithelium, bladder distension, urethral spasm, and irritation all contribute to pain.*

Hypovolemia
- *Obstructed animals suffer from severe fluid and electrolyte disturbances. IV fluids commonly selected include 0.9 percent NaCl, Normosol-R, and PLASMA-LYTE.*

Electrolyte Imbalance
- *Electrolyte disturbances most commonly associated with obstruction include metabolic acidosis, increased PCV/TP, hyperkalemia, increased BUN, and creatinine.*
- *Abnormalities seen on the ECG as a result of a severe hyperkalemia include bradycardia, changes in T wave (tented or high-peaked), decreased heart rate, flattened P wave, prolonged PR, widening of QRS intervals, and increased incidence of ventricular premature complex (VPC).*

Vomiting
- *Emesis is secondary to triggering the chemoreceptor trigger zone (CRTZ) caused by elevated BUN and electrolyte disturbances.*

Altered Urinary Production
- *Normal urine production is 1 cc/lb/h. Anticipate 4 cc/lb/h for animals undergoing diuresis.*
- *Obstructed animals typically have an indwelling urinary catheter for 48 hours. During this time, the animal should be placed on an elevated grate to prevent contact with urine. Alternatively, a continuous collection system can be attached to collect all urine and provide exact urine production volumes. Remove the litter box from the cage, as indwelling urinary catheters can be obstructed by litter.*
- *Application of an E-collar will reduce incidence of unintended removal of the urinary catheter.*

- Reobstruction upon removal of an indwelling urinary catheter occurs in 14 percent of cats. Careful monitoring is critical.
- Struvite crystals (triple phosphate) are most commonly associated with feline uroliths (75 percent)/obstruction.

Risk for Infection
- Urine stasis and damage to urinary tract epithelium predispose animals to urinary tract infections. Adhere to strict aseptic technique when placing and caring for urinary catheters. A percentage of animals will have concurrent UTI and uroliths.

Client Knowledge Deficit
- Separation of a single animal for feeding in a multianimal household is often difficult. The number of litter boxes and their cleanliness often poses problems in a multicat household. Because of these issues, many owners disregard suggestions for long-term prevention. Discuss various options that can be employed to resolve these issues.
- *General prevention measures*:
  - Monitor for signs of cystitis and eradicate if present.
  - Ensure adequate water intake. Use of moist foods (> 60 percent moisture) is highly recommended in cats with prior urinary tract conditions. Moving/fountain water encourages consumption. Providing multiple feedings, adding ice cubes to the water dish, top-dressing food with broth, and placing multiple water bowls in the area also may facilitate increased water consumption.
  - Provide dogs with frequent opportunities to void. Ensure that cats have access to a clean litter box at all times. (Refer to the inappropriate elimination interventions.)
  - Monitor urine sediment and pH as directed by the veterinarian.
  - Provide diet determined by the veterinarian. Common prescription diets include Hill's s/d, c/d, u/d, and x/d; Eukanuba Low pH/S and Moderate pH/O; Purina UR Urinary and Urinary St/Ox; and Royal Canin URINARY SO, CONTROL FORMULA, and DISSOLUTION FORMULA.
  - Reduce stress via environmental enrichment. Consider use of pheromones.

CONDITIONS INVOLVING THE MUSCULOSKELETAL AND NEUROLOGIC SYSTEMS

Arthritis

Arthritis is an inflammation of the joint. Arthritis etiologies are numerous and frequently classified as traumatic, immune-mediated, or septic (infectious) in origin. Bacteria represent the most common infectious agent, although septic arthritis also has been associated with mycoplasma, rickettsia, and fungal and viral infections. The term *arthritis* is often used synonymously with *degenerative joint disease (DJD)*. Despite virtually identical clinical manifestations seen in these conditions, DJD differs in that it is a progressive noninflammatory disorder resulting in articular cartilage damage and degeneration. DJD has no known cure.

- Diagnostic Tests: Physical examination, radiographs, synovial fluid analysis (joint tap) and culture, biopsy, serology, CBC, and serum chemistry.
- Potential Treatments: Variable dependent upon inciting cause. Antibiotics and antifungal and antiviral agents. Symptomatic: NSAIDS, steroids, dietary management, chondroprotectants, and fatty acid supplementation.

Technician evaluations commonly associated with this medical condition:

Acute Pain or Chronic Pain
- *Arthritis is a very painful condition. Pain control for acute arthritis is often achieved through medications such as NSAIDs and analgesics or through diet. Pain relief from chronic arthritis can be addressed through the following measures:*
  - *Maintain animal at ideal body weight. An increased body mass increases mechanical stress, thereby increasing pain in the affected joints.*
  - *Use NSAIDs judiciously.*
  - *Employ diets specifically designed for management of osteoarthritis, such as Hill's j/d, Royal Canin MOBILITY SUPPORT JS or MOBILITY SUPPORT JS LARGE BREED, or Purina JM Joint Mobility.*
  - *Improve general joint health through use of chondroprotectants and supplements.*
  - *Modify environment to provide warm padded areas for resting.*
  - *Encourage moderate exercise to maintain joint range of motion (ROM) and to optimize weight.*

Overweight
- *Animals suffering from arthritis often gain weight as a result of a sedentary lifestyle. Weight reduction can be accomplished through use of foods such as Hill's w/d, Royal Canin CALORIE CONTROL CC HIGH FIBER or Purina OM. Conversely, animals that develop arthritis are at greater risk for future obesity.*

Client Knowledge Deficit
- *Owners must be made aware of the importance of maintaining the animal's ideal body weight. In addition, owners should be encouraged*

to follow the veterinarian's orders for exercise regimes, supplements, or other medications that must be administered on a daily basis.

Reduced Mobility
- Animals suffering from arthritis experience a reduced ROM and pain in affected joints. These factors can combine to limit the animal's ability to exercise, thus contributing to obesity. Continual low-impact movement is critical to maintaining remaining joint health. Swimming or other aquatic activity is ideal for maintaining joint health.

Self-Inflicted Injury
- Animals often lick repeatedly at affected joints. Carpal joints in particular are at risk for repetitive licking. If pain remains uncontrolled despite all medical therapies, animals may require the use of E-collars or other devices to minimize self-traumatization.

Aggression
- Previously even-tempered animals can exhibit aggressive tendencies when they anticipate or experience pain. Owners with small children should ensure that the children do not climb on or pull an arthritic animal's extremities, hips, etc. Encourage owners to provide a designated "dog-safe/child-free" area where the dog can rest comfortably without being disturbed.

Ataxia

Ataxia is a nonspecific clinical sign associated with many disease conditions. Characterized by lack of muscular coordination, ataxia is often described as a "drunken movement" in which the animal moves with a hyper- or hypometric gait, falls easily, demonstrates weakness of movement, shuffles, and has difficulty maintaining direction. Ataxia can be seen whenever the sensory pathways responsible for proprioception are disrupted. Examples of inciting causes include intoxication, infectious disease, congenital lesions, vestibular abnormalities, and trauma to the central nervous system (CNS).

- Diagnostic Tests: Extremely variable depending on inciting cause. CBC, serum chemistry, serology, toxicology tests, radiographs, myelogram, MRI, and CT.
- Potential Treatments: Extremely variable depending on inciting cause. Surgery, specific intoxication therapies, antibiotics, antiparasitics, and anti-inflammatories.

Technician evaluations commonly associated with this medical condition:

Hyperthermia
Hypovolemia
Risk of Aspiration

Vomiting
Reduced Mobility
Acute Pain
Fear

Brachial Plexus Avulsion

The brachial plexus is a nerve bundle formed by the ventral roots of C6, C7, C8, T1, and T2 nerves. Together these nerves function to innervate the front limb. An avulsion is essentially the tearing away or pulling apart of a structure. Avulsion of the brachial plexus occurs when the thoracic limb is severely abducted (moved away from midline). This type of movement is often associated with automobile injuries and other scenarios in which the animal is suspended by a forelimb. Clinical signs associated with this condition include lameness, Horner's syndrome (exophthalmos, elevation of lower eyelid, ptosis of upper eyelid, and miosis), and loss of limb sensation. Muscle atrophy is associated with chronic cases.

- Diagnostic Tests: Physical examination.
- Potential Treatments: Prevent further damage to limb, monitor for return to function for 4–6 months, and amputate the affected limb. Neurosurgery as a treatment option has not been well documented in animals.

Technician evaluations commonly associated with this medical condition:

Impaired Tissue Integrity
- *Loss of motor innervation can result in dragging of the front limb. This leaves the skin highly susceptible to ulceration, laceration, and trauma. A variety of bandaging techniques can be used to elevate the limb to prevent dragging. Preventative bandaging is highly recommended as restoration of normal integument postulceration can be clinically unrewarding.*

Reduced Mobility
- *Denervated muscles undergo severe atrophy in a relatively short period of time. To prevent atrophy and joint contracture, physical therapy should be instituted early in the course of treatment. Heat, ultrasound, massage, and passive ROM exercises are acceptable methods. Whirlpool therapy t.i.d. for 15 minutes is ideal. Care should be taken to prevent overheating of the skin or overstretching of muscles.*

Self-Inflicted Injury
- *A tingling or itching sensation associated with early nerve regeneration may prompt some animals to self-mutilate the limb. Standard methods of preventing self-mutilation can be employed.*

Altered Mentation
- Animals suffering from brachial plexus avulsion have typically experienced a blunt trauma such as HBC. Altered mentation can be associated with injuries that occurred simultaneous to the avulsion.

Decreased Perfusion (Peripheral)
- Damage to the vascular system also can occur during the avulsion. Monitoring of the limb for edema, color changes, temperature, and pulse strength is critical during the acute period.

Coma

Coma is a state of unconsciousness characterized by lack of awareness and inability to respond to environmental stimuli. Animals often appear to be in a state of deep sleep with intact reflex activity. Coma can occur secondary to inflammatory diseases (bacterial, fungal, or viral infection) or to neoplastic, traumatic, toxic (EG, lead, carbon monoxide, or drugs), or metabolic disturbances (diabetes mellitus, hepatic encephalopathy, hypoglycemia, or uremia).

- Diagnostic Tests: Extremely variable depending on suspected cause. Physical examination, CBC, serum chemistries, antibody titers, bile acids, blood ammonia, urinalysis, coagulation profiles, CSF analysis, radiographs, CT, and MRI.
- Potential Treatments: Extremely variable depending on cause.

Technician evaluations commonly associated with this medical condition:

Altered Mentation
Risk of Infection
- Comatose animals are severely debilitated; therefore, they are at much greater risk for infection. Pneumonia is of particular concern.

Hypovolemia
- Unresponsive animals are unable to consume fluids or nutrients. To avert dehydration and cachexia, IV catheters are used to provide appropriate fluids, electrolytes, and TPN. Fluids are limited to maintenance volumes as they can contribute to cerebral edema.

Risk of Aspiration
- Comatose animals do not possess a normal swallow reflex and are very susceptible to aspiration pneumonia. Oral administration of drugs should be avoided. Comatose animals should be monitored closely for signs of pneumonia secondary to aspiration or pulmonary edema. Comatose animals should be examined every 15–60 minutes depending on vital sign status.

Impaired Tissue Integrity
- Reduced tissue perfusion and lack of movement predispose comatose animals to decubital ulcers. Unresponsive animals should be rotated from

side to side every 2–3 hours. Frequent turning also helps prevent hypostatic lung congestion. Use of appropriate padding and cage racks prevents wetness and ulceration.

Bowel Incontinence
- The volume of fecal material is reduced secondary to diminished intake of solid food. Altered fecal consistency and constipation also can accompany incontinence.

Urinary Incontinence
- Urinary catheters can be placed to facilitate continual emptying of the bladder and to monitor urine production. Manual expression of the urinary bladder should be performed 2–4 times per day in noncatheterized animals.

Reduced Mobility
- Nonmobile animals are susceptible to muscle atrophy and joint contracture. Passive ROM exercises, massage, and ultrasound therapy can be instituted to minimize such secondary complications.

Self-Care Deficit
- Coma renders self-care impossible. Daily brushing helps maintain hair coat quality and appearance in addition to facilitating circulation. Reduced blinking increases susceptibility to corneal ulceration. Daily application of an intraocular sterile lubricant/ointment is recommended. Open-mouth breathing coupled with a diminished swallowing rate serves to rapidly dry the tongue and gums. Application of glycerin to oral surfaces reduces dryness and discomfort.

Decreased Perfusion
- Coma can be associated with reduced perfusion of kidneys, intestines, and integument.

Client Coping Deficit
- Owners often require quiet, undisturbed bonding time with the animal. If medically appropriate, owners can assist with brushing the animal's coat or with passive ROM exercises. Coma is often associated with a grave prognosis.

Degenerative Joint Disease (DJD)

DJD is a progressive, painful deterioration of the articular cartilage and constitutes the most common joint disease of dogs. Clinical signs typically develop over time with animals initially "warming out" of the lameness or exhibiting signs only during cold or damp periods. As the disease progresses, the severity of the lameness increases and loss of mobility or muscle atrophy can ensue.

- Diagnostic Tests: Radiographs and arthrocentesis with synovial fluid analysis.

- Potential Treatments: Symptomatic: anti-inflammatories (steroids and NSAIDS), chondroprotective drugs, weight loss, and moderate exercise plan. Surgery may be indicated to correct joint laxity.

Technician evaluations commonly associated with this medical condition:

Chronic Pain
- *Pain control for DJD is often achieved through medications such as NSAIDs and analgesics. In addition, pain relief can be addressed through the following measures:*
  - *Maintain animal at ideal body weight. An increased body mass increases mechanical stress, thereby increasing pain in the affected joints.*
  - *Improve general joint health through use of chondroprotectants and supplements.*
  - *Apply thermal therapy (conductive or converted), which can be helpful.*
  - *Modify environment to provide warm padded areas for resting.*
  - *Encourage moderate exercise to maintain joint ROM and to optimize weight. Avoid high-impact activities such as running, chasing balls, and jumping. Encourage low-impact activities such as swimming and walking.*
  - *Employ diets specifically designed for management of osteoarthritis, such as Hill's j/d, Royal Canin MOBILITY SUPPORT JS, or Purina JM.*

Overweight
- *Animals suffering from DJD often gain weight as a result of a sedentary lifestyle.*

Client Knowledge Deficit
- *Owners must be made aware of the importance of maintaining the animal's ideal body weight. In addition, owners should be encouraged to follow the veterinarian's orders for exercise regimes, supplements, or other medications that must be administered on a daily basis.*

Reduced Mobility
- *Animals suffering from DJD experience a reduced ROM and pain in affected joints. These factors can combine to limit the animal's ability to exercise, thus contributing to obesity. Continual low-impact movement is critical to maintaining remaining joint health. Swimming or other aquatic activity is ideal for maintaining joint health.*

Self-Inflicted Injury
- *Animals often lick repeatedly at affected joints. Carpal joints in particular are at risk for repetitive licking. If pain remains uncontrolled despite all medical therapies, animals may require the use of E-collars or other devices to minimize self-traumatization.*

Aggression
- *Previously even-tempered animals can exhibit aggressive tendencies when they anticipate or experience pain. Owners with small children should ensure that the children do not climb on or pull a painful animal's extremities, hips, etc. Encourage owners to provide a designated "dog-safe/child-free" area that permits the dog to rest comfortably without being disturbed.*

Hip Dysplasia

Hip dysplasia is a developmental DJD affecting the coxofemoral joint of dogs. Affected animals have a high degree of instability or laxity in the hips that ultimately leads to osteoarthritis. Clinical signs associated with this condition include hind limb lameness, bunny hopping gait, decreased ROM, and pain upon palpation of the hips. This disease has a high degree of heritability, and breeding of affected animals is not recommended.

- Diagnostic Tests: Radiographs and specific radiographic evaluations including PennHIP and Orthopedic Foundation Association certification.
- Potential Treatments: Surgery: triple pelvic osteotomy, femoral head and neck excision, and total hip replacement. Medical symptomatic treatment for osteoarthritis: NSAIDs, chondroprotective agents, and physical therapy.

Technician evaluations commonly associated with this medical condition:

Overweight
- *Although hip dysplasia has a definite genetic component, dietary management has a significant influence on the progression of the disease. Excessive or rapid growth has been shown to increase the risk for hip dysplasia. Owners should be instructed to avoid free-choice feeding and should be encouraged to feed their dogs growth diets that are specifically formulated for large or giant breeds. Supplementation of these diets with vitamins or minerals (calcium) should be discouraged.*
- *Animals that have received a diagnosis of hip dysplasia will inevitably suffer from osteoarthritis. Pain associated with the arthritis encourages a sedentary lifestyle, which in turn favors obesity. Owners should weigh animals monthly to ensure optimal body mass. The diet selected will reflect the animal's age, weight, and activity level.*

Acute Pain or Chronic Pain
- *Hip dysplasia is a DJD. This disease can be surgically addressed; however, owners may elect to forego surgical intervention, instead opting*

to manage the disease using techniques/medications as described for arthritis/DJD.
- Pain relief can be addressed through the following measures:
  - Maintain animal at ideal body weight. An increased body mass increases mechanical stress, thereby increasing pain in the affected joints.
  - Use NSAIDs judiciously.
  - Improve general joint health through use of chondroprotectants and supplements.
  - Modify environment to provide warm padded areas for resting.
  - Encourage moderate exercise to maintain joint ROM and to optimize weight.
  - Employ diets specifically designed for management of osteoarthritis, such as Hill's j/d, Royal Canin MOBILITY SUPPORT JS, or Purina JM.

Reduced Mobility
- Excessive exercise of young at-risk breeds of dogs should be avoided until the animal has reached musculoskeletal maturity (approximately 12–14 months of age).
- Animals undergoing surgery should be postoperatively exercise-restricted. The length and restriction limits are determined by the surgical procedure performed.
- Animals undergoing surgery should receive postoperative physical therapy. This can include ROM exercises, heat therapy, massage, and other intervention. Hydrotherapy, which permits animals to exercise on a water-submersed treadmill, is of great benefit as it promotes joint movement with less mechanical stress.

Aggression
- Previously even-tempered animals can exhibit aggressive tendencies when they anticipate or experience pain. Owners with small children should ensure that the children avoid causing hip pain by climbing on or pushing the dog.

Intervertebral Disc Disease (IVDD)

Two types of IVDD are seen in animals. Type I is characterized by a rapid, sudden extrusion of ruptured disc material into the spinal cord. This type is commonly seen in chondrodystrophic breeds such as the dachshund, shih tzu, basset, and corgi. IVDD Type I is often described by owners as "suddenly going down in the back end." Type II IVDD is associated with a gradual protrusion of an intact disc into the spinal cord. Type II protrusion typically affects larger dog breeds such as the German shepherd, mastiff, and Labrador retriever. Observant owners often describe a gradual onset of a "wobbly hind end." Clinical signs of IVDD reflect the severity of spinal cord damage. Signs range from pain, loss of proprioception, reluctance to move, stiffened appearance, ataxia, mild paraparesis, and paralysis to paraplegia.

- Diagnostic Tests: Radiographs, myelogram, MRI, and CT.
- Potential Treatments: Surgical intervention (laminectomy, hemilaminectomy, and decompression), cage rest, and steroids.

Technician evaluations commonly associated with this medical condition:

Client Coping Deficit
- Owners often experience a high degree of anxiety when IVDD has caused acute paralysis of their pet. It is common for owners to anthropomorphize and express feeling of depression or despair.

Risk of Infection
- Cystitis is a common sequela of reduced bladder innervation secondary to IVDD. Animals with an indwelling urinary catheter are especially at risk for cystitis. Culture of the urine is recommended upon removal of the catheter. Atelectasis secondary to recumbency predisposes animals to respiratory infection.

Overweight
- Chondrodystrophic breeds such as the dachshund also are likely to experience excessive weight gain. This attribute, in conjunction with a conformational predisposition, places additional stress on the vertebral column. Selecting an appropriate diet to reduce weight and to maintain it within normal parameters is highly recommended.

Impaired Tissue Integrity
- Paralysis leaves animals very susceptible to decubital ulcers. Recumbent animals should be placed on a well-padded surface and turned every 3–4 hours.
- The suture line of animals undergoing surgery should be examined b.i.d.

Urinary Incontinence
Bowel Incontinence
Reduced Mobility
- Depending on the location and extent of spinal cord damage, mobility impairment can range from reluctance to ambulate to paralysis. Patient interventions selected by the technician will reflect patient need and the veterinarian's orders.
- Extreme care should be taken when moving patients with spinal cord injuries. Stabilization of the spinal cord to prevent further trauma is of utmost importance. Animals should be moved using a firm, flat surface such as plywood or a stretcher.
- Animals with Type I IVDD may suffer from permanent hind limb paralysis. Mobility carts have been specifically developed for these animals.

- Cage rest: Exercise restriction is critical for spinal cord repair. Cage rest can be combined with surgical intervention or used as a primary medical intervention. Animals must be confined in close quarters to restrict movement. Thus, cages are preferred over runs as a method of confinement. If able to ambulate, animals should remain on a leash at all times when brought outside to urinate or defecate. A harness or another alternative to a collar should be used on animals affected in the cervical area.
- Physical therapy such as passive and active ROM exercises and hydrotherapy is of benefit. IVDD patients typically experience a loss of muscle mass, which further increases susceptibility to pressure sores. Passive manipulation of joints is recommended 4–6 times per day to prevent joint contracture. Therapy should be initiated 2 weeks postoperatively.

Self-Care Deficit
- Paretic animals are unable to maintain normal hygiene and often require assistance with feeding and grooming.

Meningitis

Meningitis is an inflammation of the meninges covering the brain or spinal column. This condition often accompanies encephalitis or inflammation of the brain. Simultaneous occurrence of the two conditions is termed *meningoencephalitis*. Bacterial, viral, fungal, parasitic, rickettsial, and protozoal causes of meningitis are seen. Clinical signs associated with this condition include pain; fever; cervical hyperesthesia; and an altered state of consciousness such as stupor or depression, incoordination, seizures, and muscle rigidity.

- Diagnostic Tests: Physical examination, CSF culture and analysis, CBC, serum chemistry, and radiographs.
- Potential Treatments: Antibiotics, antivirals, antifungals, anti-inflammatories (steroidal and nonsteroidal) supportive fluids, and nutrition.

Technician evaluations commonly associated with this medical condition:

Hyperthermia
- Elevations in body temperature can result from the body's response to infectious agents, seizure activity, or muscle fasciculations.

Risk of Infection Transmission
- Animals suffering from meningitis should be isolated until the risk of transmission has been determined.

Self-Care Deficit
- Meningitis often renders animals incapable of performing basic grooming and other self-care needs. Food and water consumption should be closely monitored.

Acute Pain
- *Cervical pain is most common.*

Altered Mentation
- *Mentation alterations can include varying degrees such as depression, overt stupor, or coma. Supportive measures will reflect the severity of altered mentation. Intravenous fluids and TPN may be warranted.*

Osteomyelitis

Osteomyelitis is an infection and inflammation of the cortex, medulla, or periosteum of bone. Orthopedic surgery, bone exposure, or penetrating injuries typically precede this condition. Bacteria most commonly associated with osteomyelitis include *Staphylococcus, Streptococcus*, and *Escherichia coli. Coccidioides, Blastomyces,* and *Histoplasma* are frequently implicated in fungally induced infections. Clinical signs associated with osteomyelitis can include lameness, pain, swelling of the affected area, draining tracts, fever, anorexia, and depression.

- Diagnostic Tests: Radiographs, culture, fine needle aspirates, cytology, and CBC.
- Potential Treatments: Surgical intervention (debridement, bone grafts, sequestrum removal, and implant removal). Medical therapy: antibiotics, anti-inflammatories, and analgesics.

Technician evaluations commonly associated with this medical condition:

Acute Pain or Chronic Pain
- *Osteomyelitis can be difficult to resolve. Chronic pain is associated with infections requiring months of treatment. Chronic infections are associated with a very poor prognosis.*

Hyperthermia
- *Hyperthermia occurs secondary to infection.*

Altered Tissue Integrity
- *Infections may track to the dermal surface, resulting in localized heat, swelling, redness, or discharge of infection.*

Decreased Tissue Perfusion
- *The inflammatory reaction with associated tissue destruction, edema, and loss of vascular integrity results in loss of adequate perfusion.*

Underweight
- *Anorexia secondary to pain can lead to weight loss.*

Reduced Mobility
- *Varying degrees of lameness are noted.*

Self-Inflicted Injury
- *Animals often chew or repetitively lick at the affected area.*

Panosteitis

Panosteitis is a self-limiting condition affecting young large-breed dogs. It is characterized by shifting leg lameness with dogs showing a marked pain response upon palpation of long bones. Bones most frequently affected include the humerus, radius, ulna, femur, and tibia. In addition to lameness, dogs can exhibit anorexia, fever, and lethargy during acute episodes. The cause of this disease is unknown.

- Diagnostic Tests: Physical examination and radiographs.
- Potential Treatments: Symptomatic treatment, NSAIDS, and steroids.

Technician evaluations commonly associated with this medical condition:

Acute Pain
- *Exercise restriction should be implemented during painful episodes. NSAIDs are frequently used to control pain. Nonpharmaceutical measures also should be implemented.*

Client Knowledge Deficit
- *Caution must be exercised in selecting the appropriate diet for young large-breed dogs. Rapid bone growth may trigger episodes of this condition; therefore, dogs must be maintained on an appropriate ration that encourages an acceptable rate of gain. Suitable diets are commercially available through Hill's, Iams, Purina, and Eukanuba. Inform owners that free-choice feeding a growth diet is medically contraindicated. Owners also should be discouraged from adding vitamin and mineral supplements to balanced diets. Animals experiencing anorexia during acute episodes can be encouraged to eat via warming food or switching to a canned form of the selected diet.*

Fever
- *Mild elevations in temperature company this condition.*

Seizures

Also termed *convulsions* and *epilepsy*, seizures result from abnormal brain activity. The condition is divided into three distinct periods: preictal (period just prior to), ictus (actual seizure), and postictal (period immediately after). The type of seizure, which can be described as grand mal, petit mal, focal, partial, generalized, or tetanic, reflects the underlying cause. Seizures may be congenital or acquired in origin. Some of the numerous diseases/disorders that incite seizures include hydrocephalus, bacterial or viral encephalitis, idiopathic epilepsy, hepatic encephalopathy, hypoglycemia, lysosomal storage diseases, cerebral abscess, trauma, and intoxications. Clinical signs associated with this condition include the following:

- Preictal: acting nervous, pacing, whining, and hiding

- Ictus: stiffness, muscle spasm, paddling, vocalization, defecation, urination, salivation, and jaw chomping
- Postictal: weakness, blindness, depression, nervousness, and pacing
    - Diagnostic Tests: Extremely variable depending on inciting cause. CBC, serum chemistry, CSF analysis, serology, toxicology tests, radiographs, myelogram, MRI, and CT.
    - Potential Treatments: Extremely variable depending on inciting cause. Anticonvulsants, surgery, specific intoxication therapies, antibiotics, antiparasitics, and IV fluids.

Technician evaluations commonly associated with this medical condition:

Altered Mentation
Client Knowledge Deficit
- *Animals may suffer a seizure when the owner is not home. Instruct owners to look for evidence of seizure activity such as disturbed furniture, urine or feces on floor, and unexplained cuts or bruises on the animal.*
- *Treatment of some conditions such as idiopathic epilepsy necessitate long-term use of anticonvulsants. Ensure adequate client education regarding importance of following treatment protocol, signs of toxicity, drug interactions, and need to prevent inadvertent human ingestion.*
- *Animals receiving anticonvulsant therapy should avoid swimming.*

Hyperthermia
- *Seizure activity can induce severe hyperthermia secondary to muscle activity.*

Hypovolemia
Risk of Aspiration
- *Never administer oral medication to animals experiencing seizure activity.*

Anxiety
- *Petting the animal and providing reassurance can reduce the animal's anxiety during the pre- and postictal periods.*

Inappropriate Elimination
- *Urinary elimination occurs during the actual seizure. Once control of the seizure is obtained, the inappropriate elimination will resolve.*

CONDITIONS INVOLVING THE INTEGUMENT AND SPECIAL SENSES

Conjunctivitis

Conjunctivitis, an inflammation of the conjunctiva, is a clinical sign associated with many ocular diseases. It is characterized by hyperemia, erythema, and ocular discharge. Corneal ulceration, bacterial infections, viral infections, foreign bodies, allergies, keratoconjunctivitis sicca, distichiasis, entropion, and ectropion can all be associated with conjunctivitis.

- Diagnostic Tests: Schirmer's test, fluorescence stain, ocular culture, and serology.
- Potential Treatments: Extremely variable depending on inciting cause.

Technician evaluations commonly associated with this medical condition:

Impaired Tissue Integrity
- Monitor and record degree of hyperemia and ocular discharge.

Acute Pain
- The majority of diseases associated with conjunctivitis also cause a significant amount of ocular pain.

Self-Inflicted Injury
- Ocular pain and irritation can induce self-trauma. E-collars may be necessary to prevent the animal from rubbing its eye.

Risk of Infection
- Conjunctivitis associated with corneal ulceration should be monitored closely for secondary bacterial infection.

Corneal Ulceration

Corneal ulceration is a clinical sign associated with many ocular conditions. Trauma; foreign bodies; distichia; keratoconjunctivitis sicca; entropion; and mycotic, viral, and bacterial infection can all be associated with corneal ulceration. This clinical sign is characterized by ocular pain, blepharospasm, epiphora, and photophobia.

- Diagnostic Tests: Physical examination, fluorescence stain, Schirmer's test, and culture.
- Potential Treatments: Extremely variable depending on inciting cause. Can include surgery, third eyelid flap, tarsorrhaphy, topical or subconjunctival antibiotics, and other symptomatic treatment as required.

Technician evaluations commonly associated with this medical condition:

Risk of Infection
- Corneal ulceration should be monitored closely for secondary bacterial infection.

Impaired Tissue Integrity
- Antibiotics or other topical medication is often applied to the eye to minimize the adverse effects caused by loss of integrity.

Acute Pain
- Ulceration is an extremely painful condition. Face rubbing is the most common indicator of ocular pain.

Aggression
- Some animals may become aggressive secondary to ocular pain. Exercise caution when examining or applying medication.

Self-Inflicted Injury
- E-collars can be used to minimize self-inflicted injury.

Risk for Altered Sensory Perception

Entropion and Ectropion

Rolling of the eyelid margin can occur as an inversion (entropion) or an eversion (ectropion). The outward roll, or everting margin, of ectropion often gives dogs a "sad hound dog" appearance. Clinical signs associated with both conditions can include epiphora, blepharospasm, corneal ulceration, conjunctivitis, face rubbing, and photophobia.

- Diagnostic Tests: Physical examination and fluorescence stain.
- Potential Treatments: Surgical correction and symptomatic treatment of associated clinical signs (intraocular antibiotics or steroids).

Technician evaluations commonly associated with this medical condition:

Risk of Infection
- Chronic irritation of the conjunctiva and corneal surfaces leaves the eye susceptible to secondary bacterial infection.

Impaired Tissue Integrity
- Ulceration of the corneal surface from hair (entropion) or foreign material (ectropion) is common. Antibiotics or other topical medication is often applied to the eye to minimize the adverse effects caused by loss of integrity.

Acute Pain
- Entropion is an extremely painful condition. Ectropion can be associated with pain if foreign material enters the eye. Face rubbing or head pressing are the most common indicators of ocular pain.

Aggression
- Some animals may become aggressive secondary to ocular pain. Exercise caution when examining or applying medication to the ocular area.

Self-Inflicted Injury
- Excessive face rubbing can worsen both conditions. Use of an E-collar is highly recommended postoperatively for animals undergoing surgical correction.

FLEA ALLERGY DERMATITIS

Flea allergy dermatitis is the most common hypersensitivity skin disorder in dogs and cats. Caused by an exaggerated response to flea's saliva, this condition is evidenced by pruritis; excessive grooming; military dermatitis (cats); alopecia; skin crusting; erythema; and a classic distribution pattern involving the dorsal lumbosacral area, medial thighs, and tail head area. Given that a single flea bite can precipitate this condition, many animals do not demonstrate an overt flea infestation.

- Diagnostic Tests: Gross examination for fleas (visual, flea comb, and flea excrement), serology, intradermal skin tests, and CBC.
- Potential Treatments: Flea control, steroids, and hyposensitization therapy.

Technician evaluations commonly associated with this medical condition:

Impaired Tissue Integrity
- *Self-trauma or severe infestations can result in loss of tissue integrity and secondary superficial pyodermas.*

Risk of Infection Transmission
- *Fleas are readily transmissible between animals. Fleas do not demonstrate species specificity and will readily transfer between species (dog/cats).*
- *Fleas can act as vectors for various parasites and diseases including* Dipylidium *(tapeworm) and* Yersinia pestis *(bubonic plague). Although humans are not a natural host, people experience itching and irritation secondary to flea bites.*

Client Knowledge Deficit
- *Eliminating a flea infestation can be frustrating and expensive. Owners are often embarrassed to acknowledge this issue due to its preserved association with poor hygiene, which is an incorrect perception. Control measures must address both the animal and the environment to be successful.*
- *Educate clients regarding use of available products. The following products can be utilized. Emphasize that the majority of fleas are in the environment. Depending on the products selected, repeat treatments may be required in 2–4 weeks.*
  - *Treat the environment (house and yard). Remove animals from environment during treatment.*
    - *Use insecticide foggers and sprays indoors. Make certain to spray under beds and along baseboards.*
    - *Ensure that all bedding is thoroughly washed.*
    - *Vacuum carpets and dispose of vacuum bag immediately.*
    - *Cut grass and trim bushes in the yard to minimize environment suitable for flea development. Spray yard using over-the-counter (OTC) insecticides.*
  - *Treat the animal.*
    - *Kill adult fleas using:*
      - *Medicated shampoos (pyrethrins or carbamates).*
      - *Imidacloprid (Advantage).*

- o Fipronil (FRONTLINE Spray, FRONTLINE Plus, and FRONTLINE Top Spot).
- o Selamectin (Revolution).
- o Nitenpyram (CAPSTAR).
- Prevent development of future fleas via insect growth regulators.
  - Lufenuron (Program and Sentinel)
  - Pyriproxifen (Nylar)

Self-Inflicted Injury
- Intense pruritis can result in excessive scratching and self-trauma.

Glaucoma

Glaucoma is an elevation in intraocular pressure (IOP) resulting from impaired aqueous humor outflow. IOP values above 25–30 mm Hg are indicative of glaucoma. Normal dog and cat IOP ranges from 15–20 mm Hg. Glaucoma can occur secondary to other ocular conditions such as anterior uveitis, lens luxation, and neoplasia. Glaucoma resulting from such conditions is termed *secondary glaucoma*. Alternatively, primary glaucoma is manifested by an increase in IOP in the absence of detectable ocular disease. Both forms of glaucoma are associated with optic nerve degeneration and blindness. Clinical signs associated with this condition can include an enlarged globe, which appears as a more protruding eye; sudden onset vision loss; corneal clouding; episcleral hyperemia; ocular pain; and face rubbing.

- Diagnostic Tests: IOP measurement through indentation tonometry (Schiötz tonometer) or applanation tonometry (TONO-PEN).
- Potential Treatments: Medical treatment: administration of autonomic agents (pilocarpine, epinephrine, dipivefrin, or timolol), diuretics (mannitol), carbonic anhydrase inhibitors (dichlorphenamide), and prostaglandin analogs (latanoprost). Surgery: anterior shunts, transscleral ciliary body coagulation, cryosurgery, and enucleation.

Technician evaluations commonly associated with this medical condition:

Acute Pain or Chronic Pain
- Mild increases in IOP can occur in the absence of ocular pain; however, many animals have an advanced to chronic state in which marked elevation of IOP and associated ocular degeneration produces significant pain. Acute and significant increases in IOP also induce a pain response.
- In addition to pharmaceutical measures, nonpharmaceutical pain relief can be achieved postoperatively through application of warm moist compresses b.i.d. for 1 week.

Altered Sensory Perception
- Animals with diminished sight have a greater risk for environmental injury and may experience personality changes. Caution should be exercised

when animals are near children until the animal has adapted to sensory limitations.
- Animals experiencing glaucoma in one eye have a 50 percent chance of developing glaucoma in the unaffected eye. Dogs experiencing unilateral glaucoma should have the normal eye examined two or three times per year to facilitate early detection.
- Forty percent of dogs diagnosed with glaucoma experience vision loss within the year.

Client Coping Deficit
- Many clients experience distress upon learning that their pet will experience loss of vision. Clients should receive assurances that blind animals adapt very well to their environment.

Keratoconjunctivitis Sicca (KCS)

"Dry eye" is the common name for this ocular condition resulting from inadequate tear production. Decreased tear production can be attributed to a congenital predisposition or result from other events such as nictitating membrane eversion (cherry eye), loss of parasympathetic innervation, or any systemic metabolic disease (diabetes mellitus, Cushing's, or hypothyroidism). Clinical signs associated with KCS include face rubbing, conjunctivitis, mucopurulent ocular discharge, blepharitis, squinting, and corneal vascularization.

- Diagnostic Tests: Schirmer's test.
- Potential Treatments: Symptomatic medical treatment: lacrimostimulants (pilocarpine or cyclosporine), tear substitutes (artificial tears), antibiotics, mucinolytics (acetylcysteine), and anti-inflammatories. Surgery: parotid duct transposition.

Technician evaluations commonly associated with this medical condition:

Acute Pain
- Animals suffering from KCS frequently experience concurrent corneal ulceration and conjunctivitis, making the condition particularly uncomfortable.

Self-Inflicted Injury
- Face rubbing and other types of self-trauma typically resolve with adequate corneal lubrication. The temporary use of an E-collar may be warranted.

Impaired Tissue Integrity
- Accumulation of periocular mucoid discharge can irritate the delicate tissue around the eye. Moist warm towels should be use to remove periocular discharge. Long-haired animals should have facial hair trimmed around the eyes to improve hygiene.

Ocular Proptosis

Most cases of ocular proptosis occur secondary to trauma or dog fights. Exopthalmic dog breeds that normally have a "bulging appearance" to the eye are particularly at risk for this condition. Prognosis for vision and globe viability is determined by the degree of ocular trauma associated with the proptosis. Penetration of the globe, infection, damage to the optic nerve, blood supply, or extraocular muscles negatively affect prognosis.

- Diagnostic Tests: Physical examination.
- Potential Treatments: Surgery (ocular replacement with tarsorrhaphy or enucleation) and concurrent medical treatment.

Technician evaluations commonly associated with this medical condition:

Acute Pain
- *In addition to pharmaceutical measures, nonpharmaceutical pain relief can be achieved postoperatively through application of warm moist compresses b.i.d. for 1 week.*

Self-Inflicted Injury
- *Both pre- and postoperative discomfort associated with proptosis incites self-mutilation behaviors. It is extremely important to implement protective means to minimize further damage to the globe.*
- *Postoperative prevention of self-mutilation can be achieved via E-collars. E-collars should be removed during meals or other times of bonding when the owner is able to provide direct supervision of the animal. Dogs and cats may bump into objects in the home when adjusting to the E-collar. Owners should be advised to remove fragile objects from the immediate area.*

Client Knowledge Deficit
- *Inform owners that proptosis is an emergency that must receive immediate medical attention. Instruct owner to hold the pet to prevent rubbing against paws, walls, or other objects. Keep the eye moist and protected through use of ocular lubricants or artificial tears. If these are unavailable, sterile petroleum jelly can be used. Do not attempt to remove any penetrating objects from the eye.*

Altered Sensory Perception
- *Animals with diminished sight have a greater risk for environmental injury and may experience personality changes. Animals typically require 2–3 days to adjust to limitations induced by the E-collar. Long-term sensory (vision) loss (if any) reflects the degree of ocular damage that occurred during proptosis.*

Otitis

Inflammation of the ear is a common malady caused by a variety of etiologies. Foreign bodies, allergies (atopy and food hypersensitivity), yeast, parasites, and bacterial infection can all contribute to this condition. In addition to categorizing by etiology, otitis is classified with regard to the portion of the ear affected. Otitis externa, which is most common in dogs and cats, affects structures of the external ear (tympanic membrane, horizontal and vertical ear canal, and pinna). Otitis media and otitis interna affect structures of the middle ear (auditory tube and ossicles) and inner ear (cochlea, vestibule, and semicircular canal), respectively. Clinical signs associated with otitis externa include head shaking, head tilting, ear rubbing, ear pawing, hyperemia, pruritis, and otic discharge and odor. Otitis media also can be associated with facial nerve paralysis, Horner's syndrome, and ataxia.

- Diagnostic Tests: Otoscope examination, culture, microscopic examination, cytology, and radiographs.
- Potential Treatments: Extremely variable depending on etiology. Can include flushing, antibiotics, antifungals, antiparasitics, anti-inflammatories, foreign body removal, allergy control, and surgery.

Technician evaluations commonly associated with this medical condition:

Impaired Tissue Integrity
Acute Pain or Chronic Pain
- *Otitis is a painful disease. Acute pain often can be controlled through use of NSAIDS. Chronic yeast infections are often associated with chronic pain and associated personality changes.*

Aggression
- *Animals that have not been aggressive previously may undergo a temporary change in behavior as a result of the pain response. Owners who have young children should be advised to minimize petting of the dog's head and neck until the problem has resolved.*

Risk of Infection
- *Ear flushing helps acidify and dry the ear canal and removes microorganisms (yeast and bacteria) or purulent material.*
- *Most owners use a commercially prepared flushing solution. Alternative solutions can include chlorhexidine solution (.05 percent), saline, Docusate DSS, or 50:50 water:vinegar mix. The type of solution selected will reflect the veterinarian's preference and the etiology of the disorder. A bulb syringe is not required for flushing unless an excessive amount of purulent material or debris is noted. A volume of 20–35 cc of solution should be instilled each time the canal is flushed. If a bulb syringe is used, it should be cleaned after use with 50:50 vinegar: alcohol solution.*

Altered Sensory Perception
- *Chronic otitis or severe acute otitis can diminish hearing.*

Pyoderma

Defined as a bacterial infection of the skin, pyoderma is a common malady of small animals. Pyodermas are classified as surface (hot spots and fold pyoderma), superficial (impetigo and folliculitis), and deep (acne and furunculosis). Clinical signs can include alopecia, crusts, erythema, pustules, ulceration of the skin, and acute pain.

- Diagnostic Tests: Skin scraping with cytology, culture (bacterial, fungal), and biopsy.
- Potential Treatments: Antibiotics, anti-inflammatories, and topical therapy.

Technician evaluations commonly associated with this medical condition:

Impaired Tissue Integrity
Acute Pain

Retinal Detachment

Retinal detachment is an ocular emergency that causes acute vision loss. The retina is the innermost of the three tunics, or layers, of the eyeball. The portion of the retina that is sensitive to light is comprised of rods and cones and is responsible for vision sensory input. Both congenital and acquired causes of detachment are found in animals. Acquired conditions account for the majority of cases and can include hypertension, trauma, infectious, immune-mediated, neoplastic, and idiopathic conditions. Acute vision loss is the most common clinical sign associated with a detached retina.

- Diagnostic Tests: Examination using ophthalmoscope. Tests for underlying conditions can include serum chemistry, CBC, and measurement of blood pressure.
- Potential Treatments: Surgery; medical therapy including diuretics, antibiotics, steroids, and antihypertensives; and treatment of underlying medical condition (hypertension and neoplasia).

Technician evaluations commonly associated with this medical condition:

Altered Sensory Perception
- *Detached retina results in acute blindness.*

Self-Inflicted Injury
- *E-collars may be warranted.*

Acute Pain

Ringworm

Dermatophytoses, or "ringworm infections," are superficial fungal infections involving the skin. *Microsporum* (70–90 percent) and *Trichophyton* (10 percent) species account for the majority of canine and feline infections. Clinical signs can include alopecia (often round), scaling/crusting of skin, broken hairs, folliculitis, military dermatitis (cats), and pruritis. This fungal infection is easily transferred directly through contact with infected animals and indirectly through fomites such as brushes and bedding. Ringworm is a zoonotic disease, and humans can readily contract the condition through contact with clinically/subclinically affected animals or fomites.

- Diagnostic Tests: Skin scraping with concurrent cytology, Wood's lamp examination, and culture dermatophyte test medium (DTM).
- Potential Treatments: Antifungal shampoos, topical agents (miconazole or clotrimazole), systemic treatment (griseofulvin), lufenuron, and treatment of environment (washing blankets and discarding brushes).

Technician evaluations commonly associated with this medical condition:

Impaired Tissue Integrity
- Hair regrowth in affected areas can take several weeks.

Risk of Infection Transmission
- This condition is readily transferable between animals and from animals to humans. Infected animals should be separated until the treatment is complete. A vaccine has been approved for **Microsporum canis** *in cats.*

Client Knowledge Deficit
- Provide client education regarding zoonotic potential.
  - Disinfect (or discard) all bedding, brushes, cat carriers, and other objects using appropriate solution (diluted bleach 1:10).
  - Practice good hygiene and wash immediately after handling or treating infected animals.
  - Do not allow children to handle infected animals.
  - Wear gloves when applying topical medications or handling infected animals. Wash clothing immediately after handling animals.

Risk for Vomiting
- Treatment with griseofulvin is often associated with vomiting. Absorption of the drug is enhanced if administered concurrently with a high-fat meal.

MISCELLANEOUS NEOPLASTIC CONDITIONS

Hemangioma

Classified as benign, this tumor consists of a neoplastic proliferation of blood vessels. Hemangiomas are typically located in the cutaneous or other soft tissues (spleen) and have the capacity to metastasize to additional locations such as the spleen, liver, and heart. Hemangiomas are most commonly seen in adult canine and feline patients. Canines tend to develop tumors on their body trunk or extremities, while felines tend to develop tumors on the head, abdomen, and extremities. Clinical signs are variable and can include the presence of a soft cutaneous mass, weakness with pale muocus membranes, exercise intolerance, and sudden collapse.

- Diagnostic Tests: Fine needle aspirate, biopsy, CBC, radiographs, and ultrasound.
- Potential Treatments: Surgical excision and chemotherapy.

Technician evaluations commonly associated with this medical condition:

Impaired Tissue Integrity
Exercise Intolerance
Cardiac Insufficiency
- *Highly vascular large tumors can divert blood away from the patient's normal vascular system.*

Impaired Tissue Integrity

Hemangiosarcoma

Noted as one of the most aggressive soft tissue tumors, this highly malignant cancer of the vascular endothelium can affect the heart, liver, and spleen. Clinical appearance of visible tumors on the trunk or extremities can vary from erythematous nodules to a generalized bruising. Clinical manifestations reflect the organ system involved and can include weakness, lethargy, dyspnea, syncope, hind limb paresis, seizures, pale mucous membranes, epistaxis, melena, tachycardia, abdominal distension, and sudden collapse. Prognosis for this condition is very poor, with average survival times of 20–60 days postdiagnosis.

- Diagnostic Tests: Fine needle aspirate, biopsy, CBC, serum chemistry, coagulation profiles, radiographs, and ultrasound.
- Potential Treatments: Surgical excision, chemotherapy, and symptomatic/supportive therapy.

Technician evaluations commonly associated with this medical condition:

Hypovolemia
- *Approximately 50 percent of dogs presented for hemangiosarcoma have experienced sudden collapse secondary to rupture of the primary tumor. Massive blood loss is associated with rupture. DIC commonly accompanies this neoplasia.*

Impaired Tissue Integrity
- *Cutaneous lesions are noted.*

Altered Mentation
- *Depression, dullness, and seizures can accompany the disease.*

Exercise Intolerance
- *Associate anemia decreases oxygen carrying capacity of blood.*

Self-Care Deficit

Client Coping Deficit
- *A grave prognosis with short survival times accompanies this disease.*

Insulinoma

Insulinoma is a cancer of the beta cells (insulin-producing) in the pancreas that results in excessive insulin production. Excessive amounts of insulin induces a severe, often life-threatening hypoglycemia. Clinical signs can include fatigue, weakness, ataxia, confusion, seizures, and coma.

- Diagnostic Tests: CBC, serum chemistry, biopsy/histology, ultrasound, and glucose/insulin ratio.
- Potential Treatments: IV dextrose, surgery, and chemotherapy.

Technician evaluations commonly associated with this medical condition:

Altered Mentation
Exercise Intolerance
Self-Care Deficit

Lymphoma

Lymphoma, a malignant neoplasm originating from lymph tissue, is the most common hematopoietic malignancy of cats. Four anatomical types of lymphoma have been identified: multicentric (most common), alimentary, mediastinal, and extranodal. Multicentric lymphoma is associated with numerous internal organs, bone marrow, and extranodal areas such as the CNS and skin. Alimentary lymphoma resides in the intestinal tract. Mediastinal lymphoma is associated with the thymus and cranial mediastinal lymph nodes. Extranodal lymphoma is found in the skin, lungs, kidneys, CNS, and eyes. Known risk factors associated with this cancer include a positive FeLV/FIV status. Clinical signs associated with lymphoma are highly variable and reflect the organ systems involved. Observed abnormalities can include enlarged lymph nodes, weight loss, anorexia, vomiting, diarrhea, palpable abdominal masses, dyspnea, tachypnea, lethargy, seizures, apparent blindness, circling, behavior changes, paresis, and anemia.

- Diagnostic Tests: Biopsy, cytology, CBC, fluorescence nuclear antibody FNA, FeLV/FIV testing, radiographs, and ultrasound.
- Potential Treatments: Chemotherapy and lymphadectomy (very limited).

Technician evaluations commonly associated with this medical condition:

Underweight
- *Patients may suffer from weight loss due to anorexia and/or vomiting.*

Diarrhea
Vomiting/Nausea
- *Patients with lymphomas in the intestinal tract experience vomiting and nausea. Chemotherapy can be associated with gastrointestinal upset for several days posttreatment.*

Exercise Intolerance
- *Patients may experience anemia when the bone marrow is involved.*

Risk of Infection
- *Chemotherapy suppresses the immune system.*

Altered Mentation
- *Extranodal lymphoma affecting the CNS can result in an array of mentation alterations.*

Abnormal Mobility
- *Abnormal mobility is associated with lymphoma affecting the CNS.*

Self-Care Deficit
Client Knowledge Deficit
- *FeLV-positive animals experience a significantly shorter survival time. Monitoring of animals during and after chemotherapy is critical. Exercise restriction should be enforced with animals suffering from low platelet or WBC counts.*

Mammary Tumors

Cancers involving the mammary glands can be benign or malignant depending on the type of neoplasia involved. Common benign tumors include adenoma, cystadenoma, and "mixed benign" forms. Malignant tumors include adenocarcinoma and carcinoma. Feline mammary tumors are associated with a much higher incidence of malignancy than are canine tumors. Malignancy rates of 90 percent (feline) and 50 percent (canine) are expected. Animals typically present with a palpable mass in the mammary tissue; however, animals with metastases (lungs) also may exhibit coughing, tachypnea, and exercise intolerance.

- Diagnostic Tests: Biopsy, histology, radiographs, and CBC.
- Potential Treatments: Surgery (lumpectomy or mastectomy) and chemotherapy.

Technician evaluations commonly associated with this medical condition:

Risk of Infection
- Loss of dermal tissue integrity or surgical incision predisposes animals to infection.

Impaired Tissue Integrity
- Ulceration of the dermal surface over the tumor can occur.

Acute Pain
- Palpation of the mass may elicit a pain response. Ulcerative masses are painful.

Ineffective Nursing
- Queens and bitches suffering from mammary tumors may not permit offspring to suckle.

Client Knowledge Deficit
- Mammary tumors are the most common tumor of female dogs.
- Owners should conduct a prophylactic examination of the intact female dog every few months.
- OHE has dramatically decreased the incidence of mammary tumors. Timing of the OHE is important. If performed prior to the first estrus (< 6 months), tumors are extremely rare. Incidence increases dramatically when OHE is performed after the second or third estrus.
- In dogs, tumors found to be malignant have already metastasized in 50 percent of the cases.
- Pregnancy, pseudopregnancy, and lactation do not affect incidence of tumors.
- Castration does not appear to decrease tumor incidence in cats.
- Progesterone therapy is associated with an increased risk of tumor development.

Mast Cell Tumors

Mast cell tumors, which are neoplastic accumulations of mast cells, are a common malignant skin tumor found in dogs and cats. Many clinical signs associated with this tumor result from degranualization of the histamine-containing mast cells. Examples include local pruritis, alopecia, bruising, erythema, and gastric ulceration (vomiting, diarrhea, melena, and weight loss). Mast cell tumors are noted for a varied appearance and can be readily confused with skin tags, lipomas, or insect bites. Although the vast majority of mast cell tumors are located within the skin, this tumor may be found in visceral or splenic locations.

- Diagnostic Tests: Physical examination, cytology, histology, radiograph, ultrasound (hepatosplenomegaly), CBC, and lymph node aspirate (metastasis).
- Potential Treatments: Surgical excision, chemotherapy, radiotherapy, and supportive therapy.

Technician evaluations commonly associated with this medical condition:

Impaired Tissue Integrity
- *Delayed healing in the area of the tumor is common.*
- *Mast cell tumors that degranulate, causing a local erythema and wheal formation when manipulated, are referred to as having a positive Darier's sign.*

Vomiting/Nausea
- *Vomiting occurs secondary to histamine effects on gastrointestinal system.*

Diarrhea
- *Diarrhea occurs secondary to histamine effects on gastrointestinal system.*

Underweight
- *Loss of weight can occur as a result of vomiting/diarrhea or secondary to histamine effects on gastrointestinal system.*

Client Knowledge Deficit
- *Fifteen percent of dogs diagnosed with this condition develop additional mast cell tumors. Instruct owners to complete a monthly examination of the skin.*
- *Mast cell tumors account for 25 percent of all skin tumors in dogs.*

Osteosarcoma

Osteosarcoma, a malignant neoplasm of the bone, occurs with greater frequency in dogs than in cats. In addition, it appears to have a greater affinity for large and giant breeds. This tumor can develop in the axial or appendicular skeleton; however, the radius, tibia, and humerus represent the bones that are predominantly affected (95 percent). Metastasis to the lungs is common and is associated with a grave prognosis. Clinical signs include pain, swelling of the affected area, and lameness. Animals suffering from concurrent metastasis also may exhibit neurologic deficits, lethargy, and anorexia.

- Diagnostic Tests: Radiographs and biopsy.
- Potential Treatments: Amputation, surgical excision, chemotherapy, analgesics, and supportive therapy.

Technician evaluations commonly associated with this medical condition:

Acute Pain
- Patients with osteosarcomas experience severe acute pain at the site of tumor. Effective analgesia is imperative.

Reduced Mobility
- Patients may present with an acute onset of lameness of the affected limb. Spontaneous pathological fractures are common.

Client Coping Deficit
- This disease is associated with a grave prognosis. Median survival rates are less than one year. Survival times for animals with metastasis at time of diagnosis is less than 2 months.

Squamous Cell Carcinoma

Squamous cell carcinoma is a malignant neoplasm of the skin (squamous epithelium) that is often associated with prolonged or chronic exposure to sunlight. Lesions are typically raised and firm, irregular, and thickened; and they are often ulcerated. There are two forms of squamous cell carcinoma: cutaneous and oral.

- Diagnostic Tests: Biopsy and radiographs.
- Potential Treatments: Surgical excision, amputation, chemotherapy, and radiotherapy.

Technician evaluations commonly associated with this medical condition:

Impaired Tissue Integrity
- Patients with squamous cell carcinomas suffer from lesions of the skin and possible delayed healing postoperatively.

Reduced Mobility

Testicular Tumors

Testicular neoplasm occurs in the scrotum, inguinal area, and testicles. Sertoli cell, interstitial (Leydig) cell, and seminomas account for the majority of testicular tumors. Leydig cell tumors and seminomas typically affect descended testes whereas the majority of Sertoli cell tumors occur in undescended (cryptorchid) testes. Tumors are most commonly found in older dogs with many being incidentally identified during routine physical examination. It is rare to find testicular tumors in cats. Clinical signs can include swelling of the scrotum and/or inguinal area. Paraneoplastic signs such as alopecia and feminization can occur secondary to estrogen production by Sertoli cell tumors.

- Diagnostic Tests: Ultrasonography, fine needle aspiration, cytology, and physical examination.
- Potential Treatments: Castration, chemotherapy, and supportive therapy.

Technician evaluations commonly associated with this medical condition:

Reproductive Dysfunction
– *The affected testicle may not have full or any reproductive function.*

Client Knowledge Deficit
– *Cryptorchidism is associated with an increased incidence of Sertoli cell tumors. Owners of cryptorchid animals should be informed of the preventative benefits of castration.*

SELECT INFECTIOUS DISEASES

Rabies

Rabies is a potentially zoonotic viral disease that causes progressive, fatal encephalomyelitis in affected animals. Dogs, cats, and livestock are typically infected through a bite from an infected skunk, bat, fox, or raccoon. Two distinct clinical appearances are seen with this disease. The "dumb form," in which animals are less responsive to the environment, is seen in horses and cows. Dogs and cats are more likely to exhibit the "furious form," which is characterized by aggression. All animals can demonstrate signs of behavioral changes, hypersalivation, difficulty swallowing, cranial nerve alterations, ataxia, and paralysis.

Zoonotic potential is high. All testing of potentially infected animals and postmortem handling of tissue should be performed with extreme caution.

– Diagnostic Tests: Fluorescent antibody test, CSF antibody titers, and postmortem immunofluorescent antibody testing.
– Potential Treatments: Animals are not treated for this condition. Euthanasia is indicated. Prevention is achieved via vaccination.

Technician evaluations commonly associated with this medical condition:

Risk of Infection Transmission
– *Rabies is a fatal zoonotic disease. Human exposure has occurred as a result of:*
  - *Being exposed to virus-laden saliva through bite wounds or "apparently choked" animals.*
  - *Handling sick wildlife such as bats, raccoons, or skunks. Owners should be instructed to avoid contact with wild animals showing any signs of disease.*
  - *Handling infective tissues or fluids.*
– *Additional precautions are as follows:*
  - *Contact state veterinary boards for information on how to report instances to individual states, how to handle suspect animals, and what protocols to follow for specimen submission.*

- In addition to federal and state regulations, local counties may have quarantine regulations regarding dogs that have bitten humans.
- Most agencies recognize 10 days as the time period required for quarantine of dogs that have bitten humans.
- Given their relatively high risk of exposure, immunization of veterinary personnel is recommended.
- Periodic review of the National Association of State Public Health Veterinarians (NASPHV) and American Veterinary Medical Association (AVMA) guidelines is recommended.

Risk of Aspiration
– Inability to swallow precipitates aspiration. Saliva is highly infective.

Reduced Mobility

Aggression
– Avoid contact with potentially rabid animals. Veterinary personnel should exercise extreme caution when dealing with a potentially infected animal.

Altered Mentation
– The degree of mental changes reflects the affected nerves and stage of disease progression. Some animals may appear hyperresponsive while those nearing death are often comatose.

Self-Care Deficit
– Animals appear unkempt. No attempt should be made to treat this issue.

Feline Infectious Diseases

Feline Infectious Peritonitis (FIP)

FIP is a contagious, viral, immune-mediated disease that is induced by a feline coronavirus. Upon initial exposure to the virus, cats may remain asymptomatic. Within weeks of the infection, however, an inappropriate immune response against the virus is mounted. Immune-mediated complexes deposited in blood vessels result in severe vasculitis. Two distinct forms of this disease can then develop. The hallmark of the wet (effusive) form is abdominal distension secondary to ascites or pleural effusion. The dry (noneffusive or granulomatous) form is more insidious, with clinical signs reflecting the particular organ system involved. Generalized clinical signs of FIP include fever, depression, vomiting, diarrhea, weight loss, ataxia, seizures, dyspnea, and icterus.

- Diagnostic Tests: CBC, serum biochemistry, abdominocentesis, thoracocentesis with concomitant cytology of fluid, and antibody titers.
- Potential Treatments: Symptomatic treatment, immunotherapy (interferon), immunosuppressives (cyclophosphamide), nutritional support, steroids (catabolic and anabolic), and antibiotics (secondary infections).

Technician evaluations commonly associated with this medical condition:

Altered Gas Diffusion
- One third of cats suffering from wet FIP develop pleuritis and pleural effusion. Dyspnea is commonly noted.

Hyperthermia

Risk of Infection Transmission
- The coronavirus is transmitted to other animals through oral and respiratory secretions, feces, and urine. Appropriate aseptic technique should be utilized to prevent spread of the virus. Although readily killed by common disinfectants, the virus can survive in the environment for several weeks.
- Prevention can be attempted via:
  - Avoidance of infected cats. (All infected animals should be removed from catteries.)
  - Good sanitation practices.
  - Vaccination. (Current vaccines have a relatively low efficacy.)
- Current serologic testing cannot differentiate between vaccinated and infected cats.

Hypovolemia
- Reduced fluid intake, vomiting, diarrhea, and ascites lead to dehydration.

Diarrhea

Vomiting

Altered Mentation
- Changes in mentation are varied, ranging from overt seizures to mild depression. Neurologic involvement may be more common in the "dry" form.

Underweight
- Mortality rates for this disease approach 100 percent. Cats typically succumb to the disease within days to months of diagnosis. Animals surviving for a period of time may lose weight.

Feline Immunodeficiency Virus Infection (FIV)

This infection, caused by a lentivirus, is often compared to HIV infections in humans. (Neither of these viruses can cross species.) The virus induces an immunosuppression, making the cat susceptible to a variety of secondary infections. Following inoculation, the cat may exhibit a transient period of illness with accompanying fever and malaise. This episode is often followed by a latent period (months to years) when the cat appears clinically normal. Ultimately, the virus induces an immunosuppression that causes the cat to succumb to clinical disease. Clinical signs associated with chronic infections of the respiratory, gastrointestinal, urinary, and integumentary systems predominate. Generalized clinical signs associated with this condition include fever, lethargy, anorexia, and weight loss. Coinfection with FeLV is a significant finding.

- Diagnostic Tests: ELISA Snap test (detects FIV antibodies and is often combined with FeLV tests; all positives should be confirmed with IFA or Western blot immunoassay as ELISA has a higher number of false positives) CBC, and serum biochemistry.
- Potential Treatments: Symptomatic antibiotics for secondary infections and immunotherapy (interferon).

Technician evaluations commonly associated with this medical condition:

Hyperthermia
Vomiting
- *Vomiting is often associated with secondary infections.*

Diarrhea
- *Diarrhea is often associated with secondary infections.*

Altered Gas Diffusion
- *Altered gas diffusion is often associated with secondary infections.*

Altered Urinary Production
- *Altered urinary production is often associated with secondary urinary tract infections.*

Risk of Infection
- *Chronic, unresponsive infections (oral and respiratory) are typical.*

Client Coping Deficit
- *The chronicity and gradual debilitating nature of this disease can be difficult for clients to accept.*

Underweight
- *The chronic nature of this disease is often evidenced through gradual weight loss. High-quality diets should be fed at all times. Growth or nutrient-dense diets may be indicated to counteract effects of anorexia and repeated secondary infection. Examples of nutrient-dense diets suitable for cats include Hill's p/d, Purina CV Cardiovascular, Abbott CliniCare, and Royal Canin IVD.*

Risk of Infection Transmission
- *Large numbers of the virus are present in saliva. Transmission takes place primarily through biting.*

Client Knowledge Deficit
- *During client education, if the virus is described as "being like HIV in people," emphasize that FIV is not zoonotic and cannot be transmitted to humans. The primary similarity is that both viruses destroy the immune system of the host, making it susceptible to secondary infections.*

- Prevention Considerations
  - Vaccinate. (Current testing practices are unable to differentiate between vaccinated and infected cats. Vaccinated animals will test positive.)
  - Prevent biting. Neutering males decreases incidences of fighting. Housing cats indoors minimizes exposure.
  - FIV-test all new animals older than 6 months of age.

Feline Leukemia (FeLV)

Feline leukemia is a contagious viral disease spread via contact with saliva or nasal secretions from infected cats. Transmission typically occurs through mutual grooming, bites wounds, or shared food/water bowls. Once exposed, cats can become persistently viremic (30 percent), clear the virus (30 percent), or develop latent infections (30 percent). Nonspecific clinical signs such as anorexia, weight loss, and lethargy are associated with the many disorders associated with feline leukemia. Immunosuppression, lymphoid tumors (leukemia), enteritis, immune complex disorders, anemia, and reproductive problems are frequently seen.

- Diagnostic Tests: ELISA Snap test and immunofluorescent antibody test (IFA).
- Potential Treatments: Antivirals (AZT); immunotherapy (interferon); and symptomatic therapy including fluids, antibiotics, antiemetics, and transfusions. Preventative vaccination is recommended.

Technician evaluations commonly associated with this medical condition:

Vomiting
- *Acute and chronic vomiting can occur.*

Hypovolemia
- *Cats suffering from concurrent vomiting and diarrhea are most likely to dehydrate.*

Electrolyte Imbalance

Underweight
- *Chronic diarrhea, anorexia, and reduced nutrient absorption contribute to weight loss.*

Hyperthermia

Diarrhea
- *Chronic diarrhea is a common sequela. In addition, animals suffering from feline leukemia are more susceptible to concurrent panleukopenia infection.*

Altered Gas Diffusion
- *Secondary respiratory infections are common.*

Risk of Infection
- *Given the immunosuppressive nature of feline leukemia, care must be taken to minimize stress and exposure of feline leukemia-positive cats to secondary infections.* **Cats should receive quality nutrition, routine vaccinations against other infectious diseases, and regular deworming.**

Risk of Infection Transmission
- **Testing cats to determine immune status prior to vaccination is recommended.**
- *Persistently infected carrier cats account for the majority of virus dissemination.*
- *Infected cats shed large numbers of the virus in saliva. Tears, urine, and feces contain smaller amounts of viral particles. Both vertical (in utero and milk) and horizontal (secretions from infected cats) modes of transmission exist.*
- *Kittens are especially susceptible to infection.*
- *The following measures are recommended for prevention/control of feline leukemia:*
    - Test entire cat population; remove all positive cats if possible. If positive animals are found, retest cat population in 12 weeks to detect cats incubating the virus during the first test period.
    - If infected cats are not removed, keep infected cats indoors isolated from susceptible cats.
    - Thoroughly disinfect food and water bowls when infections are noted.

Feline Panleukopenia

Also called infectious enteritis or distemper, panleukopenia is a highly contagious viral disease caused by a parvovirus. Similar to the parvovirus infection in dogs, feline parvovirus infects rapidly, dividing cells such as intestinal epithelium, lymphoid tissue, and bone marrow. Fever, depression, anorexia, vomiting, diarrhea, abdominal pain, ataxia, and tremors are noted in affected animals. Queens infected during pregnancy may experience stillbirth, fetal mummification, abortion, or delivery of kittens suffering from cerebellar hypoplasia (ataxia and incoordination). Care should be taken to isolate all cats that test positive as the virus is present in all secretions.

- Diagnostic Tests: Physical examination, CBC, and serology.
- Potential Treatments: Supportive care including fluids with correction of electrolyte abnormalities and colloid pressure, antiemetics, parenteral nutrition, vitamin B, and antibiotics (secondary infections). Prevention through vaccination is highly recommended.

Technician evaluations commonly associated with this medical condition:

Vomiting
- Animals should be maintained on IV fluids until vomiting has resolved. Adequate hydration is fundamental to recovery. Cats should be maintained on a bland diet for 2 weeks postresolution of vomiting.

Hypovolemia
- Vomiting results in dehydration.

Electrolyte Imbalance
- Intestinal inflammation/necrosis and vomiting can result in severe protein loss and electrolyte imbalances. Potassium chloride 20–30 mEq/L is often added to fluids. In addition to crystalloid fluids such as LRS, NaCl, dextrose, and Normosol-R, colloids such as Hetastarch or plasma may be used to combat hypoalbuminemia.

Hyperthermia
- Feline panleukopenia can be associated with high fevers (103–107°F).

Acute Pain
- Abdominal pain occurs secondary to viral replication and invasion of intestinal cells.

Client Knowledge Deficit
- Client education regarding the importance of preventative vaccination programs is imperative. This disease can devastate feral and semiferal cat populations. Owners with large outdoor cat populations may neglect vaccination programs.
- Owners should be instructed to isolate the cat from susceptible animals for 2 weeks posthospital discharge. Care should be taken to dispose of fecal material properly. Dilute bleach 1:32 is an appropriate disinfectant that can be used by owners to disinfect surfaces in the home. This virus can remain patent in the environment for 1 year.

Risk of Infection Transmission
- Panleukopenia is highly contagious. Cages should be thoroughly disinfected after use. All personnel should maintain aseptic technique practices when handling infected animals.

Self-Care Deficit
- Cats are unlikely to maintain grooming efforts when suffering from panleukopenia.

Feline Upper Respiratory Disease (FURD)

Respiratory disease is common in cats. Feline calicivirus (FCV) and feline viral rhinotracheitis (FVR) account for the majority (90 percent) of cases, with *Chlamydia psittaci* and *Mycoplasma* playing lesser roles. Clinical signs associated with FURD include fever, lacrimation, serous or purulent ocular and nasal discharge, conjunctivitis, hypersalivation,

anorexia, ulcerative stomatitis, ulcerative keratitis, polyarthritis, dehydration, dyspnea, and sneezing.
- Diagnostic Tests: Physical examination, fluorescent antibody testing, virus isolation, and serum antibody titers.
- Potential Treatments: Supportive treatment, including fluids, antivirals, decongestants, ocular medications, and broad spectrum antibiotics (secondary bacterial infections). Prevention through vaccination is recommended. Vaccinations are currently available for FVR, FCV (highly recommended), and chlamydia.

Technician evaluations commonly associated with this medical condition:

Hyperthermia
- *Viral infections can be associated with very high fevers (105°F).*

Underweight
- *Oral ulceration and loss of olfactory senses predispose cats to anorexia. Modifying the consistency and texture of the diet using soft moist foods or heating the diet to increase odor and palatability can facilitate consumption.*

Hypovolemia
- *Moist foods encourage fluid intake. IV fluids are required in severely debilitated animals.*

Altered Oral Health
- *Oral ulceration, gingivitis, and stomatitis predispose cats to anorexia and secondary infections.*

Risk of Infection Transmission
- *FURD is very contagious. Cages should be thoroughly cleaned after use, with assistants taking special care to clean all surfaces (especially cage bars as cats often rub their face against these surfaces). All personnel should maintain aseptic technique practices when handling animals.*

Client Knowledge Deficit
- *Households with multiple cats should be educated regarding disease transmissibility and prevention. This syndrome can devastate catteries and outdoor semiferal cat populations. General recommendations include the following:*
  - *Vaccinate all cats. Kittens can be vaccinated as young as 4–5 weeks in problem households.*
  - *Handle cats in order of susceptibility. Handle kittens first, followed by healthy adults, then clinically ill cats.*
  - *Wash thoroughly between handling cats.*
  - *Use separate feeding, bedding, and litter supplies.*

- *Assess ventilation adequacy in indoor catteries.*
- *Increase overall resistance to disease through quality nutrition, parasite control, and a comprehensive vaccination program.*

Altered Ventilation
- *Severely debilitated cats may require supplemental oxygen therapy. Owners can be instructed to place the cat in a steamy room (bathroom) for 10–20 minutes b.i.d. to facilitate removal of respiratory secretions.*

Altered Gas Diffusion
- *Severe infections, especially those associated with secondary bacterial infections, reduce gas exchange at the alveolar level.*

Impaired Tissue Integrity
- *Corneal ulceration commonly accompanies FURD. Ophthalmic antiviral or antibacterial ointments should be applied as directed.*

Self-Care Deficit
- *Debilitated cats require grooming assistance. FURD often results in excessive ocular and nasal discharge. Excessive nasal discharge, which can accumulate and occlude nares, should be removed using warm moist towels.*

Hemobartonella

Also termed *feline infectious anemia, Hemobartonella felis* is a rickettsial parasite affecting the red blood cells of cats. It appears as a dark red-violet cocci adhering to the surface of the infected cells. Many cats remain asymptomatic until stressed by concurrent illness. Transmission typically occurs via blood-sucking arthropods such as fleas, but also can result from intrauterine infections or blood transfusions. Clinical signs reflect the underlying anemia and include fever, pale mucous membranes, jaundice, anorexia, lethargy, dyspnea, and depression.

- Diagnostic Tests: Blood smear, CBC, chemistry profile, and serology.
- Potential Treatments: Oxygen therapy, blood transfusions, and antibiotics.

Technician evaluations commonly associated with this medical condition include:

Hyperthermia
- *Fevers of 103–106°F are noted.*

Altered Gas Diffusion
- *Loss of oxygen carrying capacity results in tachypnea.*

Self-Care Deficit

Exercise Intolerance
- *Anemic animals are unable to meet additional oxygen demands of exercise.*

Risk of Infection Transmission
- *Fleas transmit the parasite from one host to another.*

Client Knowledge Deficit
- *Educate clients regarding use of flea and tick control products. The following products can be utilized. Remember to emphasize that the majority of fleas are found in the environment.*
- *Treat the environment (house and yard)*
  - *Use insecticide foggers and sprays indoors.*
  - *Ensure that all bedding is thoroughly washed.*
  - *Vacuum carpets and dispose of bag immediately.*
  - *Spray yard using OTC insecticides.*
- *Treat the animal*
  - *Kill adult fleas using:*
    - *Medicated shampoos (pyrethrins or carbamates).*
    - *Imidacloprid (Advantage).*
    - *Fipronil (FRONTLINE Spray, FRONTLINE Plus, and FRONTLINE Top Spot).*
    - *Selamectin (Revolution).*
    - *Nitenpyram (CAPSTAR).*
- *Prevent development of future fleas via insect growth regulators:*
  - *Lufenuron (Program and Sentinel)*
  - *Pyriproxifen (Nylar)*

Toxoplasmosis

Toxoplasmosis is a potentially zoonotic disease caused by the protozoan parasite *Toxoplasma gondii*. This parasite affects all warm-blooded mammals, but cats are the definitive host and primary reservoir. Infection of cats most likely occurs secondary to consumption of bradyzoites during carnivorous feeding. Cats then shed oocysts in feces for 4–21 days. Many cats remain asymptomatic; however, clinical signs can include fever, diarrhea, coughing, muscle stiffness, ataxia, paresis, seizures, stillbirth, and abortions. Clinical signs of disease are most likely to manifest in the young or immunocompromised animals.

- Diagnostic Tests: Serology (a fourfold rise in antibody titers) and CSF analysis. Fecal examination is invalid due to the intermittency of oocysts shed by cats. Definitive diagnosis is difficult.
- Potential Treatments: Antibiotics.

Technician evaluations commonly associated with this medical condition:

Diarrhea
- *Feces should be considered infective to other animals. Dispose of appropriately. Oocysts can remain in the environment from months to years.*

Hyperthermia
- Fever commonly occurs when tachyzoites (disseminating form) spread throughout the body systems.

Altered Mentation
- Animals experiencing meningoencephalomyelitis may suffer seizures.

Reproductive Dysfunction
- Queens infected during pregnancy commonly abort. Kittens infected transplacentally or early in the neonatal period are unlikely to survive.

Altered Gas Diffusion
- Respiratory involvement can occur as a result of secondary infections or extraintestinal dissemination to pulmonary tissues. Animals suffering from overt toxoplasmosis may be coinfected with FIV, FeLV, or FIP.

Risk of Infection Transmission
- Zoonotic potential is high.

Client Knowledge Deficit
- Pregnant women and immunocompromised individuals are of particular concern. Women exposed during pregnancy risk transmitting the infection to the fetus. Miscarriage, stillbirth, retinal abnormalities, blindness, hydrocephalus, and retardation are potential consequences of human fetal infection. Humans can be exposed via ingestion of fecal oocysts or bradyzoites present in muscle tissue (lambs and pigs).
- The following guidelines are recommended to minimize zoonotic potential:
  - Wear gloves when gardening.
  - Wash hands and all surfaces when handling raw meat. Thoroughly cook all meat (66°C). Freezing meat 3 days also destroys the organism.
  - Do not allow pregnant women to clean litter box.
  - Wash fresh vegetables thoroughly.
  - Clean litter box daily. Use a litter box liner. Do not dispose of fecal material in the backyard or other accessible location.
  - Wear gloves and wash thoroughly if you are a pregnant woman who must clean litter box.
  - Do not handle cats with diarrhea. Determine the cause and treat cats appropriately if diarrhea develops.
  - Cover children's sandboxes.
  - Do not acquire a new cat if pregnant or immunosuppressed.
  - Do not feed cats undercooked meat or allow them to hunt.

Canine Infectious Diseases

Adenovirus

Canine infectious tracheobronchitis (IT) has long been associated with three major pathogens. These include canine adenovirus type 2 (CAV-2), parainfluenza virus (PIV), and *Bordetella bronchiseptica*. Classic clinical signs include a dry hacking cough that is often followed by retching or gagging. Palpation or pressure on the trachea (leashes) will trigger a coughing spasm. Unless complicated by secondary infections, IT is typically self-limiting with few other clinical signs.

- Diagnostic Tests: Physical examination.
- Potential Treatments: Cough suppressants, bronchodilators, supportive nutrition, and antibiotics (only if secondary infection occurs).

Technician evaluations commonly associated with this medical condition:

Underweight
- A small percentage of animals become anorexic. Moistening food may encourage animals to maintain food consumption.

Exercise Intolerance
- Activity can trigger coughing episodes. Exercise restriction for 2–3 weeks postremission of clinical signs is recommended.

Altered Ventilation
- Avoid use of collars for 1 week postresolution of clinical signs. Head collars or harnesses can be used in place of collars.
- Humidification of air may assist in ventilation. Placing the animal in a steam-filled room such as a bathroom t.i.d. is helpful. In addition, nebulization using 4–6 mL sterile saline over 15 minutes q.i.d. (oxygen flow of 3–5 L/min) will assist with removal of tracheal secretions.

Risk of Infection Transmission
- This virus is readily transmitted in confined kennels or boarding facilities that are not well ventilated. Transmission occurs via aerosol spread and fomites. Strict isolation of infected animals is imperative. Disinfect all surfaces, food bowls, and other fomites using bleach, Nolvasan, or ROCCAL.
- Kennel ITB Prevention Recommendations
  - Vaccinate susceptible animals.
  - Match population density to facility capabilities. (Do not overcrowd.)
  - Maximize air changes to more than 12 per hour.
  - Segregate any suspect or new animals.
  - Use metal dishware that can be washed and disinfected daily.
  - Use bleach for disinfection at a 1:30 dilution (inexpensive and effective).

Bordetella Bronchiseptica

Commonly known as kennel cough, *Bordetella* is a well-known cause of infectious tracheobronchitis (IT). Classic clinical signs include a dry hacking cough that is often followed by retching or gagging. Palpation

or pressure on the trachea (leashes) will trigger a coughing spasm. Unless complicated by secondary infections, this condition is typically self-limiting with few other clinical signs.
- Diagnostic Tests: Physical examination.
- Potential Treatments: Cough suppressants, bronchodilators, supportive nutrition, and antibiotics (of limited value because they fail to reach the site of infection).

Technician evaluations commonly associated with this medical condition:

Underweight
- *A small percentage of animals become anorexic. Moistening food may encourage animals to maintain food consumption.*

Exercise Intolerance
- *Activity can trigger coughing episodes. Exercise restriction for 2–3 weeks postremission of clinical signs is recommended.*

Altered Ventilation
- *Avoid use of collars for 1 week postresolution of clinical signs. Head collars or harnesses can be used in place of collars.*
- *Humidification of air may assist in ventilation. Placing the animal in a steam-filled room such as a bathroom t.i.d. is helpful. In addition, nebulization using 4–6 mL sterile saline over 15 minutes q.i.d. (oxygen flow of 3–5 L/min) will assist with removal of tracheal secretions.*

Risk of Infection Transmission
- *This bacteria is readily transmitted in confined kennels or boarding facilities that are not well ventilated. Transmission occurs via aerosol spread and fomites. Strict isolation of infected animals is imperative. Disinfect all surfaces, food bowls, and other fomites using bleach, Nolvasan, or ROCCAL.*
- *Kennel ITB Prevention Recommendations*
  - *Vaccinate susceptible animals.*
  - *Match population density to facility capabilities. (Do not overcrowd.)*
  - *Maximize air changes to more than 12 per hour.*
  - *Segregate any suspect or new animals.*
  - *Use metal dishware that can be washed and disinfected daily.*
  - *Use bleach for disinfection at a 1:30 dilution (inexpensive and effective).*

Canine Distemper

Canine distemper is a highly contagious viral disease that affects many organ systems. Transmission occurs through contact with virally loaded respiratory exudates, saliva, feces, urine, and ocular discharge. Given the multisystemic nature of this disease, clinical signs are numerous and can include depression, fever, oculonasal discharge, cough, dyspnea, vomiting, diarrhea, dehydration, paresis, ataxia, muscle twitching, seizures, blindness, keratoconjunctivitis sicca, and hyperkeratosis of pads.

- Diagnostic Tests: Physical examination, CBC, serum biochemistry, serology (antibody titers), and virus detection (FA of conjunctival scrapes).
- Potential Treatments: Supportive treatments such as IV fluids, anticonvulsants, intestinal protectants, antiemetics, B vitamins, and antibiotics (secondary infections).

Technician evaluations commonly associated with this medical condition:

Vomiting
Dehydration
Electrolyte Imbalance
Diarrhea
Hyperthermia
Underweight
- *Lack of adequate nutrient consumption is typical and can last several days. Dogs with anorexia/vomiting lasting for more than 48 hours are considered candidates for TPN or PPN.*

Obstructed Airway
- *Respiratory secretions impeded oxygen flow. Humidification of airways will benefit animals. Nasal discharge (encrusted) should be removed regularly and a light petroleum product used to protect nasal integument.*

Sleep Disturbance
- *Hospitalized animals, especially those receiving continual intervention, risk losing periods of sleep.*

Altered Mentation
- *Neurologic signs associated with distemper are varied (depression and coma) and numerous. When selecting a nutrient support plan, care should be taken to assess the degree of mental involvement. Owners should be informed that neurologic signs can develop up to 2 months postacute infection.*

Reduced Mobility
- *Ataxia, paresis, hypermetria, muscle spasm, and other CNS-induced signs can reduce mobility.*

Impaired Tissue Integrity
- *Recumbent animals are at high risk for decubital ulcers. KCS/ocular involvement is damaging to corneal tissues.*

Risk of Infection Transmission
- *The canine distemper virus survives a limited time in the environment (20 minutes in exudates) and is easily destroyed by heat or routine disinfectants. Infected dogs should be maintained under quarantine. This virus is capable of infecting most carnivores, including dogs, foxes, ferrets, and cats. Infection of ferrets results in mortality approaching 100 percent.*
- *Vaccination of dogs is highly recommended.*

Risk of Infection
- *Secondary bacterial infections of the respiratory and gastrointestinal systems are common.*

Reproductive Dysfunction
- *In utero infections can induce abortions or stillbirths. All reproductive tissues and fluids are highly contagious and should be disposed of appropriately.*

Ineffective Nursing
- *If they survive, puppies infected in utero often exhibit clinical signs of the disease. "Fading puppy syndrome" is frequently noted during the neonatal period.*

Hepatitis

Infectious canine hepatitis, which is inflammation of the liver, is caused by the Canine Adenovirus Type 1 (CAV-1). (Type 2 adenovirus [CAV-2] induces infectious tracheobronchitis.) Clinical signs associated with infectious hepatitis range from subclinical infection to death. Clinical signs include lethargy, anorexia, thirst, jaundice, serous ocular discharge, vomiting, abdominal pain, fever, petechial hemorrhage, and corneal edema (opacity).

- Diagnostic Tests: CBC, serum biochemistry, clotting times, and virus isolation.
- Potential Treatments: IV fluids, blood transfusion, antibiotics, antiemetics, intestinal protectants, and analgesics.

Technician evaluations commonly associated with this medical condition:

Hyperthermia
- *Fevers of 104°F are common and can be biphasic.*

Hypovolemia
- *IV fluids are usually required.*

Vomiting
- *Once vomiting has been controlled, animals should receive frequent meals to avoid hypoglycemia.*

Acute Pain
- *Palpation of the abdomen, particularly the cranial quadrants, often elicits pain.*

Underweight
- *Severely debilitated animals may require several months to regain weight. Use of nutritionally dense foods is recommended. Examples include Hill's p/d, Royal Canin EARLY CARE PUPPY, and CliniCare.*

Risk of Infection Transmission
- *Large numbers of the canine distemper virus are present in urine, feces, and saliva of infected animals. Recovered dogs can shed the virus in urine for up to 6 months. Although readily inactivated by most disinfectants, the virus can remain viable in the environment for weeks to months.*
- *Isolation of infected animals is highly recommended. Adhere to strict aseptic technique.*
- *Vaccination is recommended.*

Leptospirosis

Leptospirosis is an infectious, zoonotic, spirochete bacterial disease that affects many animal species. Dogs are most commonly affected by *Leptospira interrogans*. Transmission occurs through direct contact, indirect contact (fomites), venereal and placental transfer, bite wounds, and ingestion of infected meat. Clinical signs of infection can include fever, anorexia, muscle soreness, conjunctivitis, rhinitis, petechial hemorrhage, melena, epistaxis, vomiting, dehydration, tachypnea, oliguria, jaundice, DIC, vascular collapse, and death. Acute kidney or liver failure can occur secondary to infection with leptospirosis.

- Diagnostic Tests: CBC, serum chemistry, urinalysis, antibody titers, and culture of urine or blood.
- Potential Treatments: Supportive treatments such as antibiotics, IV fluids, exercise restriction, and nutritional support.

Technician evaluations commonly associated with this medical condition:

Hyperthermia
- *Fevers reach 103–104°F.*

Underweight
- *Animals suffering from chronic leptospirosis develop chronic nephritis or hepatitis. This is associated with PU/PD, weight loss, ascites, and hepatic encephalopathy.*

Hypovolemia
- *Lack of fluid intake, vomiting, and diarrhea contribute to severe dehydration.*

Altered Urine Production
- *Infection can result in ARF with decreased urine production.*

Exercise Intolerance
- Strict cage rest should be enforced during acute episodes. Animals are anemic and unable to withstand high oxygen demands.

Risk for Impaired Tissue Integrity
- Care should be taken to provide adequate bedding for debilitated animals.

Acute Pain
- Abdominal and muscular pain is frequently noted.

Self-Care Deficit
- Debilitated animals are unlikely to maintain good hygiene. Wear gloves at all times when handling these animals. Urine and feces are infective.

Risk of Infection Transmission
- Strict quarantine and aseptic technique should be employed when handling infected dogs.
- Recovered animals can shed bacteria in urine for months to years postrecovery.
- Vaccination is associated with a higher risk of allergic reaction. Common vaccine reactions include fever, anorexia, and pain at the vaccination site. Allergic reaction can be evidenced by vomiting, diarrhea, facial swelling, whole body itching, urticaria, dyspnea, or sudden collapse.
- To avoid zoonotic disease, exercise extreme caution when handling animals or their secretions.

Client Knowledge Deficit
- Clients should be educated about zoonotic potential and proper handling of animal.

Parainfluenza

Canine infectious tracheobronchitis has long been associated with three major pathogens: parainfluenze virus (PIV), Canine Adenovirus Type 2 (CAV-2), and *Bordetella bronchiseptica*. Parainfluenza is the most common virus isolated from the upper respiratory tract. Classic clinical signs include a dry hacking cough that is often followed by retching or gagging. Palpation or pressure on the trachea (leashes) will trigger a coughing spasm. Unless complicated by secondary infections, parainfluenza is typically self-limiting with few other clinical signs.

- Diagnostic Tests: Physical examination, CBC, transtracheal wash, and virus isolation.
- Potential Treatments: Cough suppressants, bronchodilators, supportive nutrition, and antibiotics (only if secondary infection occurs).

Technician evaluations commonly associated with this medical condition:

Underweight
- A small percentage of animals become anorexic. Moistening food may encourage animals to maintain food consumption.

Exercise Intolerance
- Activity can trigger coughing episodes. Exercise restriction for 2–3 weeks postremission of clinical signs is recommended.

Altered Gas Diffusion

Altered Ventilation
- Avoid use of collars for 1 week postresolution of clinical signs. Head collars or harnesses can be used in place of collars.
- Humidification of air may assist in ventilation. Placing the animal in a steam-filled room such as a bathroom t.i.d. is helpful. In addition, nebulization using 4–6 mL sterile saline over 15 minutes q.i.d. (oxygen flow of 3–5 L/min) will assist with removal of tracheal secretions.

Risk of Infection Transmission
- This virus is readily transmitted in confined kennels or boarding facilities that are not well ventilated. Transmission occurs via aerosol spread and fomites. Strict isolation of infected animals is imperative. Disinfect all surfaces, food bowls, and other fomites using bleach, Nolvasan, or ROCCAL.
- Kennel ITB Prevention Recommendations
  - Vaccinate susceptible animals.
  - Match population density to facility capabilities. (Do not overcrowd.)
  - Maximize air changes to more than 12 per hour.
  - Segregate any suspect or new animals.
  - Use metal dishware that can be washed and disinfected daily.
  - Use bleach for disinfection at a 1:30 dilution (inexpensive and effective).

Parvovirus

Canine parvovirus is a highly contagious immunosuppressive virus that induces acute gastroenteritis in affected animals. Dogs under 1 year of age are most likely to experience clinical signs of the disease. Classic clinical signs include vomiting, bloody diarrhea, dehydration, depression, painful abdomen, and pale mucous membranes.

- Diagnostic Tests: ELISA Snap test and CBC.
- Potential Treatments: Supportive therapies such as IV fluids, antiemetics, nutritional support, anti-inflammatories, intestinal protectants, and probiotics. Routine antibiotic use is discouraged.

Technician evaluations commonly associated with this medical condition:

Hypovolemia
- *Maintaining adequate fluid volumes is critical to recovery. Common fluid choices include lactated Ringer's and 5 percent dextrose with added potassium chloride (10–20 mEq/L).*

Electrolyte Imbalance
- *Electrolyte monitoring is highly advisable.*

Vomiting
- *Once vomiting has resolved, a very bland diet should be reintroduced. Maintain the bland diet for 1–2 weeks postrecovery. Examples include Hill's i/d, Royal Canin INTESTINAL HE or DIGESTIVE LOW FAT LF, and Purina EN. Success also has been reported with recovery-type diets used early in the recovery, such as Hill's a/d and Royal Canin RECOVERY RS.*

Acute Pain
- *Enteritis causes abdominal pain. Care should be exercised when palpating the abdomen.*

Diarrhea
- *Diarrhea resulting from parvovirus is often voluminous and bloody. Abnormal gastrointestinal motility predisposes these dogs to intussusception. Palpate abdomen b.i.d. to monitor for intussusception.*

Self-Care Deficit

Altered Mentation
- *Depression often accompanies dehydration and acute pain.*

Sleep Disturbance
- *Animals suffering from gastroenteritis are often hospitalized for several days. Care should be taken to ensure that the animal is not continually disturbed.*

Cardiac Insufficiency
- *A rare complication of parvovirus can occur in young dogs. Puppies infected during the early neonatal period can develop an acute myocarditis. This typically leads to cardiopulmonary failure manifested by pulmonary edema, cyanosis, and sudden death.*

Risk of Infection Transmission
- *Infected dogs should be isolated. Utilizing equipment and clothing that have been reserved for quarantine is highly recommended. Strict aseptic technique should be maintained.*
- *The parvovirus is very stable, is resistant to many disinfectants, and can survive for months in the environment.*
- *Transmission is fecal-oral. Infection occurs directly through contact with infected animals and indirectly through feces.*
- *The virus can be shed in feces for several weeks postinfection. Isolate dogs 3 weeks postinfection to decrease environmental contamination.*

Client Knowledge Deficit
- Prevention Recommendations
  - Vaccinate.
  - Complete vaccination series prior to taking immunologically naive animals to public locations such as parks.
  - Isolate infected animals 3 weeks postinfection.
  - If an infection has occurred, thoroughly clean all surfaces and fomites using dilute bleach (1:30) or Parvocide.

Bowel Incontinence
- Owners may report house soiling in previously housetrained animals. This is most likely secondary to the profuse diarrhea and will resolve with correction of the diarrhea.

BIBLIOGRAPHY

Aiello, S. E. (Ed.). (1998). *The Merck veterinary manual* (8th ed.). Whitehouse Station, NJ: Merck & Co.

Battaglia, A. (2001). *Small animal emergency and critical care: A manual for the veterinary technician.* Philadelphia: W. B. Saunders.

Bistner, S. I., Ford, R. B., & Raffe, M. R. (1995). *Kirk and Bistner's handbook of veterinary procedures & emergency treatment.* (6th ed.). Philadelphia: W. B. Saunders.

Bonagura, J. D. (2000). *Kirk's current veterinary therapy XIII: Small animal practice.* Philadelphia: W. B. Saunders.

Cohan, M. (2006, September). Portosystemic shunts. *Veterinary Technician,* 544–552.

Cote, E. (2007). *Clinical veterinary advisor: Dogs and cats.* St. Louis, MO: Mosby.

Cox, M. (2006, September). Unraveling the mystery of feline cholangiohepatitis. *Veterinary Technician,* 538–542.

Doenges, M. E., Moorhouse, M. F., & Murr, A. C. (2006). *Nurse's pocket guide: Diagnoses, prioritized interventions, and rationales* (10th ed.). Philadelphia: F. A. Davis.

Ettinger, S. J., & Feldman, E. C. (2005). *Textbook of veterinary internal medicine.* Philadelphia: W. B. Saunders.

Gfeller, R. F., & Messonnier, S. P. (1998). *Handbook of small animal toxicology and poisonings.* St. Louis, MO: Mosby.

Greene, C. E. (2006). *Infectious diseases of the dog and cat* (3rd ed.). Philadelphia: W. B. Saunders.

Hand, M. S., & Novotony, B. J. (2002). *Pocket companion to small animal clinical nutrition* (4th ed.). Topeka, KS: Mark Morris Institute.

Hendrix, C. M. (2002). *Laboratory procedures for veterinary technicians* (4th ed.). St. Louis, MO: Mosby.

Jack, C. M., Watson, P. M., & Donovan, M. S. (2003). *Veterinary technician's daily reference guide. Canine and feline.* Baltimore: Lippincott Williams & Wilkins.

Merial EDU. (2005). *A four course continuing education workbook for veterinarians and clinic staff.* Lenexa, KS: Advanstar Healthcare Communications.

Nelson, R. W., & Couto, C. G. (1999). *Manual of small animal internal medicine.* St. Louis, MO: Mosby.

Norkus, C. (2006, September). Feline urethral obstructions. *Veterinary Technician,* 556–564.

Plumb, D. C. (2005). *Plumb's veterinary drug handbook* (5th ed.). Stockholm, WI: Blackwell Publishing.

Romich, J. A. (2008). *Understanding zoonotic diseases.* Clifton Park, NY: Delmar Cengage Learning.

Wingfield, W. (1997). *Veterinary emergency medicine secrets.* Philadelphia: Hanley & Belfus.

Zoran, D. (2006). Pancreatitis in cats: Diagnosis and management of a challenging disease. *Journal of the American Animal Hospital Association, 42,* 1–9.

Chapter 5
Surgical Procedures and Associated Technician Evaluations

INTRODUCTION

This chapter describes selected surgical interventions of dogs and cats. A brief description of each procedure is followed by technician evaluations that are commonly associated with the surgery. Additional comments are included with certain technician evaluations; this information typically relates to the specific procedure under discussion. However, the student technician still must refer to Chapter 3 for general information regarding each technician evaluation.

It is assumed that the technician will identify pre- and postoperative compliance as an evaluation relevant to all surgical interventions; therefore, these evaluations have not been identified for each surgery. Additional technician evaluations considered relevant to all surgical interventions include acute pain, risk of infection, risk of hypothermia, client knowledge deficit, and risk of self-inflicted injury.

This chapter is for reference purposes only; students wanting a broader, more detailed description of any surgical interventions are referred to standard veterinary technician or veterinary medical textbooks.

ANAL GLAND REMOVAL

Anal glands are located ventral to the anus in the five and seven o'clock positions. The malodorous substance expelled by these glands during defecation is believed to play an important role in territorial marking. Dogs can experience an array of problems associated with anal glands, the most common being recurrent perianal fistulas, neoplasia, and chronic impaction. Removal of the glands provides a permanent resolution to the medical problems and is frequently the treatment of choice that frustrated canine owners choose.

Technician evaluations commonly associated with this surgical procedure:

Acute Pain
– *Application of warm compresses or heating pads is the most appropriate nonpharmacological method of postoperative pain relief. Many dogs will willingly sit on a heating pad next to their owner.*

Constipation
– *Postoperative constipation is typically associated with pain upon defecation. Adequate control of postoperative pain is imperative. Stool softeners are often used to prevent constipation.*

Bowel Incontinence
– *Incontinence is typically associated with excessive surgical trauma or damage to the pudendal nerve. Monitoring for impaired pudendal nerve*

function is critical. Incontinence, decreased anal sphincter tone, and laxity in the perianal area can be associated with impaired function. The appearance of fecal material in the cage of housetrained dogs or elimination that is not associated with typical defecation behaviors warrants notifying the veterinarian immediately.

Self-Inflicted Injury
- Due to the sensitive nature of the perianal area, excessive postoperative chewing and licking is likely. Prophylactic application of E-collars is recommended. Dogs unable to lick the surgical area may resort to scooting along the floor. If this behavior is noted, appropriate environmental limitations should be instituted.

Risk of Infection
- Surgical incisions adjacent to the anus are exposed to high levels of bacteria. Maintaining good perianal hygiene reduces bacterial exposure. E-collars restrict the grooming and self-care abilities of canine patient. Gentle removal of perianal fecal material using a warm water wash or spritz is recommended.

AMPUTATION

Defined as the surgical removal of a body part, amputation most commonly involves the extremities. Amputation is typically regarded as a "salvage" procedure and is performed only when successful repair of the affected limb is highly improbable. Severe traumatic injuries, cancer, neurologic dysfunction, peripheral vascular disease, and congenital disorders are common reasons for amputation. Patients with an amputated limb tend to adjust quickly and ambulate with reasonable ease. Forelimb amputation is typically performed at the middiaphysis level of the humerus or through removal of the scapula and limb. Hind limb amputation is performed at the middiaphysis of the femur or through the coxofemoral joint.

Technician evaluations commonly associated with this surgical procedure:

Acute Pain
- Incisional pain may be expected. Severance of the nerve pathway may contribute to painful muscle spasms.

Reduced Mobility
- In the initial postoperative stage, the patient may require assistance in ambulation, such as use of a sling. The patient should be kept on a leash or in a confined space until balance is reestablished.

Altered Perfusion, Peripheral
- Decreased perfusion at the distal end of an amputated limb can lead to delayed healing and possible tissue necrosis. The surgical site should

be closely examined at each bandage change to ensure that tissue is adequately perfused.

Risk of Infection

Client Coping Deficit
- *Many owners have anthropomorphic views of their pets and fear that a pet may suffer psychological damage following amputation. Clients should be reassured that most animals adjust well to amputation.*

ARTHROSCOPY

Arthroscopy is the use of an endoscope to visualize a joint for diagnostic or surgical procedures. Repair of ligaments, meniscus, and cartilage and removal of bone fragments/excess tissue are common indications for arthroscopy. Although multiple incisions are required (one for the endoscope; others for the placement of surgical implements), the procedure remains significantly less invasive than open surgery.

Technician evaluations commonly associated with this surgical procedure:

Acute Pain
- *Patients may experience postsurgical pain due to the incision, increased fluid pressure within the joint, or edema.*

Risk of Infection
- *Joints are not well vascularized, with a consequent decrease in the ability to counter invading pathogens. Postsurgery, the affected joint should be closely monitored for signs of infection, including heat, erythema, and swelling.*

Impaired Mobility
- *Restricted/controlled exercise is indicated following arthroscopic surgery. Range of motion (ROM) in the affected joint may be maintained through passive ROM exercises.*

Client Knowledge Deficit
- *The owner should be educated regarding signs of infection and the prescribed exercise/rehabilitation program.*

AURAL HEMATOMA

Aural hematoma, which results from a rupture of an auricular vessel, is the accumulation of blood between the two layers of cartilage in the ear. Shaking or scratching of the ear secondary to bacterial/yeast otitis, ear mites, foreign bodies, or allergies remains the most common means of inducing auricular vessel leakage. Treatment of the underlying or predisposing cause is fundamental to recovery, and surgical intervention without concurrent resolution of head shaking inevitably results in treatment failure. Correction options for this condition include:

- Aspiration using a large-gauge needle (usually a temporary measure).
- Placement of a Penrose drain: A drain is inserted via two stab incisions.
- Insertion of a teat cannula: A cannula is inserted into a stab incision and secured via suture or tissue glue.
- Incision with mattress sutures: An S-shaped incision is made on the ventral pinna, and skin is secured via parallel mattress sutures.
- Use of laser: Several small holes are made in the ventral pinna followed by a slightly larger incision on the distal aspect. No sutures are placed.

Technician evaluations commonly associated with this surgical procedure:

Impaired Tissue Integrity
- *Postoperative contracture can result in a permanent cosmetically undesirable condition known as cauliflower ear. This condition results in shriveling and distortion of the pinna.*

Acute Pain
- *Pain can be associated with the hematoma, with the underlying cause of irritation, and with the surgical procedure.*

Risk of Infection
- *Infection occurs postoperatively due to contamination of the surgical site or as a result of the predisposing otitis.*

Self-Inflicted Injury
- *Head shaking and scratching occur secondary to discomfort. Application of an E-collar or a bandage can reduce the incidence of self-trauma. Use of an E-collar postoperatively is critical.*

Altered Sensory Perception
- *E-collars and bandages impair perception.*

Client Knowledge Deficit
- *Education of the client regarding home care and reoccurrence prevention is vital.*

BLADDER SURGERY (CYSTOTOMY)

A cystotomy, an incision into the bladder, is performed to treat congenital or acquired disorders of the urinary bladder. Common surgical indications include uroliths, a ruptured bladder (trauma or secondary to obstruction), and neoplasia. Most cystotomies entail a ventral midline incision as this approach provides optimal exposure of the bladder. An indwelling urinary catheter is indicated postoperatively. Complications associated with this surgery can include peritonitis, failure to flush calculi through the bladder neck, and occlusion of the ureters secondary to inflammation.

Technician evaluations commonly associated with this surgical procedure:

Acute Pain
- In addition to pain from the surgical incision, animals may experience pain during urination due to inflammation of the bladder or urethra.

Urinary Incontinence
- Surgical trauma, medications, and anesthetics can cause decreased muscle control, resulting in incontinence. Placement of a urinary catheter can lead to irritation of the bladder; following removal, animals may experience a frequent urge to urinate. This problem is usually self-limiting.

Risk of Infection

Client Knowledge Deficit
- Owners should be educated regarding signs of UTIs and methods used to prevent the formation of uroliths, such as prescribed diets and increased fluid intake.

CESAREAN SECTION

A cesarean section (c-section) is the surgical removal of the fetus(es) from the uterus. Cesarean sections may be planned due to breed predisposition or the dam's previous history of dystocia, or it may be performed on an emergency basis due to obstructive dystocias or non-progressing labors. Dogs and cats undergoing cesarean sections are often dehydrated and physiologically stressed. Rehydration and stabilization of the bitch/queen prior to surgery is fundamental to a successful outcome. In addition, oxygen therapy of the female and clipping prior to induction minimize the likelihood of inadequate oxygenation during surgery. If possible, avoid placing the female in dorsal recumbency until performing final surgical preparations.

Preoperative tests can include radiographs, complete blood count (CBC), and serum chemistry profile. Clients may choose to have an elective ovariohysterectomy performed upon removal of the fetus(es).

A recovery station is recommended for the care of newly delivered puppies and kittens. Supplies should include warmed towels/incubator, bulb syringe, Kelly forceps, suture, epinephrine, naloxone, and doxapram. Newborns should remain in an incubator until the mother is fully recovered from anesthesia.

Technician evaluations commonly associated with this surgical procedure:

Acute Pain
- The dam may experience pain due to both the surgical incision and uterine contractions following birth. Administration of oxytocin (to facilitate

uterine involution and to stimulate milk production) may lead to stronger and therefore more painful contractions.

Hypovolemia
- Animals are often slightly anemic at the conclusion of gestation. Prolonged labor predisposes females to dehydration. Postoperative uterine hemorrhage can induce rapid hypovolemia and sudden hypotension. Close monitoring during the first few hours postsurgery is warranted.

Risk of Infection
- Special attention needs to be paid to the sutures postoperatively since the newborns will be nursing next to the incision site. Also, the mammary glands may cover the incision site, presenting ideal growth conditions for yeast and bacteria. Antibiotics are often administered postoperatively.

Electrolyte Imbalance
- Hypocalcemia and hypoglycemia commonly occur with dystocia.

Anxiety
- The dam should be recovered from anesthesia and fully aware of her surroundings before the newborns are placed with her. Dams tend to be protective of and/or anxious about their young; it is important to provide privacy.

Risk of Ineffective Nursing
- Dams that undergo a cesarean section are at risk of rejecting their young. Pain from the surgical incision may cause the dam to limit or prohibit infant nursing. Females undergoing a c-section may experience a temporary agalactia. Normal milk production should be noted within 24 hours of surgery. While females are recovering from anesthesia, all residual surgical preparation solutions should be removed from the mammary area.

Client Knowledge Deficit
- Owners who have not previously bred animals may need to be educated regarding normal maternal and newborn behavior. They also should be instructed to monitor the nutritional status of all newborns.

Altered Ventilation
- Gravid females placed in dorsal recumbency can experience inadequate oxygenation due to compression of the vena cava and pressure on the diaphragm. Preoxygenation prior to surgical induction will facilitate adequate oxygenation of both mother and fetuses.

DECLAW (ONYCHECTOMY)

An onychectomy (declaw) is the removal of the third phalynx (P3). This procedure is typically performed on the front paws of cats. Owners usually request this elective procedure to prevent furniture damage or to reduce intercat aggression injuries. A nail clipper, small scalpel blade, and/or laser may be used to amputate the P3 while a tourniquet is placed to control bleeding. Surgeons can close the incisions using an absorbable suture or tissue adhesive. Alternatively, the surgical site may be allowed to close via second intention healing. Cats are routinely bandaged postoperatively. For this procedure to be successful, the germinal epithelium must be removed; otherwise, the claw may regrow.

Technician evaluations commonly associated with this surgical procedure:

Acute Pain
Risk of Infection
- *Care needs to be taken to make sure the paws are kept bandaged and clean. For 5–7 days following surgery, shredded paper instead of clay litter should be used in the litter box to prevent contamination/infection of the surgical site.*

Risk of Self-Inflicted Injury
Reduced Mobility
- *Cats, especially older and overweight animals, refuse to engage in normal physical activities for several days postoperatively.*

Owner Knowledge Deficit
- *The owner needs to be aware that the patient can no longer defend itself with its front paws or climb to escape predators. The owner should be advised that the animal must remain indoors at all times.*
- *Performing an onychectomy on species such as the dog (to control digging or destructive behavior) or the rabbit (to control scratching) is medically contraindicated as these species utilize P3 as a primary weight-bearing structure. Removal of P3 in these animals results in chronic pain and lameness.*

DENTAL PROPHYLAXIS

Dental prophylaxis involves cleaning the teeth as well as evaluating the oral cavity for disease through visual inspection and radiographs. The plaque found on unclean teeth harbors bacteria that cause gingivitis, tooth loss, and halitosis. In addition, advanced periodontal disease can lead to systemic infections. General anesthesia is required for the ultrasonic scaling and polishing of teeth. The patient is placed in lateral recumbency with the head lower than the neck to prevent aspiration.

Technician evaluations commonly associated with this surgical procedure:

Altered Oral Health
Acute Pain

Risk of Aspiration
 – An endotracheal tube should be placed to prevent aspiration; in addition, the patient's head is placed lower than its body.

Client Knowledge Deficit
 – Many clients are unaware that daily brushing is beneficial to their pet's health. Clients should be educated regarding the technique and use of appropriate animal toothpaste.

Noncompliant Owner
 – Owners may find daily brushing tedious. In addition, pets may be extremely resistant to brushing.

DIAPHRAGMATIC HERNIA

A diaphragmatic hernia is a hole or tear in the diaphragm that allows abdominal organs to protrude into the thoracic cavity; it may be congenital or acquired. Clinical signs associated with this condition can include dyspnea, shock, a "tucked-up" abdomen, muffled heart and lung sounds, and abdominal sounds noted in the chest cavity. Radiograph and ultrasound evaluation are the primary tools of diagnosis.

Traumatic diaphragmatic herniation typically results from injuries associated with automobile accidents. Preoperative stabilization of these patients remains critical to successful surgical intervention. Animals undergoing surgery within 24 hours posttrauma traditionally experience a much higher mortality rate than those stabilized for longer periods of time. Cardiopulmonary dysfunction, organ entrapment, and pulmonary compromise secondary to pleural space filling make up the primary assessment parameters for this condition. Ultimately, the veterinarian's decision to proceed with surgery reflects the relative risk of anesthesia on these parameters against the benefit of surgical intervention.

Although some diaphragmatic hernias are immediately life-threatening, others present with milder, vague symptoms, sometimes of many months' duration. A history of past trauma is often reported by owners.

Technician evaluations commonly associated with this surgical procedure:

Acute Pain
 – Uncontrolled pain can lead to shallow breathing and inadequate ventilation, both predisposing the patient to developing pneumonia.

Hypovolemia
 – Shock is often associated with traumatic hernias. Appropriate fluid therapy is vital.

Altered Ventilation
- Dyspnea associated with traumatic herniation can result from hypovolemic shock, chest wall trauma, pleural space filling, pulmonary contusion, and cardiac dysfunction.
- Hypoventilation associated with herniation often results from voluntary restriction of chest motion secondary to pain, mechanical dysfunction (rib fractures and flail chest), inadequate filling of lungs (due to filling of thoracic cavity by organs or air) and atelectasis. Animals should be closely monitored for hypoventilation both pre- and postoperatively.
- Preoxygenation of all surgical patients is highly recommended. Placing animals in an oxygen-rich environment (40 percent) is advised postoperatively.
- Twenty-four-hour supervision and extreme care must be taken to ensure safety and patency of indwelling chest tubes. Petroleum-impregnated gauze and Kelly clamps should be kept at hand to clamp the tube and occlude the incision in the event the chest tube develops an air leak.

Decreased Perfusion Cardiac
- Traumatic herniation is often associated with myocardial contusion and dysfunction, which causes a decreased cardiac output. Arrhythmias also are associated with this condition, with ventricular tachycardia being most common.

Exercise Intolerance
- Animals may go undiagnosed for months or years. Owners often report that the animal tires easily with exercise or refuses to engage in activities.

Risk of Infection

DISLOCATIONS

Dislocations are the displacement or separation of a joint. The most common joints affected are the hip, elbow, and stifle; but all joints have the potential for dislocation. Reduction of a joint should be performed promptly to minimize damage to surrounding soft tissues. Reductions may be surgical or nonsurgical (closed). Both procedures require that the patient be under general anesthesia. Radiographs are taken to ensure proper placement of the joint. A bandage splint or an Ehmer sling may be applied to limit mobility postoperatively.

Technician evaluations commonly associated with this surgical procedure:

Acute Pain or Chronic Pain
- Previously dislocated joints are at a high risk of arthritis.

Client Knowledge Deficit
- The owner needs to be aware of and understand the importance of exercise restriction for the patient.

Risk of Self-Inflicted Injury
- The patient must be kept calm during recovery from anesthesia as thrashing may cause further injury.

Reduced Mobility
- The patient may require assistance ambulating while wearing the sling/bandage.

EAR CROPPING

Ear cropping is an elective cosmetic procedure performed to conform animals to breed standards. The amount of pinna removed reflects individual breed specifications, with ranges from one-third (schnauzer) to one-eighth (Great Dane) of the pinna being removed. Ears are typically cropped at 10–12 weeks of age while the puppy is under general anesthesia. Follow-up care is extensive and necessitates a high degree of owner compliance. Most breeds require "racking" of the ears, which entails the use of an aluminum frame to hold ears upright for 3 weeks postoperatively. After removal of the rack, ears are taped in position for an additional 1–8 weeks.

Technician evaluations commonly associated with this surgical procedure:

Acute Pain
- Puppies are typically placed on nonsteroidal anti-inflammatories to minimize postoperative discomfort.

Self-Inflicted Injury
- The animal can easily reach and scratch the pinna using the hind limbs. Trauma to the ears will result in treatment failure. E-collars are often used to prevent self-inflicted trauma. Application of vitamin A and D ointment to the incision line after suture removal can help reduce itching.

Client Knowledge Deficit
- Owner compliance is critical. Prior to surgery, owners should receive education about the extensive aftercare that this procedure requires.
- Once the rack has been removed, retaping of the ears must be performed every 3–10 days. In addition, owners should be aware that surgery and bandaging do not guarantee a cosmetically acceptable appearance and that the ears may not stay erect.
- Ears should be standing erect by 9-10 months of age, and interventions after this time are unlikely to succeed.

Impaired Tissue Integrity
- Inadequate cartilage development, scarring with wound edge contracture, and irritation to the suture line constitute the primary reasons for ears failing to stand erect. Tape used for stabilization purposes should never come in contact with the suture line.

ENDOSCOPY

Endoscopy is the use of an endoscope to visualize the internal organs or body cavity for diagnostic or surgical procedures. Gastric ulcers, removal of foreign bodies, biopsy, tumors, and exploration of the gastrointestinal (GI) tract are common reasons to use an endoscope.

Technician evaluations commonly associated with this surgical procedure:

Risk of Aspiration

FRACTURES

Pelvic Fracture

The pelvis, made up of the ilium, ischium, and pubis, is a common site of fracture in veterinary patients. Given the boxlike configuration of the fused sacrum and pelvis, multiple fractures with more than three bones involved are the norm rather than the exception. Fractures typically occur in response to blunt trauma such as automobile accidents and are classified as surgical or nonsurgical. Nonsurgical candidates, which are treated via cage rest, anti-inflammatories, analgesics, and good nursing care, are characterized by an intact acetabulum, minimal displacement of fracture segments, and a viable pelvic canal. Surgery is indicated in the presence of a reduced pelvic canal diameter or hip joint instability. Fixation options are numerous and can include bone plates, screws, external fixators, intramedullary pins, cerclage wire, or a combination thereof. The surgical approach selected will reflect the fracture location.

Technician evaluations commonly associated with this surgical procedure:

Acute Pain
Chronic Pain
– *Osteoarthritis is a potential sequela of pelvic fracture.*
Constipation
– *Pain in the pelvic area may make the patient reluctant to defecate. Administration of stool softeners can facilitate defecation.*
Urinary Incontinence
– *Concurrent injury of the bladder or urethra can occur with pelvic fracture.*
Risk of Reproductive Dysfunction
– *Females suffering from severe fractures may experience dystocia secondary to reduced pelvic diameter.*
Reduced Mobility
– *The pelvis is the primary weight-bearing structure of the hind end. Animals suffering from severe pelvic fracture may be unable or unwilling*

to bear weight on the hind limbs. Damage to the spinal cord or sciatic nerve can result in neurologic dysfunction.
- Animals suffering from acetabular fractures are often placed in an Ehmer sling, while those experiencing difficulty adducting the hind limbs may be placed in a hobble bandage. The degree of exercise restriction ordered by the attending veterinarian will reflect fracture severity and the degree of stability achieved during surgery.

Risk of Impaired Tissue Integrity
- Pain may make the patient unwilling to change position; skin should be monitored for early signs of decubital ulcers, with particular attention being paid to pressure points.

Self-Care Deficit
- Animals in pain may be reluctant to clean the perineal area adequately or may urinate in a recumbent position. Maintaining a clean and dry hair coat is fundamental in preventing decubital ulcers.

Risk of Infection
- Delayed emptying of the bladder can predispose animals to urinary tract infection (UTI). Suture lines exposed to pressure, urine, or feces are at a higher risk of complications.

Client Knowledge Deficit
- The client should be educated regarding exercise restrictions and physical therapy.

Limb Fractures

Fractures of the limbs are typically associated with blunt trauma and can range in severity from mild (phalynx fracture) to severe (femoral fracture). Terminology used to describe and classify limb fractures include the following:

- Transverse: Fracture line is at right angle to long axis of bone (straight across).
- Oblique: Fracture line is diagonal to long axis of bone.
- Spiral: Fracture line spirals along long axis of bone.
- Comminuted: Splintering or fragmentation is present (shattered china).
- Avulsion: Portion of bone upon which muscle inserts is detached.
- Greenstick: Fracture is a crack in the bone with only one side of the cortex affected.
- Physeal: Fracture involves growth plate. Further classified according to Salter-Harris system.
- Open: Fracture communicates with outside.

Surgical fixation options for limb fractures can include intramedullary pins, external fixators, plates, screws, and wires. Postoperatively, the limb may be placed in a sling or bandaged (lightly or heavily). Repair of limb fractures also can be achieved via fiberglass casts or splints.

Technician evaluations commonly associated with this surgical procedure:

Acute Pain

Reduced Mobility
- *Fracture location, severity, and stabilization determine the degree of lameness. Immature animals suffering from fractures of the growth plate may experience lack of normal bone growth with resultant "shortening" of the affected limb.*

Impaired Tissue Integrity
- *Open fractures involve damage to soft tissue surrounding the bone. External fixators require additional care of integument near areas of pin placement.*

Risk of Infection
- *Open fractures provide a pathway for invading pathogens.*

Client Knowledge Deficit
- *Owners must be educated regarding exercise restrictions and physical therapy.*

Facial Fractures

Fractures of the mandibular and maxillary bones are commonly open and contaminated. Fixation includes the use of interfragmentary stainless steel wire, intramedullary pins, external fixation, acrylic splints, and/or bone plates. Most commonly the animal's ability to eat and drink is compromised. In severe cases, an esophagostomy tube is placed for postoperative feeding and watering. Healing time for mandibular fractures averages 3–5 weeks.

Technician evaluations commonly associated with this surgical procedure:

Acute Pain

Self-Care Deficit
- *Due to pain or inability to open its mouth, the patient will lack the ability/desire to groom itself.*

Risk of Infection
- *Bacteria can become entrapped around fixation devices. Ulcers of the mucous membranes may provide a pathway for invading pathogens.*

Altered Oral Health
- *Oral mucous membranes may be damaged or ulcerated from the initial trauma or from the presence of the fixation devices.*

Risk of Aspiration
- *The patient may have a decreased ability to expel saliva, vomit, or blood.*

Abnormal Eating Behavior
- *The patient's diet will need to be changed to accommodate facial structure stability. The attending veterinarian will prescribe the appropriate diet.*

Risk of Obstructed Airway
- Posttrauma and postsurgical edema may block the patient's airway. Similarly, the patient's decreased ability to expel saliva, blood, or vomit can lead to an obstructed airway.

Client Knowledge Deficit
- The owner must be instructed to observe the animal for signs of infection or obstructed airway.

Underweight
- Animals unwilling or unable to consume adequate nutrition may require an esophagostomy tube.

GASTRIC DILATATION VOLVULUS (GDV)

Commonly referred to as canine bloat, GDV is a veterinary emergency that typically affects large and giant breed dogs. Clinical signs of this condition can include abdominal distension, retching, depression, labored breathing, and shock. Treatment for shock, gastric decompression and correction, or fluid deficits/electrolyte disturbances should precede surgery. Surgical intervention involves decompression of the stomach, correction of the gastric/esophageal torsion, correction of splenic displacement, and gastropexy. Animals typically require constant monitoring during the postoperative period. (See Chapter 4.)

Technician evaluations commonly associated with this surgical procedure:

Acute Pain
- Control of postoperative pain is critical to recovery.

Hypovolemia

Risk of Aspiration
- Retching and increased thoracic pressure due to gastric distension predispose animals to aspiration.

Cardiac Insufficiency
- Gastric torsion results in compression of the vena cava, which leads to decreased venous cardiac return and ischemia. Reduced cardiac output and arrhythmias (ventricular premature contractions and ventricular tachycardia) are frequently noted. Ventilation perfusion mismatch can occur.

Decreased Perfusion Cardiovascular, Intestinal
- Gastric torsion reduces blood flow to the portal vein and impairs perfusion of abdominal organs. Severe compression of gastric vessels can result in tissue necrosis and subsequent toxemias. Displacement of the spleen from the left to the right side of the abdomen obstructs splenic vessels and induces splenomegaly/ischemia.

Electrolyte Imbalance
- Common abnormalities include hypokalemia and hypomagnesemia.

Vomiting
- Postoperative gastritis can induce nausea and vomiting. Prior to surgery, animals may exhibit signs of retching but are unable to move injesta through the torsed esophagus. Dogs are typically maintained n.p.o. for 12 hours postsurgery.

Risk of Infection
- Decompression and resection of necrotic tissues predispose animals to infection.

Client Knowledge Deficit
- Educate owners regarding prevention techniques such as feeding smaller, more frequent meals; reducing aerophagias; and prohibiting exercise following a meal. Stressed or fearful dogs may be predisposed to this condition due to inhibition of gastric motility.

Altered Ventilation
- Gastric distension compresses the diaphragm, resulting in dyspnea, decreased tidal volume, and impaired oxygenation. Supplemental oxygen may be indicated.

HERNIAS

A hernia, which is a protrusion of tissue through a rent (hole) in surrounding musculature, is a common malady of small animals. Hernias encountered in small animal medicine include diaphragmatic (discussed previously), hiatal, umbilical, and inguinal. Umbilical and inguinal hernias are typically congenital, but they can occur secondary to trauma. Uncomplicated, nonemergency hernia repair surgery is often performed in conjunction with routine spaying or neutering. Surgical reduction of hernias entails returning herniated structures to the abdominal cavity, obliterating the peritoneal sac, and closing the rent. Closure of excessively large rents may necessitate the use of a polypropylene mesh.

Technician evaluations commonly associated with this surgical procedure:

Acute Pain
- Pain is associated with surgical repair; however, uncomplicated hernias such as small umbilical hernias may not induce a pain response.

Risk of Infection
- Abdominal organs that protrude through a small opening may become strangulated. The resulting tissue ischemia leads to necrosis and sepsis.

Vomiting
- Animals that have developed an intestinal blockage at the site of a hernia will exhibit vomiting and refuse food once the digestive tract is emptied.

Decreased Perfusion, Intestinal
- Incarceration of viscera within the hernia can impede blood flow to the entrapped tissues. Complete strangulation and necrosis of viscera is a surgical emergency with an associated grave prognosis.

Reproductive Dysfunction
- Intact males can be affected by inguinal hernias, leading to reproductive dysfunction.

INTESTINAL RESECTION AND ANASTOMOSIS

Intestinal resection and anastomosis is the removal of a portion of the intestine followed by a reattachment of the healthy ends. Neoplasia, mechanical obstruction (intraluminal or extraluminal), strangulating obstruction (thrombosis, hernias, or volvulus), and intussusception are common reasons for intestinal resection.

Patients that have experienced complete obstruction may be hypovolemic, with dehydration severe enough to cause decreased renal perfusion, electrolyte imbalances, and shock. In the immediate postoperative period, food and water are withheld until the determination has been made that peristalsis has returned.

Technician evaluations commonly associated with this surgical procedure:

Acute Pain

Risk of Infection
- Destruction of the normal mucosal barrier permits bacteria/endotoxins to pass transmurally into the bloodstream, leading to septic shock. Leakage of intestinal contents before, during, or after surgery can lead to peritonitis.

Vomiting

Decreased Perfusion, GI
- Distension of the lumen due to pressure from trapped air, gas, and fluid can lead to ischemia.

Hyperthermia
- Fever indicates peritonitis and/or necrosis.

Hypovolemia
- Increased capillary permeability allows intravascular fluid to move into the lumen of the intestine, leading to third spacing and consequent dehydration.

Electrolyte Imbalance
 – *Metabolic acidosis and hypokalemia are common.*
Constipation

LACERATIONS

Defined as wounds or cuts through the skin that are not created for the purpose of surgery, lacerations are a common condition requiring surgical intervention. Lacerations can range from superficial, involving only the dermis, to deep, exposing underlying tissues and structures. Fresh, clean superficial lacerations may be managed via primary closure through suture or tissue adhesive. Deeper lacerations may necessitate placement of drains, while contaminated lacerations may require debridement and secondary intention healing. Ultimately, the surgical intervention selected by the veterinarian will reflect the laceration's location, depth, degree of contamination, wound edge tension, and damage to underlying structures.

Technician evaluations commonly associated with this surgical procedure:

Impaired Tissue Integrity
Acute Pain
Risk of Infection
 – *Openings through the skin provide a pathway for invading pathogens.*
Hypovolemia
 – *Deep lacerations associated with penetration of arteries can result in significant blood loss.*
Decreased Perfusion, Peripheral
 – *Loss of vascular integrity can induce necrosis and sloughing of affected tissues.*

OCULAR INJURY/SURGERY

Entropion repair, foreign body penetration, proptosis, meibomian gland prolapse, and lacerations of the eyelid or the cornea represent common reasons for ocular surgery. Surgical preparation of the eye will reflect the intended surgical procedure. The following preoperative measures are applicable to the majority of eye injuries:

- Use only sterile eyewash or saline to flush the eye.
- Keep eye moist at all times.
- Unless the cornea has been penetrated, use a nonsteroidal ophthalmic ointment to keep the eye moist. Avoid petroleum-based ointments in cases of corneal penetration.
- Use cold packs to minimize periorbital swelling.

Technician evaluations commonly associated with this surgical procedure:

Acute Pain
- Cold packs may be used to alleviate periorbital pain and swelling.

Altered Sensory Perception Visual
- Due to the nature of the surgery, bandaging, and/or E-collar placement, the patient's vision may be altered or impaired.

Risk of Self-Inflicted Injury
- Irritation from the surgery or initial trauma may cause the patient to rub or scratch the eye. An E-collar or a bandage may be warranted.

Risk of Infection

PROLAPSES

Defined as a movement or a sliding away of a tissue from its normal location, the rectum and vagina represent the organs most commonly prolapsed. Rectal prolapse, which is seen primarily in younger patients, is often associated with tenesmus secondary to colitis. Vaginal prolapses frequently follow dystocias, hormone surges, and/or tenesmus. The surgical intervention selected for prolapse repair will vary depending on the severity and tissue involved. However, reduction and placement of the affected organ followed by a purse-string suture is typical of many repairs. Prior to surgery, general care of any prolapsed organ centers on monitoring prolapse size and preventing further damage to the exposed tissue.

Technician evaluations commonly associated with this surgical procedure:

Acute Pain

Risk of Infection
- Exposed mucosal tissue is very vulnerable to trauma and infection.

Reproductive Dysfunction
- Reproductive dysfunction is associated with prolapses of the vagina or uterus.

Decreased Perfusion
- Compression of the venous vasculature impairs blood flow and leads to edema in the affected tissue.

Impaired Tissue Integrity
- Monitor mucosal lining for ulceration/necrosis. A warm saline solution may be used to clean and moisten the tissue. Application of a petroleum product such as Vaseline helps retain moisture when repair is delayed.

Self-Inflicted Injury
- E-collars help prevent licking and chewing of sutures and exposed tissue.

RENAL SURGERY

Treatment of trauma, neoplasia, and nephroliths account for many of the performed renal surgeries. Surgical interventions are varied and can range from nephrectomy, which is the removal of the kidney, to nephrolithotomy, which involves the removal of kidney stones.

Technician evaluations commonly associated with this surgical procedure:

Risk of Infection
- Urinary catheters are often utilized postoperatively. Use of sterile technique and careful monitoring for UTIs is important.

Urinary Incontinence
- Incontinence can occur secondary to inflammation or irritation.

Acute Pain

SPAYING/NEUTERING

Spaying entails removing the ovaries/uterus in the female; neutering is the removal of the testicles in the male. These elective surgeries are performed to prevent breeding and to minimize the incidence of future reproductive disease. Sterilization also can reduce/prevent gender behavior traits from developing (marking, traveling, yowling, and being aggressive). Dogs and cats can be spayed/neutered as early as 8 weeks, but the general recommendation remains 6 months of age. Removal of reproductive organs also may be recommended as a treatment component for other diseases such as pyometra and prostatitis.

Technician evaluations commonly associated with this surgical procedure:

Acute Pain
- Cold packs may be used to alleviate swelling or discomfort in the dog's scrotum.

Self-Inflicted Injury
- E-collars can be used to prevent self-injury.

Risk of Infection
- Prophylactic antibiotics are not routinely used for this elective procedure.

SPINAL CORD/VERTEBRAE SURGERY

Surgical procedures involving the spinal cord are typically performed to decompress or stabilize the spinal cord tissue. Decompression techniques include the following:

- Hemilaminectomy: Bone (lamina) on one side of the vertebrae is removed, thereby providing ventral and lateral access to the spinal cord.

- Dorsal laminectomy: Bone is removed from the dorsal aspect of the vertebrae allowing for dorsal, lateral, and ventral exposure of the spinal cord.
- Ventral slot decompression: Bone is removed from the ventral aspect of the vertebrae. This procedure is reserved for cervical lesions.
- Fenestration: In this prophylactic procedure, the nucleus pulposus is removed from the intervertebral disc. The vertebral canal is not entered. The procedure is reserved for animals experiencing pain secondary to mild disc protrusion.

Common indications for surgery of the spinal column include intervertebral disc protrusion, trauma, spondylomyelopathy (Wobbler syndrome), cauda equina syndrome, and neoplasia.

Technician evaluations commonly associated with these surgical procedures:

Acute Pain or Chronic Pain
Bowel Incontinence
- *Damage to the spinal cord can result in loss of anal innervation.*

Urinary Incontinence
- *Damage to the spinal cord can result in loss of bladder innervation.*

Client Coping Deficit
- *Spinal cord injury is a severe and often grave condition.*

Impaired Tissue Integrity
- *In addition to loss of skin integrity at the surgical site, animals may experience decubital ulcers or abrasions secondary to loss of mobility and innervation. Recumbent animals should be placed on a well-padded surface and turned every 3–4 hours.*

Reduced Mobility
- *Depending on the location and extent of spinal cord damage, mobility impairment can range from a reluctance to ambulate to paralysis. Patient interventions selected by the technician will reflect patient need and the veterinarian's orders.*
- *Extreme care should be taken when moving patients with spinal cord injuries. Stabilization of the spinal cord to prevent further trauma is of utmost importance. Animals should be moved using a firm, flat surface such as plywood or a stretcher.*
- *Cage rest: Exercise restriction is critical for spinal cord repair. Animals must be confined in close quarters to restrict movement. Thus, cages are preferred over runs as a method of confinement. If able to ambulate, animals should remain on a leash at all times when brought outside to urinate or defecate. A harness or another alternative to a collar should be used on animals affected in the cervical area.*
- *Physical therapy such as passive and active ROM exercises and hydrotherapy is of benefit. Spinal cord patients typically experience a loss*

of muscle mass, which further increases susceptibility to pressure sores. Passive manipulation of joints is recommended four to six times per day to prevent joint contracture. Therapy should be initiated 2 weeks postoperatively.

Risk of Infection
- The suture line of all surgical animals should be examined at least b.i.d.
- Cystitis is a common sequela of reduced bladder innervation secondary to spinal cord injury. Animals with a urinary catheter are especially at risk for cystitis. Culture of the urine is recommended upon removal of the catheter. Atelectasis secondary to recumbency predisposes animals to respiratory infection. Antibiotics are routinely administered.

Self-Care Deficit
- Pain and loss of mobility predispose animals to discontinuing appropriate grooming behaviors.

Overweight
- Chondrodystrophic breeds suffering from intervertebral disc disease (IVDD) are likely to be overweight. Selection of an appropriate diet to reduce weight and maintain it within normal parameters is highly recommended.

TAIL DOCKING

Tails are most commonly docked for breed standards at 3–5 days of age with the amount of tail removed reflecting breed requirements and the owner's personal preference. The tail is amputated using scissors/scalpel blade, and closure is achieved via a suture or skin adhesive. Once the surgery is completed, the newborn is immediately returned to the dam. When performed at 3–5 days of age, anesthesia is seldom used. Dewclaw removal is typically performed in conjunction with tail docking.

Removal of an adult animal's tail may be required for treatment of traumatic or neurologic injuries. General anesthesia and standard aseptic technique are required for amputation.

Technician evaluations commonly associated with this surgical procedure:

Acute Pain
- Adult animals often exhibit signs of severe pain and experience difficulty sitting.

Ineffective Nursing
- Observe puppies for nursing behaviors after they are returned to the bitch. Monitor dam for anxiety or inappropriate grooming of the puppies.

Bowel Incontinence
- Amputation close to the base of the tail may lead to peripheral nerve damage and incontinence.

Risk of Infection
- Risk of infection is substantial due to the proximity to fecal material and propensity for the dam's licking of the surgical area on the puppy.

Self-Inflicted Injury
- Adult animals may self-traumatize the tail end. Application of an E-collar is advised. Bitches should be monitored closely for grooming behaviors that traumatize the puppy's surgical site.

THORACIC SURGERY

Made up of the ribs, sternebrae, and respiratory muscles, the thoracic cavity is often entered to perform surgery on the heart, lungs, esophagus, caudal vena cava, diaphragm, thoracic duct, and trachea. Treatment of pneumothorax, lung lobectomy, esophageal obstruction/stricture, chylothorax, neoplasia, trauma, and diaphragmatic hernias are common reasons to open and enter the thoracic cavity. The lateral intercostal thoracotomy is the most common approach. However, various conditions may necessitate alternative approaches such as the lateral approach with rib resection, lateral approach with rib pivot thoracotomy, median sternotomy, and transsternal thoracotomy. Regardless of the inciting condition/surgical approach, restoration and maintenance of negative intrathoracic pressure is of obvious importance with all surgeries involving the thorax.

Technician evaluations commonly associated with this surgical procedure:

Acute Pain
- Pain significantly impairs postoperative ventilation. Animals experiencing postoperative pain are unlikely to fully expand the chest wall during inspiration, causing a reduced tidal volume and contributing to atelectasis.

Altered Ventilation
- Impaired ventilation can result from pneumothorax, pain, poor cough reflex, reluctance to breathe deeply, or ventilation depressants such as anesthetics and medications. Secretions (mucus and edema) caused by the primary disease or a poor cough reflex, pain, or weakness may obstruct the airway.
- Always place the animal in recovery with the surgical side down as this permits the lung that has been dependent during surgery to fully reinflate. Turn the animal every 15 minutes postoperatively.
- Thoracostomy tubes are routinely placed during thoracic surgery and require 24-hour monitoring. Tube removal typically occurs 6–24 hours postoperatively when air or fluid accumulation is no longer significant.
- Loss of negative intrathoracic pressure during surgery necessitates assisted ventilation during the procedure. A ventilatory pressure of 15–20 cm H_2O at 12–15 bpm is typical.

– In addition to decreasing tidal volume (lungs collapse), a positive intrathoracic pressure reduces cardiac venous return.

Altered Gas Diffusion
– Atelectasis secondary to thoracic surgery or underlying disease processes impair oxygen exchange.

Risk of Infection
– Surgical entry into the chest cavity provides a potential entry for invading pathogens. Atelectasis secondary to thoracic surgery predisposes the animal to respiratory infections. Infections associated with a thoracotomy have a poor prognosis.

Reduced Mobility
– An incision of the latissimus dorsi muscle during surgery will cause the animal to experience pain while walking for several weeks postoperatively. Exercise restriction to leash-walking several times per day for 3 weeks postoperatively is recommended.

TUMOR REMOVAL

Tumor removal and the surgical intervention selected will reflect the type, size, and location of the mass. Malignant tumors are often removed to prevent the neoplasia from metastasizing. Benign tumors may be removed to alleviate pressure, obstruction, or ambulation interference. Owners also may request the removal of a benign tumor for aesthetic purposes.

Technician evaluations commonly associated with this surgical procedure:

Decreased Tissue Perfusion
– Removal of large tumors can result in loss of vascularity to surrounding tissues.

Impair Tissue Integrity
– Depending on the size of the tumor removed, tissue apposition may be difficult to achieve. In addition, the underlying disease may impair incision healing.

Risk of Infection

Acute Pain
– Surgical incisions are associated with pain.

WOUNDS

Physical injuries caused by the tearing, piercing, or laceration of tissues, wounds are a common malady of animals. Wounds are classified using a variety of parameters, with the most common being degree of contamination.

- Clean: Nontraumatic, surgically created wound with no break in aseptic technique
- Clean-contaminated: Minimal break in aseptic technique or involvement of the alimentary, urogenital, or respiratory tracts
- Contaminated: Moderate amount of contamination with major break in aseptic technique or an open traumatic wound of more than 6 hours' duration (fresh wound)
- Dirty: High degree of contamination such as with abscess or an open wound of more than 12 hours' duration (old wound)

Severity and treatment protocol selected will reflect the degree of contamination, location, size, depth (superficial, full thickness, or deep), amount of exudates, degree of pain, and origin of the wound.

Wounds that are mildly contaminated can be lavaged with a cleansing solution (iodine or chlorhexidine) diluted with sterile saline. More severely contaminated wounds may require general anesthesia for cleaning and surgical debridement of nonviable tissue. Closure may be primary, may be delayed, or may incorporate the use of surgical drains. Regardless of the treatment protocol selected, all wounds should be closely monitored for evidence of infection such as localized redness, swelling, heat, pain, and discharge. The animal should be monitored for decreased appetite or activity, hyperthermia, leukocytosis, and increased serum fibrinogen levels.

Technician evaluations commonly associated with this surgical procedure:

Impaired Tissue Integrity
- *Crushing or degloving wounds cause extensive tissue damage and are associated with prolonged recovery times.*

Acute Pain

Reduced Mobility
- *Wound location can affect mobility.*

Risk of Infection
- *Openings through the skin provide a pathway for invading pathogens.*

Hypovolemia
- *Deep wounds associated with penetration of arteries can result in significant blood loss.*

Decreased Perfusion, Peripheral
- *Loss of vascular integrity can induce necrosis and sloughing of affected tissues.*

Client Knowledge Deficit
- *Wounds requiring surgical drains or continual bandaging necessitate extensive client education.*

BIBLIOGRAPHY

Aiello, S. E. (Ed.) (1998). *The Merck veterinary manual* (8th ed.). Whitehouse Station, NJ: Merck & Co.

Battaglia, A. (2001). *Small animal emergency and critical care: A manual for the veterinary technician.* Philadelphia: W. B. Saunders.

Bistner, S. I., Ford, R. B., & Raffe, M. R. (2000). *Kirk and Bistner's handbook of veterinary procedures & emergency treatment* (7th ed.). Philadelphia: W. B. Saunders.

Brinker, W. O., Piermattei, D. L., & Flo, G. L. (1990). *Handbook of small animal orthopedics & fracture treatment* (2nd ed.). Philadelphia: W. B. Saunders.

Crow, S. E., & Walshaw, S. O. (1997). *Manual of clinical procedures in the dog, cat, & rabbit* (2nd ed.). Philadelphia: Lippincott Williams & Wilkins.

Doenges, M. E., Moorhouse, M. F., & Geissler-Murr, A. C. (1997). *Nursing care plans: Guidelines for individualizing patient care* (5th ed.). Philadelphia: F. A. Davis.

Ettinger, S. J., & Feldman, E. C. (2000). *Textbook of veterinarian internal medicine* (5th ed.). Philadelphia: W. B. Saunders.

Fenner, W. R. (1991). *Quick reference to veterinary medicine* (2nd ed.). Philadelphia: J. B. Lippincott.

Harari, J. (1996). *Small animal surgery (the national veterinary medical series).* Baltimore: Williams & Wilkins.

Hickman, J., Houlton, J., & Edwards, B. (1995). *An atlas of veterinary surgery* (3rd ed.). Oxford: Blackwell Science.

McCurnin, D. M. (1998). *Clinical textbook for veterinary technicians* (4th ed.). Philadelphia: W. B. Saunders.

McFarland, G. K., & McFarlane, E. A. (1993). *Nursing diagnosis & intervention: Planning for patient care* (2nd ed.). St. Louis, MO: Mosby.

Nelson, R. W., & Couto, C. G. (1999). *Manual of small animal internal medicine* (2nd ed.). St. Louis, MO: Mosby.

Norkus, C., & Juda, H. (2005, April). Gastric dilatation volvulus. *Veterinary Technician, 26*(4), 269.

Pratt, P. (1994). *Medical, surgical and anesthetic nursing for veterinary technicians* (2nd ed.). Goleta, CA: American Veterinary Publications.

Pratt, P. W. (1998). *Principles and practice of veterinary technology.* St. Louis, MO: Mosby.

Tracy, D. L. (2000). *Small animal surgical nursing* (3rd ed.). St. Louis, MO: Mosby.

Chapter 6
Therapeutic Procedures and Associated Technician Evaluations

INTRODUCTION

This chapter addresses selected therapeutic procedures employed in canine and feline care. A brief description of each procedure is followed by technician evaluations that pertain specifically to the procedure under discussion. Additional comments are included with some technician evaluations; this information typically includes rationale or data unique to the specified medical procedure and associated technician evaluation. However, the student technician should refer to Chapter 3 for appropriate interventions specific to each technician evaluation. This chapter is for reference purposes only; students requiring more detailed descriptions of therapeutic procedures are referred to standard veterinary technician or veterinary medical textbooks.

BEHAVIORAL COUNSELING

Behavioral counseling consists of training an animal in order to prevent unwanted behaviors and to encourage desired behaviors. Education of the owner is a necessary concomitant part of behavioral counseling.

Although most veterinarians and technicians are not animal behaviorists, owners often turn to them for advice when their pet is exhibiting undesirable behavior. The technician must listen carefully to the owner's description of the behavior for two reasons: (1) because some undesirable behaviors are the result of reversible medical conditions (e.g., a cat suffering from urinary tract infections that urinates on its owner's rug) and (2) because, more importantly, uncorrected "bad" behavior by a pet is a primary reason that owners choose to surrender or euthanize their pets. Owners are often unaware that certain behaviors are normal for a particular species (e.g., a cat scratching on furniture to sharpen its claws) or are in fact behaviors that the owners have initiated (e.g., a dog allowed to play "tag" with its owner that later bites a running child). Medical conditions that underlie undesirable behavior should be explained thoroughly so that the owner does not "blame" the animal. If the behavior does not have a medical basis, the veterinary staff should offer suggestions for reversing or managing the behavior. If warranted, the owner may be referred to a veterinary behaviorist, a professional animal trainer, or the National Association of Dog Obedience Instructors.

Technician evaluations commonly associated with this procedure:

Client Knowledge Deficit
- *Providing written instructions that address methods for correction and benchmarks will facilitate client education.*

Owner Noncompliance
- Altering an undesirable behavior requires 100 percent compliance/commitment from the owner and all family members.

BLOOD TRANSFUSION

A blood transfusion is the transfer of blood or blood products (red blood cells, plasma, or platelets) from one animal directly into the circulatory system of another animal. Fresh blood and blood products can be transfused, or they can be transfused after they have been properly stored. Blood/blood product transfusions are used for a variety of medical and surgical conditions, including hypovolemia secondary to trauma, surgical blood loss, anemia, thrombocytopenia, and/or coagulation disorders. While blood transfusions can be life-saving, they present risks to the recipient that range from harmless and transitory to life threatening. Recipients must be monitored for fever, urticaria, chills, dyspnea, cyanosis, tachycardia, arrhythmias, and vomiting during transfusion. To minimize reactions, it is imperative that donor and recipient be typed and cross-matched. Although dogs have 11 different blood types, A1 negative is considered a universal donor. Cats have only 3 blood types; and although none is considered a universal donor, more than 90 percent of cats in the United States are type A. The ideal donor is free of blood-borne pathogens and/or parasites.

Technician evaluations commonly associated with this procedure:

Hypothermia
- To avoid hypothermia, warm all blood products prior to administration. NEVER place blood in a microwave to warm. Place it in a warm water bath and gently agitate. Avoid overzealous agitation as it can result in clot formation or cell disruption. The administration line also should be run through a warm water bath during the transfusion.

Decreased Perfusion
- Decreased packed cell volume (PCV) often leads to decreased perfusion of oxygen. A PCV less than 20 percent is associated with a significant reduction in oxygen transport capacity.
- The Rule of One: 1 mL of transfused whole blood per 1 lb of body weight is expected to raise the PCV by 1 percent. To determine the necessary volume of transfused blood, use the following formula: BW (kg) × 90 mL/kg × Desired PCV − Patient PCV/Donor PCV.
- Complications of the transfusion such as embolization and disseminated intravascular coagulation also can reduce perfusion. To avoid transferring emboli, use an administration set equipped with a 170 μm microfilter.

Cardiac Insufficiency
- Blood transfusions can induce cardiac arrhythmias.

Impaired Tissue Integrity
- Urticaria is a relatively common transfusion reaction. Antihistamines can be administered prior to transfusion.

Altered Ventilation
- Dyspnea and coughing are clinical signs of an adverse reaction to the transfused blood.

Vomiting/Nausea
- Vomiting and nausea are clinical signs of an adverse reaction to the transfused blood.

Hyperthermia
- Fever is a clinical sign of an adverse reaction to the transfused blood.

Risk of Infection
- Various pathogens and/or parasites can be transmitted through infected blood. Contamination also can occur through improper storage or handling of blood products.

Hypovolemia

Hypervolemia
- Overly aggressive administration of fluids can result in fluid overload. Extreme caution should be exercised when transfusing animals in congestive heart failure.

BURN THERAPY

A burn is an injury to the skin caused by thermal or electrical exposure. Burns are classified according to the depth of tissue damage: first-degree burns involve only the epidermis, second-degree burns destroy the entire epidermis and part of the dermis, and third-degree burns destroy the epidermis and the entire dermis. Treatment varies according to the degree and extent of the burns. First-degree burns may require only mild cleaning, bandaging, and analgesics, while extensive second- and third-degree burns can be life-threatening and require advanced medical support. The metabolic/physiological cascade that begins soon after an extensive burn can lead to hypovolemic shock, electrolyte imbalances, kidney and liver failure, metabolic acidosis, and acute pain. Longer term, the patient is at risk of infection and sepsis.

Technician evaluations commonly associated with this procedure:

Obstructed Airway
- Inhalation burns are likely to result in an obstructed airway as burned tissues become inflamed and edematous. Extensive burns on remote parts of the body also can lead to an obstructed airway if massive systemic edema results. Airway protection is of priority.

Altered Gas Diffusion
- Inhalation burns destroy tissues inside the lungs, resulting in reduced carbon dioxide/oxygen exchange at the alveolar level. Supplemental oxygen may increase cellular respiration.

Acute Pain
- Most burns are extremely painful. Although no pain is felt in the area of a third-degree burn due to destruction of nerves, most third-degree burns have a surrounding area of second-degree burn. Cleaning and debridement of burns may require sedation or general anesthesia.

Hypovolemia
- Increased systemic capillary permeability secondary to burns can result in massive peripheral edema. This fluid shift reduces intravascular fluid levels, which can lead to hypovolemic shock. Both crystalloids and colloids are used for fluid replacement.

Electrolyte Imbalance
- Increased secretion of aldosterone is a common sequela to a major burn. This results in sodium retention and potassium loss. Electrolyte balance must be monitored and maintained throughout the course of treatment.

Cardiac Insufficiency
- Cardiac output is often decreased immediately following a major burn injury. Later, hypovolemia can contribute to decreased cardiac output.

Impaired Tissue Integrity
- Second-degree burns destroy the epidermis and cause damage to the dermis. Often, debridement is required before the burns can be bandaged. However, since the dermis is still present, healing of the affected area is possible if infection can be prevented.
- Third-degree burns destroy the entire dermis; therefore, the skin over these areas cannot heal. A black leathery covering called eschar will form over third-degree burns that have not received skin grafts. Grafting is the only method to restore skin over the area of a third-degree burn.

Altered Perfusion, Peripheral
- Massive edema that follows major burns can result in decreased peripheral perfusion.

Altered Perfusion, Renal
- Catecholamine release results in vasoconstriction, which in turn reduces blood flow through the kidneys. Prolonged reduced glomerular perfusion rate can lead to permanent kidney damage.

Risk of Infection
- Destruction of the epidermis and dermis provides a pathway for invading pathogens. Intact blisters should not be opened as the intact (although damaged) skin prevents entrance of pathogens. Open wounds may be cultured to identify pathogens, and topical or systemic antibiotics may be ordered.

Altered Oral Health
- Inhalation burns can cause damage to oral tissues. Electrical burns caused by chewing electric cords also can cause extensive damage. (Puppies are especially prone to electrical burns in the mouth.) Animals with oral pain can experience anorexia. Nasogastric or esophagostomy tubes may be used to provide enteral nutrition, or the patient may be placed on peripheral parenteral nutrition (PPN).

Underweight
- After the initial period of shock has passed, the patient's metabolism is greatly increased. Nausea and pain can induce anorexia.

Reduced Mobility
- Pain often reduces voluntary movement. Analgesics are necessary to permit the animal to move. Burns over a joint may contribute to reduced mobility as scar tissue develops. (Scar tissue is not as elastic as healthy skin.) Passive range of motion exercises during recovery will help maintain mobility in affected joints.

Self-Care Deficit

Anxiety
- Animals that are hospitalized for significant periods of time may develop anxiety. Digestive upset is often a sequela of chronic anxiety.

Client Coping Deficit
- Euthanasia may be recommended for a severely burned animal, or the owner may be unable to afford the extensive medical treatment that is required.

CARDIOPULMONARY CEREBROVASCULAR RESUSCITATION (CPCR)

CPCR is initiated in response to cardiopulmonary arrest (CPA), which is evidenced by the absence of a palpable pulse or auscultable heartbeat, dilated pupils, and the absence of ventilation with cyanosis. Causes of CPA are numerous and can include infection, metabolic disorders, trauma, anesthetics, primary heart conditions, autoimmune diseases, and malignancy. Designed to substitute for the oxygenation, circulation, and ventilation of the cardiopulmonary system, CPCR should be maintained until cardiopulmonary function is reestablished or until the animal is pronounced dead by the veterinarian. The mnemonic ABCDEF (ABC: basic life support; DEF: advanced life support) is a useful tool for reminding inexperienced individuals of the appropriate steps.

- A: Airway. Establish a patent airway via endotracheal intubation.
- B: Breathing. Use an Ambu bag or an anesthetic machine to deliver 100 percent oxygen at a rate of 12–20 breaths per minute.
- C: Cardiac. Initiate external cardiac compression at a rate of 80–120 compressions per minute.

- D: Drugs. Consider use of fluids and the following drugs during CPCR: atropine, epinephrine, 2 percent lidocaine, naloxone, sodium bicarbonate, and bretylium.
- E: Evaluation/Electrocardiography. Assess for peripheral pulse, mucous membrane color, and end tidal CO_2. Diastolic pressure should be 40 mm Hg or greater.
- F: Fibrillation control. Initiate defibrillation protocol.

Technician evaluations commonly associated with this procedure:

Altered Ventilation
- *CPA is characterized by an absence of ventilation.*

Altered Gas Diffusion
- *Oxygen and carbon dioxide exchange is not possible due to the absence of ventilation and cardiac function.*

Cardiovascular Insufficiency
- *Palpable arterial pulses require a systolic pressure greater than 60 mm Hg.*
- *External cardiac compression is typically performed by placing the animal in lateral recumbency and placing both hands at the costrochondral junction of the 4th-5th intercostal space. The animal must be on a firm surface. Extending the arms, compress the thorax by 25–30 percent of its normal diameter. The thumb and first two fingers can be used to compress the chest of animals weighing less than 10 lb. CPCR can be performed with the animal in dorsal recumbency; however, this position is difficult to maintain.*
- *When evaluation of the patient demonstrates ineffective results, the technician can increase the rate, duration, or depth of compressions or ventilation; change hands; change personnel; or change the patient's position.*
- *Internal cardiac compression has been shown be a more effective means of perfusing tissues than external compression; however, its discussion is outside the scope of this text.*

Decreased Perfusion
- *Increasing abdominal pressure by bandaging or manually pressing on the abdomen (abdominal counterpressure technique) can facilitate venous return and decrease arterial runoff. Shock trousers (application of elastic bandaging around hind limbs and pelvis) can reduce peripheral runoff and help increase central pressures.*
- *Common fluids selected include lactated Ringer's, Hetastarch, and Dextran 70. Hetastarch and Dextran 70 are selected in cases of hypoproteinemia. Fluid rates for dogs are 40 mL/kg; for cats, 20 mL/kg.*

Client Coping Deficit
- *The prognosis for a recovery even with successful CPCR typically remains poor due to the underlying disease process that incited the CPA. In an otherwise healthy animal, acute anesthetic incidences resulting in CPA are the exception to this generalization.*

Electrolyte Imbalance
- Severe electrolyte imbalances can induce a CPA episode. CPA will induce metabolic acidosis.

Altered Mentation
- Animals are unresponsive to stimuli. Irreversible cerebral injury secondary to cerebral ischemia can occur secondary to CPA.

CASTS

Casts, as a form of external coaptation, are designed to provide rigid support during the healing process. Fractures most amendable to this procedure include closed, minimally displaced, and interlocking fractures of the radius, tibia, metatarsal, metacarpal, or tarsal bones. The ability to immobilize the joints proximally and distally ("above and below") to the fracture site is critical to successful casting. Therefore, the vast majority of casts are applied distally to the humerus or femur.

Guidelines for proper application and care of casts include the following:

- Anesthetic considerations: Heavily sedate the patient or use general anesthesia as manipulation of the affected limb is very painful.
- Application Procedure: Reduce the fracture and hold the limb in a neutral position with joint angles approximating the contralateral limb. Apply a stirrup, a minimum of cast padding over and around pressure points, and a stockinette. Submerse casting material in water 15–20 seconds; then begin rolling "up the leg" (distal-to-proximal direction), overlapping by 1/3–1/2 of the strip each time. Apply three or four layers. (Note: Rubbing K-Y Jelly on gloves will prevent them from sticking to cast tape.)
- Monitoring: Instruct owners to examine the animal and cast b.i.d. for heat, pain upon palpation, swelling of digits, odor, cast damage, or softening. Weekly veterinary rechecks are recommended for young, rapidly growing animals.
- Removal: Remove the cast using an oscillating saw. How long the animal wears the cast varies depending on the animal's age and the severity and location of the fracture. Four weeks (for dogs less than 6 months old) to 12 weeks (for middle-age dogs) can be expected.

Technician evaluations commonly associated with this procedure:

Decreased Perfusion, Peripheral
- A cast that is applied too tightly or that becomes tight as a result of continued swelling secondary to an initial trauma can result in decreased

perfusion. Cold or separated toes, swelling, and acute pain are indicators of reduced perfusion. Common sequelae of reduced perfusion include nerve damage, tissue necrosis, and paralysis.
- Correction of limb edema prior to cast application is fundamental to ensuring that the cast does not slip downward once the edema has resolved and the limb is reduced in size. Cast slipping can result in pressure point necrosis.

Acute Pain
- Pain may result from the initial injury or from tissue necrosis secondary to cast pressure.

Risk of Self-Inflicted Injury
- The presence of a cast induces many animals to start chewing the cast or the limb. Apply an E-collar as necessary.

Reduced Mobility
- Restricted exercise is usually ordered for an animal wearing a cast.

Client Knowledge Deficit
- Owners should be instructed to monitor the cast b.i.d. (as described previously), keep the cast clean and dry, and restrict the animal's activities to leash walks.

CHEMOTHERAPY

Chemotherapy is the use of cytotoxic drugs to treat a medical condition, usually cancer. Cytotoxic drugs tend to target rapidly dividing cells such as those in neoplasias. However, since the gastrointestinal (GI) tract, bone marrow, and hair follicles also consist of rapidly dividing cells, these areas are often negatively impacted by chemotherapy.

Due to the toxic nature of these drugs, an intravenous (IV) catheter should be placed prior to administration. In addition, the technician should exercise precautionary measures and utilize gloves/gown/mask/safety glasses when reconstituting or handling chemotherapeutics. Immunocompromised or pregnant individuals are advised to avoid handling these chemicals.

Technician evaluations commonly associated with this procedure:

Risk of Infection
- Bone marrow suppression caused by chemotherapeutic drugs can result in panleukopenia.
- Blood tests can be used to monitor the degree of immune system impairment. Reverse isolation may be required.

Vomiting
- Chemotherapeutic drugs often induce nausea and/or vomiting. Vomiting that occurs during administration is likely secondary to the activation of

the chemoreceptor trigger zone (CRTZ). Vomiting occurring 2–5 days posttreatment is most likely related to loss of GI epithelium.

Diarrhea
- Chemotherapeutic drugs may induce diarrhea.

Hypovolemia
- Animals that are nauseous may not have adequate PO intake of fluids. Vomiting and diarrhea contribute to a fluid deficit.

Underweight
- Animals on a prolonged course of chemotherapy often become anorexic.

Client Knowledge Deficit
- Owners need to be educated regarding the side effects of chemotherapy as well as the increased nutritional requirements of an animal receiving chemotherapy.

Impaired Tissue Integrity
- Urticaria, erythema, and facial swelling are noted during allergic reaction to chemotherapeutic drugs.
- Extravasation injury with licking at the injection site can occur up to 7–14 days posttreatment.

Anxiety
- Animals often become anxious during or prior to treatment. Administering chemotherapy in a quiet, low-traffic area is advised.

ENTERAL FEEDING TUBES

A variety of techniques are used to enable delivery of enteral nutrition. Regardless of the technique chosen, the underlying purpose of enteral feeding is to provide nutrition to animals unwilling or unable to consume food. In addition to providing the calories required for normal cellular metabolism, the presence of food within the GI tract has been shown to prevent bacterial translocation and the absorption of endotoxins. Following are brief descriptions of various enteral feeding options.

- Nasogastric (NG) intubation: This type of intubation involves the placement of an NG tube through the nares into the esophagus. Discomfort and the restricted use of liquid diets remain the primary drawbacks of this procedure. NG intubation should be avoided in cases associated with severe facial trauma, uncontrolled vomiting, reduced level of consciousness, and severe coagulopathy.
- Esophagostomy tube: Also termed an *E-tube* or *esophageal feeding tube*, this procedure involves placing a tube through the upper esophagus into the lower esophagus or stomach. Esophagostomy tubes are better tolerated than NG tubes and do not hinder an animal's ability to eat and drink. Esophagostomy tubes should be avoided in cases of prolonged vomiting or esophageal disease.

- Gastrostomy tube: A gastrostomy tube (G-tube) or a percutaneous endoscopic gastrostomy (PEG) tube allows for direct placement of food into the stomach. This technique allows for placement of relatively large diameter tubes, making food administration easier. G-tubes should be avoided in cases of pancreatitis and intestinal obstruction. G-tubes are selected only when enteral feeding is anticipated for more than a month's duration.

Technician evaluations commonly associated with this procedure:

Acute Pain
- Acute pain is associated with passage of an NG tube through the nares. This pain can be controlled via use of topical lidocaine. Secured NG tubes are associated with a significant degree of discomfort. This limitation often leads practitioners to utilize esophagostomy or G-tubes instead.
- All tubes secured for long-term placement (NG, maximum 7 days; gastrostomy and esophagostomy, several weeks) should be capped to avoid a painful distension of the stomach due to gas. Rapid distension of the stomach caused by feeding excessive amounts or feeding too quickly also should be avoided.

Risk of Aspiration
- Misplacement or movement of the feeding tube can lead to aspiration. Proper placement of the tube can be ascertained by flushing 1–3 mL of saline through the tube prior to administration of each feeding. If the patient coughs, do not administer food until contacting the veterinarian.

Risk of Infection
- Cellulitis (surrounding the ostomy site) can occur secondary to placement of the esophagostomy and gastrostomy tubes.
- To allow the formation of a tight fibrinous seal and to avoid peritonitis upon removal, G-tubes must remain in place for a minimum of 3 weeks.
- Attention to aseptic technique is essential to tube management. Wiping the tube port with alcohol after every use is imperative; in addition, the surrounding skin should be cleaned as needed and allowed to air-dry. Bandages, when used, must remain clean and free of food debris and moisture.

Decreased Perfusion
- Loss of tissue perfusion secondary to splenic laceration or gastric hemorrhage during placement of a G-tube disrupts vasculature and impairs blood flow.

Impaired Tissue Integrity
- Esophageal trauma and gastric irritation can occur secondary to tube placement. Gastric reflux can be irritating to esophageal tissues.
- Epistaxis is an associated complication of NG tube placement.

Vomiting
Self-Inflicted Injury
– E-collars are advised for animals with secured NG tubes or G-tubes.
Self-Care Deficit
– Animals with secured NG tubes rarely perform normal grooming behaviors.
Underweight
– Anorexic animals experience weight loss. Most of the commercially available liquid diets used in enteral feeding have a caloric density of approximately 1 kcal/mL. The required total daily volume of food should be divided into four to six feedings.

EUTHANASIA

Defined as an "easy" or "good" death, euthanasia is the humane destruction of an animal. Although the American Veterinary Medical Association (AVMA) has approved several methods for euthanasia, the most common method used in small animal practice remains an IV injection of sodium pentobarbital. Given the terminal nature of this procedure, a written consent form is legally required in all states. Since drugs used for euthanasia purposes are lethal, they should be handled with extreme caution. In addition, euthanasia drugs are regulated by the Drug Enforcement Administration (DEA)/State Board of Pharmacy and therefore must be recorded/logged.

Technician evaluations commonly associated with this procedure:

Client Coping Deficit
– Refer to Chapter 3 for information about assisting owners throughout the process of euthanasia. Grief-stricken owners should be referred to grief counselors or pet loss hotline numbers. See the appendices for contact information.
Client Knowledge Deficit
– Educating the owner prior to the procedure can help lessen stress.

PARENTERAL NUTRITION (PN)

PN, which involves placing nutrients directly into the bloodstream, is used when the GI tract must be bypassed. Conditions necessitating the use of PN can include intestinal obstruction, ileus, chronic vomiting, hepatitis, pancreatitis, or any condition resulting in a reduced level of consciousness. The majority of veterinary patients receive partial parenteral nutrition (PPN) as opposed to the use of total parenteral nutrition (TPN) seen in human medicine. PPN is designed to meet the energy, amino acid, electrolyte, and B vitamin requirements of the animal. Taurine, fatty acids, and many vitamins/minerals are lacking in PPN formulas. Ideally, PPN solutions are administered through a peripheral IV catheter for a maximum duration of 3–4 days. In contrast, TPN solutions can be administered over longer periods and must be delivered through a central line. (The tip of the catheter reaches the vena cava.) Catheters used for PPN or TPN must be "dedicated" and should not be used for any other purpose such as drug administration or blood sampling.

Technician evaluations commonly associated with this procedure:

Risk of Infection
- *Parenteral solutions must be kept sterile. Replace solution bags and IV lines every 24 hours. Examine catheter site q.i.d. for signs of infection or phlebitis.*

Electrolyte Imbalances
- *Disease conditions necessitating the use of PN often induce electrolyte abnormalities. All electrolyte imbalances should be corrected prior to initiating PN. Hyperglycemia is of concern during PPN administration. Monitoring blood work b.i.d. is recommended.*

Underweight
- *PPN is not all-inclusive and must be administered concurrently with enteral nutrition to meet all nutritional requirements. Patients receiving PPN or TPN should be weighed daily.*
- *Exclusive use of PN has been shown to result in gastric and intestinal atrophy.*

Decreased Perfusion
- *Thrombophlebitis is the most common complication of PN. Appropriate care of the IV catheter will minimize this risk. To minimize the risk of thrombophlebitis, use aseptic techniques during placement, reserve the catheter for PN use, change bandages daily, and use a central line whenever possible.*

Hypothermia
- *PPN solutions that are stored in a refrigerator should be warmed to room temperature prior to administration. Never microwave solutions.*

Self-Care Deficit
- *Animals receiving PN are typically debilitated and do not engage in grooming behaviors.*

PERITONEAL DIALYSIS

Peritoneal dialysis is a treatment/procedure used to remove the toxins, nitrogenous waste, or drugs that are normally excreted by the kidney. Dialysis can be useful in treating conditions such as acute renal failure, uremia, acidosis, and hyperkalemia. The procedure is very labor-intensive, and patients undergoing dialysis must receive constant monitoring. The basic procedural steps of dialysis are as follows:

- An insertion site midway between the umbilicus and the pelvis is clipped and prepared. (Dialysis catheters can be temporary, or they can be surgically placed. Surgically placed catheters are associated with fewer complications such as omental obstruction.) A peritoneal dialysis catheter is then placed into the abdomen.
- Dialysate is injected until the abdomen is slightly distended.
- The dialysate remains in the abdomen for an hour and is then removed.
- The removed dialysate is measured to allow for comparison between the volume of dialysate injected to the amount removed.

Technician evaluations commonly associated with this procedure:

Hypothermia
- *The dialysate should be warmed 2–3 degrees above body temperature prior to infusion. To ensure that the dialysate solution remains warm during the infusion, the dialysate bag should be wrapped in a warm water blanket.*

Acute Pain
- *Marked abdominal distention and the introduction of cold fluid cause acute pain.*

Risk of Infection
- *A contaminated dialysate and/or injection site can lead to peritonitis. Recovered dialysate should be cultured and cytologically examined s.i.d.*

Electrolyte Imbalance
- *The peritoneum is a permeable membrane that permits diffusion of electrolytes from the bloodstream into the dialysate and from the dialysate into the bloodstream. Electrolyte imbalances can result from the dialysis or from the underlying condition that necessitates the dialysis.*
- *Monitoring for hypoalbumemia and the associated secondary limb edema is recommended.*
- *Serum chemistries and electrolytes, PCV, and TP should be monitored at least s.i.d.*

Altered Gas Diffusion
- *Increased abdominal pressure due to the presence of the dialysate solution can induce compression of the thoracic cavity, which leads to atelectasis and pneumonia.*

Hypervolemia
- *Dialysate fluid volumes must be closely monitored. All infused and recovered volumes should be recorded on an hourly basis. Also, record IV fluid volumes and urine output on an hourly basis. Weigh animal b.i.d.*
- *Typically, patients are infused with a dialysate volume of 30–40 mL/kg.*

Hypovolemia
– *Fluid movement into the dialysate can induce hypovolemia.*
Self-Care Deficit
– *Animals requiring peritoneal dialysis are often severely debilitated.*
Nausea/Vomiting
– *Vomiting is often associated with the underlying condition.*
Underweight
– *Animals are often anorexic.*

RADIOACTIVE IODINE THERAPY

Radioactive iodine therapy (iodine-131, or I-131) is a nonsurgical treatment option used for feline hyperthyroidism. This procedure, which allows animals to avoid the risks associated with anesthesia and surgery, uses a radioactive form of iodine to destroy the abnormal thyroid tissue. With cats, a return of normal thyroid function can be expected within 1–3 months of treatment. After receiving the I-131, cats must remain quarantined in the treatment facility until radioactivity has dissipated to a state-approved level. This typically requires hospitalization of 7–14 days. During this time, contact with the cat is limited to approved personnel and handling of the animal by the designated personnel is kept to a minimum. Protective clothing consisting of a lab coat, latex gloves (double-glove), and a dosimeter is worn at all times. The primary disadvantages of this treatment include the initial cost and need for isolation as well as the potential for inducing hypothyroidism. Animals that become hypothyroid secondary to I-131 treatment are subsequently placed on thyroid hormones.

Technician evaluations commonly associated with this procedure:

Underweight
– *Hyperthyroidism induces a hypermetabolic state that often results in weight loss. Animals receiving treatment may experience difficulty swallowing, which also can contribute to weight loss.*
Client Knowledge Deficit
– *Quarantine/isolation times vary with each state. The following are general recommendations that apply to the first 2 weeks after hospital release. All issues should be discussed and provided in written form to the client prior to the animal's discharge.*
  - *Keep the cat indoors.*
  - *Wash hands after handling the animal. (I-131 is excreted in the urine and saliva; therefore, cats have the potential to accumulate radioactivity on the hair coat after coming in contact with urine or during grooming behaviors.)*
  - *Use a plastic litter box liner with flushable, scoopable litter and place the box in a low-traffic room in the house.*

- Wear gloves when handling litter and dispose of litter in the toilet.
- Avoid close contact with the cat for extended periods.
- Do not handle the cat if you are pregnant or are a child.

Vomiting
– Anorexia, nausea, and vomiting can occur secondary to I-131 treatment.

Diarrhea
– Diarrhea is often associated with the underlying hyperthyroidism.

Altered tissue integrity
– Alopecia and thinning of the hair posttreatment is common.

Anxiety
– Restlessness, anxiety, and excessive grooming behaviors can be associated with the underlying hyperthyroidism. In addition, cats accustomed to human interaction may experience stress and anxiety during the isolation period.

VASCULAR ACCESS VIA CATHETERIZATION

Catheterization can provide secure access to venous and arterial portions of the vascular system. Reasons for catheterization are as diverse as they are numerous.

- Arterial catheters: direct blood pressure monitoring and repeat arterial blood gases analysis
    - Vessels utilized: dorsal metatarsal artery (pedal), femoral artery, and auricular artery
- Venous catheters: administration of fluids, blood/blood products, medications, nutrition, diagnostic contrast media, monitoring of central venous pressure, and drawing of blood samples
    - Vessels utilized: cephalic vein, accessory cephalic vein, medial and lateral saphenous veins, auricular vein, jugular vein, and vena cava (accessed via jugular)

Potential complications associated with vascular catheterization include thrombophlebitis, infection, hematoma formation, pyrogenic reaction, allergic reaction, and air embolism.

The type of catheter selected will reflect the purpose of the catheter, intended length of use, condition of the animal, and personal preference of the medical team. Catheter types can include the following:

- Over-the-needle catheter: Most common catheter used for short-term (3 days) venous access. It also is used for arterial access of the dorsal pedal artery.
- Through-the-needle catheter: Very long catheter that is placed percutaneously into a central vein. The catheter is inserted through the needle.

- Multi-lumen central catheter: Multiple ports allow for increased access with a single catheter. It is most commonly used in the ICU, and use is limited to central veins.
- Peripherally induced central line: An extremely long catheter that is placed in a peripheral vessel and threaded into the vena cava. It is used primarily when jugular vein is not available. A through-the-needle catheter, a multi-lumen catheter, or catheters placed using the Seldinger technique are used for this purpose.
- Swan-Ganz catheter: a balloon-tipped catheter used to measure central venous pressure.
- Arterial catheter: A sturdier catheter that is made specifically for arterial access and that uses a guide wire. This catheter, which is much longer than other catheters, allows for more secure placement.
- Catheter with vascular access port: Can be used when long-term fluid therapy/chemotherapy is required. The port is placed SQ; and the catheter is fed through the port and subcutaneous tissues into the vein.

Technician evaluations commonly associated with this procedure:

Decreased Perfusion
- *Use of arterial catheters are prone to producing thrombosis.*

Impaired Tissue Integrity
- *Hematoma formation at the site of entry can occur.*

Self-Inflicted Injury
- *Catheters placed in the cephalic vein are most likely to be disturbed by the animal.*

Risk of Infection
- *Monitor for development of cellulitis, which can be evidenced by swelling, heat, or pain at the catheter site. A potential for bacteremia/septicemia also exists. Catheters should be monitored b.i.d.-q.i.d. depending on the status of the individual animal.*

BIBLIOGRAPHY

Aiello, S. E. (Ed.) (1998). *The Merck veterinary manual* (8th ed.). Whitehouse Station, NJ: Merck & Co.

Battaglia, A. (2001). *Small animal emergency and critical care: A manual for the veterinary technician.* Philadelphia: W. B. Saunders.

Bistner, S. I., Ford, R. B., & Raffe, M. R. (2000). *Kirk and Bistner's handbook of veterinary procedures & emergency treatment* (7th ed.). Philadelphia: W. B. Saunders.

Cote, E. (2007). *Clinical veterinary advisor: Dogs and cats.* St. Louis, MO: Mosby.

Chalmers, H. J., Scrivani, P. V., Dykes, N. L., Erb, H. N., Hobbs, J. M., & Hubble, L. J. (2006, September/October). Identifying removable radioactivity on the surface of cats during the first week after treatment with iodine 131. *Vet Radiology & Ultrasound, 47*(5), 507–509.

Crow, S. E., & Walshaw, S. O. (1997). *Manual of clinical procedures in the dog, cat, & rabbit* (2nd ed.). Philadelphia: Lippincott Williams & Wilkins.

Doenges, M. E., Moorhouse, M. F., & Geissler-Murr, A. C. (1997). *Nursing care plans: Guidelines for individualizing patient care* (5th ed.). Philadelphia: F. A. Davis.

Ettinger, S. J., & Feldman, E. C. (2000). *Textbook of veterinarian internal medicine* (5th ed.) Philadelphia: W. B. Saunders.

Fenner, W. R. (1991). *Quick reference to veterinary medicine* (2nd ed.). Philadelphia: J. B. Lippincott.

Harari, J. (1996). *Small animal surgery (the national veterinary medical series)*. Baltimore: Williams & Wilkins.

Marks, S. (n.d.). *Enteral feeding devices: What's old, what's new*. Retrieved September 15, 2007, from http://www.vin.com/VINDBPub/SearchPB/Proceedings/PR05000/PR00169.htm

McCurnin, D. M. (1998). *Clinical textbook for veterinary technicians* (4th ed.). Philadelphia: W. B. Saunders.

McFarland, G. K., & McFarlane, E. A. (1993). *Nursing diagnosis & intervention: Planning for patient care* (2nd ed.). St. Louis, MO: Mosby.

Nelson, R. W., & Couto, C. G. (1999). *Manual of small animal internal medicine* (2nd ed.). St. Louis, MO: Mosby.

Pratt, P. (1994). *Medical, surgical and anesthetic nursing for veterinary technicians* (2nd ed.). Goleta, CA: American Veterinary Publications.

Pratt, P. W. (1998). *Principles and practice of veterinary technology*. St. Louis, MO: Mosby.

Tracy, D. L. (2000). *Small animal surgical nursing* (3rd ed.). St. Louis, MO: Mosby.

Chapter 7

Sample Cases with Documentation

INTRODUCTION

This chapter contains 10 case scenarios with appropriate documentation by the veterinary technician. The reader should keep in mind that additional documentation is done by the veterinarian and that surgeries are normally documented through anesthesia and surgery records. Case scenarios are listed in alphabetical order.

CASE SCENARIOS

Case 1: Anal Gland Abscess

Scenario

An 8-year-old, 9-lb castrated male toy poodle presented with a history of scooting and licking the perineal area. The dog had a history of chronic anal gland impaction and had been treated for this problem 10 months prior. In an attempt to minimize the recurring episodes, the owners had begun having the dog's anal glands expressed during routine grooming appointments. Physical examination revealed the following results: TPR 101.1°F, 110, panting. An open abscess approximately 0.5" × 0.25" involving the right anal gland was apparent. The entire perineal area was hyperemic and damp from licking. A pain response of 2/5 was noted. All other body systems were within normal limits.

Upon examining the dog, the veterinarian diagnosed an anal gland abscess and requested that the technician clip around the affected area and clean the wound using a diluted chlorhexidine solution. The veterinarian prescribed enrofloxacin 5 mg/kg p.o. b.i.d. for 7 days and meloxicam 0.1 mg/kg p.o. s.i.d. for 3 days, which were both filled by the technician. Upon veterinary order, the technician administered the first doses and placed an E-collar on the dog.

The veterinarian recommended a recheck in 7 days after explaining the condition and prognosis to the client. The technician demonstrated how to apply the E-collar and how to administer medications.

Anal Gland Abscess Documentation

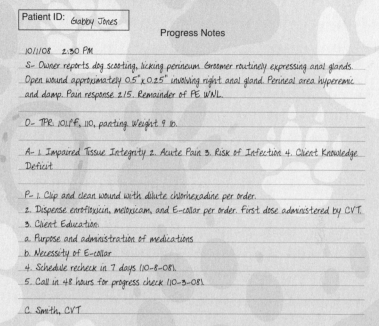

Figure 7-1

Case 2: Canine Neuter

Scenario

8 AM

A 45-lb, 6-month-old male Labrador mix presented for routine castration. The dog was current on vaccinations, deworming, and heartworm preventative. There was no history of any problems or concerns on behalf of the owner; but prior to leaving the hospital, the owner specifically requested that the dog's toenails be trimmed. The owner was provided educational material regarding the surgery, surgical consent forms, and a cost estimate. Preoperative blood tests were accepted, and the owner confirmed that the dog had been n.p.o. since 8 PM the previous evening.

Physical examination revealed TPR 101.9°F, 110 bpm, panting. The dog was BAR and very excited. Thoracic auscultation and abdominal palpation were normal, and both testicles were within the scrotum. All physical parameters appeared WNL. Following blood collection, the technician placed the dog in a surgical ward under n.p.o. status (standing order). Preoperative blood screening revealed the following:

PCV 45%
TP 6.7 g/dl

ALP 40 IU/l
AST 10 IU/l
BUN 17 mg/dl
Creatinine 0.75 mg/dl
Glucose 98 mg/dl
Amylase 200 IU
Ca 10.2 mg/dl
Na 145 mEq/L
K 4.5 mEq/L
Cl 100 mEq/K

9 AM

After administration of an Acepromazine-atropine-butorphanol premedication, an IV catheter was placed. Anesthesia was induced using ketamine-valium, and the dog was maintained on isoflurane. A standard prescrotal approach surgical castration was performed, and 200 cc of LRS were administered prior to IV catheter removal. Prior to an uneventful recovery from anesthesia, the dog's nails were trimmed and he was placed on a warmed blanket.

12 PM

The dog appeared somewhat subdued but was ambulating normally in the kennel. No swelling, discharge, or hyperemia was noted on the incision line, and the scrotum appeared normal in size. Per order, the dog was offered water. A pain response of 1/5 was noted. Carprofen 2.2 mg/kg p.o. b.i.d. 3 days was ordered, and the first dose was administered.

6 PM

Examination revealed no changes in the surgical site, the dog was alert, and half the water had been consumed. Carprofen was administered, Hill's p/d was offered, and the water was refilled.

Day 2

8 AM

Examination of the surgical site revealed extreme hyperemia with apparent clipper burn of the scrotal/prescrotal area, and the scrotum was swollen to two times the normal size. Dampness of the hair near the surgical site and inner thighs suggested that the dog had been licking the area extensively. TPR 102.2°F, 104 bpm, panting. All food and water had been consumed, and the dog urinated and defecated when brought to the exercise area. Per order, cryotherapy was administered via a cold pack (apply 10 minutes t.i.d. for 1 day), the area was sprayed lightly with DERMACOOL HC (spray hyperemic areas t.i.d. for 5 days), and an E-collar was applied. Carprofen was administered, and food and water were provided.

12 PM

The dog had consumed all of the offered food and half the water. Hyperemia of the area had decreased by one third, and scrotal size remained unchanged. Cryotherapy and topical hydrocortisone spray were administered.

6 PM

A brief examination revealed no changes in physical status. Carprofen, cryotherapy, and the hydrocortisone spray were administered. Prior to the animal's discharge, the owner was instructed on E-collar usage, medication administration (carprofen and DERMACOOL), and monitoring of the surgical complication. Appointments for a recheck in 48 hours and suture removal in 10 days were scheduled.

Canine Neuter Documentation

Patient ID: Scout Jones

Progress Notes

10/1/08 8 AM
S- Six-month-old Labrador current on vaccinations, deworming, and heartworm prevention presented for castration. Owner states patient has been NPO since 8 PM last night. Patient BAR, excited. PE WNL. Both testicles palpated in scrotum.
O- T 101.9°F, P 110 bpm, panting. Weight 45 lb.
A- 1. Preoperative Compliance.
P- 1. Initiate MAOR: NPO.
2. Owner consents to preoperative blood test. Results WNL.
3. Owner given estimate; signed consent.

10/1/08 12 PM
S- Patient ambulating normally but subdued (postanesthesia). Gag reflex present. Scrotum normal. No swelling, discharge, or hyperemia noted.
O- No data.
A- Preoperative Compliance resolved. 1. Acute Pain 2. Risk of Infection 3. SWAP
P- 1. Modify MAOR: provide water, add carprofen.
2. Monitor pain status.
3. Monitor for signs of infection.
C. Smith, CVT

10/1/08 6 PM
S- Surgical site intact, scrotum normal. BAR. Half of water consumed.
O- No data.
A- Continue TE 1, 2, 3.
P- 1. Modify MAOR: water and moist food.
2. Continue plan.
C. Smith, CVT

Figure 7-2a

Patient ID: Scout Jones

Progress Notes

10/2/08 8 AM

S- Scrotum swollen (2 times normal), hyperemic, and damp. Patient urinated and defecated when brought to exercise area. All food and water consumed.

O- TPR 102.2°F, 104 bpm, panting.

A- 1. Impaired Tissue Integrity 2. Acute Pain 3. Risk of Infection

P- 1. Modify MAOR: a. Treatments: cryotherapy, E-collar; b. Medication—DERMACOOL HC, carprofen.

2. Apply E-collar.
3. Monitor pain status.
4. Monitor for signs of infection.

C. Smith, CVT

10/2/08 6 PM

S- Scrotum remains swollen, hyperemic. Remainder PE WNL.

O- No data.

A- Continue TE 1, 2, 3. add 4. Client Knowledge Deficit.

P- 1. Dispense carprofen and hydrocortisone spray.

2. Client education:
   a. purpose and aministration of medication
   b. E-collar usage
   c. monitoring surgical site
3. Schedule recheck in 48 hours.
4. Schedule suture removal in 10 days.

C. Smith, CVT

Figure 7-2b

Medication Administration/Order Record

Patient ID	Allergies
Scout Jones	none

Initials	Signature	Title
CS	Cathy Smith	CVT

Medication Administration

Date of Order	Medication	Time	10/1	10/2	10/3	10/4	10/5	10/6	10/7
10/1	Carprofen 2.2 mg/kg p.o. b.i.d. for 3 days	8A	12P CS	CS					
		6P	CS	CS					
10/2	DERMACOOL topical spray to scrotum t.i.d. for 5 days	8A		CS					
		12P		CS					
		6P		CS					

Fluids and IV Drips

Figure 7-3a

Chapter 7

Patient ID: Scout Jones
Allergies: none

Order Record

Treatments

Date	Treatment	Time	10/1	10/2	10/3	10/4	10/5	10/6	10/7
10/2	Apply cold pack to scrotal area 10 minutes t.i.d. for 1 day	8A	╳	CS		╳			╳
		12P	╳	CS		╳			╳
		6P	╳	CS		╳			╳
10/2	Place E-collar	8A	╳	CS					

Tests

Date	Test	Time	10/1	10/2	10/3	10/4	10/5	10/6	10/7
10/1	Blood test	8A	CS	╳	╳	╳	╳	╳	╳

Diet

Date	Diet	Time	10/1	10/2	10/3	10/4	10/5	10/6	10/7
10/1	n.p.o.	8A	CS	DISCONTINUED					
10/1	Provide water	12P	CS	DISCONTINUED					
10/1	Provide water and Hill's p/d	8A	6P CS	CS					

Figure 7-3b

Case 3: Cesarean Section

Scenario
Day 1
3 PM

A 5-year-old, 22-lb female dachshund presented for suspected dystocia. The dog had been intermittently straining for the previous 3 hours, and a vaginal discharge was noted. Ultrasound examination conducted on Day 25 of pregnancy confirmed five puppies, and the dog was 63 days postbreeding.

Physical examination revealed TPR 99.8°F, 120 bpm, 28 bpm. Mucous membranes were pink and slightly tacky, and CRT was 2 seconds. No puppies were palpable in the birth canal, and the cervix was fully dilated. A pain response of 3/5 was assigned. Blood work was performed and revealed:

PCV 50%
TP 7.2 g/dl
Alb 3.7 g/dl
ALP 36 IU/I
ALT 45 IU/I
BUN 20 mg/dl
Creatinine 1.0 mg/dl
Glucose 95 mg/dl
Lipase 50 IU
Ca 9.2 mg/dl

Following IV catheter placement, LRS was administered at a rate of 50 cc/h to deliver 600 cc total volume. A single IM dose of 0.5 units of oxytocin was administered. Vaginal examination 30 minutes postinjection revealed no progression in the labor, and the dog was prepared for cesarean section. The dog was placed on n.p.o. status. Surgical consent forms and cost estimate were obtained.

4 PM

After an atropine-buprenorphine premedication and prior to induction with propofol, oxygen was administered for 10 minutes via face mask. The majority of clipping for surgery also was performed prior to induction. A puppy recovery station equipped with a heating pad, towels, mosquito forceps, oxygen delivery, and emergency drugs was placed near the surgery table. General anesthesia was maintained using isoflurane.

Five viable puppies were delivered via cesarean section and placed in the prewarmed puppy station. Prior to placing each puppy in the station, the placenta was removed, fluid was cleared from the airway, the umbilical cord was clamped, and the puppy was gently stimulated. The bitch was placed in a separate kennel. During recovery from anesthesia, a portion of the placenta was rubbed around her muzzle and vulvar area. Postoperative pain was controlled via an order for carprofen 2.2 mg/kg p.o. b.i.d. for 3 days.

6 PM

After full recovery from anesthesia, the mother was introduced to the puppies and closely monitored for any evidence of neonatal rejection. All puppies had demonstrated nursing behaviors within 4 hours of delivery, and the mother was demonstrating appropriate bonding behavior. Carprofen was administered.

Day 2

12 AM

Moist Hill's p/d and water were offered.

3 AM

IV fluids were discontinued upon administration of 600 cc.

8 AM

Physical examination of the bitch revealed TPR 100.6°F, 124 bpm, 28 bpm. Mucous membranes were pink and moist, and CRT was 2 seconds. Thoracic auscultation was normal, but the abdomen was not palpated. A pain response of 1/5 was assigned based on abdominal guarding behaviors. The incision line was clean and dry with a slight amount of redness. A small amount of vulvar discharge was noted. Mammary tissue palpated normally, and milk could be expressed. The dog appeared BAR and was exhibiting normal maternal behaviors. Although no food was consumed, the dog had drunk half the water offered. After removal of the IV catheter, carprofen was administered and fresh food/water offered.

All puppies were examined briefly and found to be within normal developmental and physical parameters.

12 PM

No changes in physical parameters were noted. The bitch had consumed half the offered food and most of the water. Fresh food and water were offered.

6 PM

A brief examination showed TPR 100.2°F, 130 bpm, 30 bpm, mucous membranes pink and moist, and CRT 2 seconds. Nipples and mammary tissue were normal, and evidence of nursing was apparent. The incision line remained dry with a slight redness. Vaginal discharge was not apparent, but the bitch had been seen engaging in grooming behaviors throughout the day. Most of the water and half the food had been consumed.

All puppies appeared within normal limits. Nursing behaviors and appropriate movements were noted throughout the day.

The client was provided postoperative care instructions regarding medication, incision monitoring, normal maternal/neonatal behaviors, and diet. All animals were discharged into the owner's care.

Cesarean Section Documentation

Patient ID: Alice Jones

Progress Notes

10/1/08 3:00 PM
S- Owner states dog has been in apparent labor, intermittently straining for 3 hours; vaginal discharge noted. Ultrasound Day 25 confirmed pregnancy with 5 puppies. Bitch is 63 days postbreeding. Mucous membranes pink and slightly tacky, CRT 2 seconds. No puppies palpable in birth canal; cervix fully dilated. Pain response 3/5. Remainder PE WNL.
O- T 99.8°F, P 120 bpm, R 28 bpm. Ca 4.4 mEq/L, glucose 95 mg/dl (low normal). Remainder blood values WNL. Weight 22 lb.

A- 1. Reproductive Dysfunction 2. Acute Pain
P- 1. Initiate MAOR to include: placement of IV catheter, blood test, fluids, medication, NPO.
2. Monitor progression of labor in response to oxytocin.
3. Monitor pain response.
C. Smith, CVT

10/1/08 3:30 PM
S- No progression of labor noted.
O- No data.
A- 1. Preoperative Compliance 2. Reproductive Dysfunction 3. Acute Pain
P- 1. Provided estimate to owner; consent signed.
2. Schedule surgery.
3. Prepare equipment for delivery of puppies.
C. Smith, CVT

10/1/08 5:00 PM
DAM
S- Sutures intact; animal still lethargic from anesthesia.
O- No data.
A- Preoperative Compliance resolved. 1. Postoperative Compliance 2. Acute Pain
3. Risk of Infection. Recovering well from anesthesia.
P- 1. Modify MAOR: additional medications: carprofen.
2. Monitor recovery from anesthesia. Withhold puppies from dam until she is fully recovered.
3. Rub placenta around mouth and vulvar area prior to introducing puppies.

Figure 7-4a

Patient ID: Alice Jones

Progress Notes

PUPPIES
S- WNL for neonates (all 5 puppies).
O- delivered via c-section.
A- 1. Risk of Impaired Ventilation 2. Risk of Hypothermia 3. Risk of Impaired Nursing
P- 1. Cleared airway, stimulated each puppy upon delivery.
2. Clamped umbilical cord; removed placenta.
3. Placed puppies in prewarmed puppy station.
4. Monitor status of neonates, especially respiratory effort and temperature, q 5 minutes x6, then q 15 minutes x4, then q 30 minutes until reunited with dam.
5. Introduce to dam after her recovery from anesthesia.
C. Smith, CVT

10/1/08 6 PM
DAM
S- Bitch recovered from anesthesia; mouth and vulvar area rubbed with placenta. Puppies introduced; dam exhibits appropriate bonding behavior.
O- No data
A- Postoperative Compliance resolved. 1. SWAP 2. Acute Pain 3. Risk of Infection
P- 1. Continue medications.
2. Monitor dam's acceptance of litter.
3. Monitor surgical incision for signs of infection.
C. Smith, CVT

PUPPIES
S- All puppies demonstrate effective nursing behavior.
O- No data.
A- 1. SWAP
P- 1. Continue to monitor nursing efforts, temperature, and general appearance.
C. Smith, CVT

10/2/08 8 AM
DAM
S- Mucous membranes pink and moist, CRT 2 seconds. Thoracic auscultation normal. Incision clean, dry, slight erythema, sutures intact. Small amount of vulvar discharge. Mammary tissue normal, milk expressed. Exhibiting normal maternal behavior. No food eaten during night.
O- T 100.6°F, P 124 bpm, R 28 bpm.
A- 1. SWAP 2. Acute Pain 3. Risk of Infection

Figure 7-4b

Patient ID: Alice Jones

Progress Notes

P- 1. Continue medication, diet per MAOR.
2. Monitor surgical incision for signs of infection.
C. Smith, CVT

PUPPIES
S- PE WNL
O- No data
A- 1. SWAP
P- Continue to monitor physical status, nursing efforts.
C. Smith, CVT

10/2/06 6 PM
DAM
S- Mucous membranes pink and moist, CRT 2 seconds. Nipples and mammary tissue normal. Incision clean, dry, slight erythema. No vulvar discharge noted. Eating and drinking well.
O- T 100.2°F, P 130 bpm, R 30 bpm.
A- 1. SWAP 2. Acute Pain 3. Client Knowledge Deficit
P- 1. Dispense caprofen.
2. Client education:
   a. Purpose and administration of medications
   b. Signs/symptoms of infections
   c. Normal maternal behavior
   d. Diet
3. Discharge to owner.
4. Call in 48 hours for progress report.
C. Smith, CVT

PUPPIES
S- PE WNL. All puppies nursing effectively.
O- No data
A- 1. SWAP 2. Client Knowledge Deficit
P- 1. Client education:
   a. Normal neonatal behavior
   b. Monitoring of nutritional status of neonates
2. Discharge to owner.
3. Call in 48 hours for progress report.
C. Smith, CVT

Figure 7-4c

Medication Administration/Order Record

Patient ID	Allergies
Alice Jones	none

Initials	Signature	Title
CS	Cathy Smith	CVT
NR	Nancy Reed	CVT

Medication Administration

Date of Order	Medication	Time	10/1	10/2	10/3	10/4	10/5	10/6	10/7
10/1	0.5 units oxytocin IM one dose	3P	CS	✕	✕	✕	✕	✕	✕
10/1	Carprofen 2.2 mg/kg p.o. b.i.d. for 3 days	8A	✕	CS			✕	✕	✕
		6P	CS			✕	✕	✕	✕

Fluids and IV Drips

10/1	LRS @ 50cc/hr total volume 600 cc	Start 10/1 3P Stop 10/2 3A	CS	CS	✕	✕	✕	✕	✕

Figure 7-5a

Patient ID: Alice Jones

Allergies: none

Order Record

Treatments

Date	Treatment	Time	10/1	10/2	10/3	10/4	10/5	10/6	10/7
10/1	Place IV catheter	3P	CS						

Tests

Date	Test	Time	10/1	10/2	10/3	10/4	10/5	10/6	10/7
10/1	Blood test	3P	CS	✕	✕	✕	✕	✕	✕

Diet

Date	Diet	Time	10/1	10/2	10/3	10/4	10/5	10/6	10/7
10/1	n.p.o.	3P	CS	DISCONTINUED					
10/2	Provide water and moist Hill's p/d	12A	✕	NR					

Figure 7-5b

Case 4: Corneal Ulceration

Scenario

A 5-month-old, 4-lb intact male DSH cat presented for an ocular discharge of 3 days' duration. The owner reported that the cat had begun squinting after playing with a littermate. During the 2 days following the incident, the squinting did not resolve and a watery discharge began. The cat had been tested negative for FeLV and was current on vaccinations and deworming.

Upon physical examination, the cat appeared BAR, TPR 101.4°F, 140 bpm, 20 bpm. Thoracic auscultation and abdominal palpation were within normal limits. Blepharospasm, conjunctivitis, and a watery discharge were noted in the right eye. The left eye appeared unaffected, and all other physical parameters were within normal limits.

The veterinarian requested a Schirmer's test of both eyes and a fluorescein stain of the right eye. Results of the Schirmer's test included 17 mm/min left eye, 20 mm/min right eye. The fluorescein stain demonstrated a small ulcer on the ventral right aspect of the cornea and a patent nasolacrimal duct. The veterinarian diagnosed corneal ulceration secondary to trauma and ordered a triple antibiotic ointment o.d. t.i.d. for 7 days. After the veterinarian explained the diagnosis and recommended reexamination in 7 days, the technician demonstrated the technique for administering the eye ointment.

Corneal Ulceration Documentation

Patient ID: Austin Jones

Progress Notes

10/1/08 10 AM

S- Five-month-old intact male DSH. Owner states patient began squinting after play session with a littermate; has had ocular discharge for 3 days. Current on vaccinations, deworming, and testing. Blepharospasm, conjunctivitis, and watery discharge noted in right eye. Left eye normal. Remainder of PE WNL.

O- T 101.4°F, P 140 bpm, R 20 bpm. Weight 4 lb.

A- 1. Impaired Tissue Integrity 2. Altered Sensory Perception 3. Client Knowledge Deficit

P- 1. Schirmer's test both eyes per order. Results: 17 mm/min left eye, 20 mm/min right eye.
2. Fluorescein stain right eye per order. Results: Small ulcer on ventral right aspect of cornea; nasolacrimal duct patent.
3. Dispense eye ointment. First dose administered by CVT while owner observed.
4. Client education:
   a. Purpose and administration of medication
5. Schedule recheck in 7 days.
6. Call in 48 hours for progress report.
C. Smith, CVT

Figure 7-6

Case 5: Feline Upper Respiratory Disease (FURD)

Scenario

A 2-month-old, 2.5-lb intact male kitten presented for severe wheezing and ocular and nasal discharge of an unknown duration. The kitten was believed to be the offspring of a semiferal barn cat currently residing on the owner's property. Two other kittens from the same litter had recently developed similar symptoms, although none were as severe as the presented kitten's. None of the animals had been vaccinated, dewormed, or tested for any diseases.

Physical examination revealed TPR 103.6°F, 150 bpm, 36 bpm. The kitten was moderately depressed and had copious amounts of purulent ocular and nasal discharge. The discharge and crusting prevented the left eye from opening and semioccluded both nares; oral ulceration of the tongue and hard palate were present. Thoracic auscultation revealed rhonchi and occasional crackles. Abdominal examination revealed a moderate distention and no pain upon palpation. All diagnostic tests including a CBC, FeLV, FIV, and fecal examination were declined by the owner.

Failure to test resulted in an unconfirmed diagnosis and primary rule outs of severe feline calicivirus (FCV) and feline viral rhinotracheitis (FVR) accompanied by a secondary bacterial infection. After explaining the medical repercussions of test refusal and reiterating the need for testing, vaccinating, deworming, and neutering/spaying the outdoor cat population, the veterinarian ordered amoxicillin 10 mg/kg p.o. b.i.d. 7 days and TriOptic-P o.u. t.i.d. 7 days. Ancillary treatment recommendations included feeding moist prewarmed food, removing ocular and nasal discharge using warm saline/towels, and placing the cat in a steamy room (bathroom) for 10–20 minutes b.i.d. to facilitate removal of respiratory secretions. Home-care techniques were demonstrated to the owner by the technician, and a 7-day recheck was recommended.

FURD Documentation

Patient ID: Little Man Jones

Progress Notes

10/1/08 2 PM

S- Two-month-old, intact male kitten. Owner states semiferal litter of 3 all demonstrate similar symptoms of wheezing and ocular/nasal discharge. None of the litter vaccinated, dewormed, or tested. PE: copious amounts of purulent ocular and nasal discharge; crusting of discharge prevents left eye from opening and has semioccluded both nares. Ulceration of tongue and hard palate. Thoracic auscultation reveals rhonchi and occasional crackles. Abdominal palpation reveals moderate distention with no pain. Patient is moderately depressed. All diagnostic tests declined by owner.

O- T 103.6° F, P 150 bpm, R 36 bpm. Weight 2.5 lb.

A- 1. Obstructed Airway 2. Altered Cellular Respiration 3. Risk of Infection Transmission 4. Underweight 5. Impaired Tissue Integrity 6. Altered Mentation 7. Altered Sensory Perception 8. Client Knowledge Deficit 9. Noncompliant Owner

P- 1. Remove ocular and nasal discharge with warm saline and towel.
2. Dispense Amoxicillin and Trioptic-P.
3. Client education:
   a. Purpose and administration of medications. First dose given while owner observed.
   b. Removal of ocular/nasal discharge with warm moist towel.
   c. Use of steam for removal of respiratory secretions.
   d. Diet and nutrition
   e. Benefits of testing, vaccinations, and spaying/neutering of outdoor cat population
4. Schedule recheck in 7 days.
5. Call in 24 hours for progress report.
C. Smith, CVT

Figure 7-7

Case 6: Laceration

Scenario

A 46-lb, 10-year-old spayed female German shorthaired pointer presented for a laceration over the right carpus. The owner saw the dog become tangled in barbed wire while quail hunting that morning. The laceration occurred as the dog struggled to free herself from the wire. After examining the wound, the owner decided that the dog would still be able to hunt and continued to do so for several hours. The injury had occurred 6 hours prior to examination. The dog was current on vaccinations and deworming and was taking a heartworm preventative. The dog had last eaten at 6 AM.

Physical examination revealed TPR 101.8°F, 126 bpm, panting. A 1.5 in laceration was located on the dorsal surface of the right carpus. The wound was heavily contaminated with grass and other organic debris; the extensor tendons appeared unharmed. Mild swelling of the area was noted, the dog had a slight limp, and a pain response of 2/5 was assigned. All other physical parameters were within normal limits.

The veterinarian recommended copious flushing and primary closure of the wound. The client was provided consent forms and a cost estimate. Preoperative blood work was declined, and the dog was placed in a kennel under n.p.o. status.

The dog was premedicated with atropine and induced using IV medetomidine (Domitor) at 3 PM After administration of a local lidocaine block, the wound was prepped, explored, and copiously flushed. Edges were freshened, and the wound was closed with 3 nonabsorbable simple interrupted sutures. A modified Robert Jones bandage was applied to minimize postsuture swelling. Anesthesia was reversed using IM atipamezole (Antisedan). The dog recovered uneventfully and was ambulatory by 3:45 PM.

An order for carprofen 2 mg/kg p.o. b.i.d. for 3 days and amoxicillin 10 mg/kg p.o. b.i.d. for 7 days was filled by the technician, who also administered the first doses and placed an E-collar on the dog. The client was instructed to remove the bandage the following morning, administer medications as directed, monitor for signs of infection, and return in 10 days for suture removal. The dog was discharged at 5 PM.

Laceration Documentation

Patient ID: Sally Jones

Progress Notes

10/1/08 2 PM
S- Owner states dog got tangled in barbed wire 6 hours ago. Laceration approximately 1 1/2" long on dorsal surface of right carpus; contaminated with grass and debris. Mild swelling, slight limp, pain 2/5. PE otherwise WNL. Owner states dog current on vaccinations, deworming, and heartworm prevention; last meal 6 AM today.
O- T 101.8° F, P 126, panting.
A- 1. Preoperative Compliance 2. Impaired Tissue Integrity 3. Acute Pain 4. Risk of Infection 5. Client Knowledge Deficit
P- 1. Provide estimate of costs; obtain signed consent. Preoperative blood work declined.
2. NPO.
3. Schedule surgery.
C. Smith, CVT

10/1/08 5 PM
S- PE WNL except wound. Bandage intact, wound not visible. Slight limp, pain 1/5.
O- No data
A- Preoperative Compliance resolved. Continue TE 2, 3, 4, 5.
P- 1. Fit E-collar.
2. Dispense carprofen 2 mg/kg PO BID for 3 days and amoxicillin 10 mg/kg PO for 7 days. First dose of both administered by CVT.
3. Owner education:
   a. Bandage removal post-op day 1
   b. Purpose and administration of medications
   c. Signs and symptoms of infections
   d. Necessity of E-collar
4. Schedule appointment for suture removal in 10 days (10-11-08).
5. Call in 24 hours for progress report (10-2-08).
C. Smith, CVT

Figure 7-8

Case 7: New Puppy Exam

Scenario

A 4-month-old, 18-lb intact male border collie mix puppy presented for examination and vaccinations. The puppy had been recently abandoned on the client's property; therefore, the exact date of birth and medical history were unknown. The owner reported that the puppy had been eating voraciously, acted afraid of people, and was not housetrained.

Physical examination revealed TPR 101.2°F, 140 bpm, panting. Thoracic auscultation and abdominal palpation were normal. The puppy's ribs were easily palpable, the hair coat was dry, and the skin was flaky. A BSC of 2 was assigned. A single testicle was palpable in the scrotum. All other physical parameters were within normal limits (WNL).

The veterinarian requested a fecal float, which was collected and examined by the technician. Fecal examination revealed *Toxocara canis*. The veterinarian discussed with the client the medical repercussions of reduced weight, cryptorchidism, roundworm infection, zoonotic potential, and recommended vaccines.

Nemex and vaccinations against canine parvovirus, distemper, adenovirus, leptospirosis, bordetella, parainfluenza, and rabies were ordered by the veterinarian and administered by the technician. (The veterinarian administered the rabies vaccine.) The technician also provided additional client education regarding diet and feeding regimes, housetraining techniques, and the importance of administering the booster vaccinations.

New Puppy Exam Documentation

Patient ID: Tank Jones

Progress Notes

10/1/08

S- Puppy abandoned on owner's premises. Owner states puppy eating voraciously, acts afraid of people, not housebroken. PE: ribs easily palpable, BSC 2; skin dry, flaky. Single testicle palpable. PE otherwise WNL.

O- T 101.2°F, P 140 bpm, panting.

A- 1. Underweight 2. Risk of Infection 3. Reproductive Dysfunction 4. Client Knowledge Deficit

P- 1. Fecal float per order. Results: Toxocara canis. Nemex administered per order.
2. Vaccinate CPV/DHL-A, B-P13 by CVT. Rabies by DVM. Rabies tag and certificate given to owner.
3. Client education:
   a. Zoonotic potential of Toxocara canis
   b. Recommended diet
   c. House training and socialization
   d. Benefits of vaccination, castration
   e. New puppy booklet given
   f. Schedule appointment in 4 weeks for boosters and castration (11-1-08).

C. Smith, CVT

Figure 7-9

Case 8: Ringworm

Scenario

A 7-month-old spayed female DSH cat presented for alopecia and pruritis. The owner stated that the area of hair loss had doubled in size during the previous week and that she had applied cortisone to the affected area in an attempt to control the constant itching. The cat was current on vaccinations and deworming.

Results of physical examination revealed TPR 100.4°F, 160 bpm, 30 bpm, and a 1" × 1" area of alopecia on the dorsal surface of the neck. Several smaller areas of alopecia also were noted on the ventral abdomen and right thoracic wall. No fluorescence was noted when the areas of alopecia were examined using the Wood's lamp. Thoracic auscultation and abdominal palpation were normal. All body systems except the integument were within normal limits (WNL).

The veterinarian ordered a skin scraping that was obtained and examined by the technician. Microscopic examination revealed *Microsporum*. The veterinarian ordered griseofulvin, ultramicrosized at 12 mg/kg p.o. to be administered with a fatty meal or corn oil b.i.d. for 2 weeks. The order was filled by the technician. The veterinarian then informed the owner of the diagnosis and zoonotic potential. The technician demonstrated techniques for administering medication, reiterated methods to avoid zoonotic spread, and described potential side effects of the medication.

Ringworm Documentation

Patient ID: Minnie Jones

Progress Notes

10/1/08 11 AM

S- Alopecia on dorsal surface of neck, approximately 1" x 1". Smaller area noted on ventral abdomen and right thoracic wall. No fluorescence with Wood's lamp. Owner states area of alopecia has doubled in last week; owner applies cortisone cream for pruritis. PE otherwise WNL.

O- T 100.4°F, P 160 bpm, R 30 bpm.

A- 1. Impaired Tissue Integrity 2. Risk of Infection Transmission 3. Client Knowledge Deficit

P- 1. Skin scraping per order. Results: microsporum.
2. Dispense grisofulvin, ultramicrosized at 12 mg/kg PO BID for 14 days.
3. Owner education:
    a. Zoonotic potential of microsporum, methods of preventing transmission
    b. Purpose, side effects, and administration of medication
C. Smith, CVT

Figure 7-10

Case 9: Tumor Removal

Scenario
Day 1
4 PM

A 6.8-lb, 2-year-old spayed female Chihuahua presented for a growth on the dorsal surface of the right forelimb. The mass was first noted 6 months prior to presentation and had gradually enlarged to the current size. The dog was current on vaccinations, deworming, and heartworm preventative.

Physical examination revealed TPR 100.2°F, 110 bpm, panting. The mass was firm on palpation, was nonadhered to underlying tissue, was nonpainful, measured 0.20" × 0.25", and was located on the dorsal surface of the right radius. No alopecia or abrasion of the overlying dermis was noted. Thoracic auscultation and abdominal palpation were within normal limits. All other physical parameters were WNL. Surgical excision and histopathology of the mass was recommended.

The owner was provided surgical consent forms and a cost estimate. Preoperative blood tests were accepted. Following sample collection, the technician placed the dog in the surgical ward under n.p.o. status (standing order) and scheduled the surgery for the following AM Results of the blood work were as follows:

PCV 42%
Hb 14.3 g/dl
TP 6.5 g/dl
ALP 13 IU/l
AST 33 IU/l
BUN 17 mg/dl
Creatinine 0.7 mg/dl
Glucose 102 mg/dl
Amylase 180 IU
Lipase 45 IU
Ca 9.8 mg/dl
Na 150 mEq/L
K 4.2 mEq/L
Cl 101 mEq/L

Day 2
8 AM

A brief physical revealed TPR 99.9°F, 110 bpm, 30 bpm, mucous membranes pink, and CRT 1 second. The dog was brought to an exercise area where she proceeded to urinate and defecate. A premedication (atropine-butorphanol) was administered at 8:30 AM, and an IV catheter was placed in the left cephalic vein. The dog was induced at 9 AM (propofol), and general anesthesia was maintained with sevoflurane. LRS was administered during the surgery at a rate of 25 mL/h and continued until 50 cc had been administered. The tumor was excised, placed in formalin, and submitted for histopathology. The incision was closed using a nonabsorbable suture in a simple interrupted pattern. To minimize postoperative swelling, a modified Robert Jones bandage was placed, with instructions to be removed at 8 AM the following morning. Recovery from anesthesia was uneventful, and the dog was ambulatory by 11:30 AM.

12 PM

Brief examination revealed an intact bandage and absence of any swelling above or below the wrap. A pain response of 1/5 noted. Meloxicam was ordered 0.1 mg/kg p.o. s.i.d. 3 days, and the first dose was administered at 12 PM.

4 PM

Per order, the dog was offered water.

6 PM

Physical examination revealed TPR 100.5°F, 133 bpm, 28 bpm, mucous membranes pink, and CRT 1 second. The dog had drunk 1/3 of the water offered, appeared alert and responsive, and urinated when brought to the exercise area. A slight limp was noted during ambulation, and a pain response of 1/5 was assigned. The bandage was intact, and no swelling was apparent. Moist Hill's p/d and water were offered per order.

Day 3
8 AM

Brief examination revealed TPR 100.1°F, 120 bpm, panting. The dog had consumed half the water and all of the food offered. Upon bandage removal, the incision site appeared clean and dry with no discharge or hyperemia. It was noted, however, that once the bandage was removed, the dog immediately began to lick the sutures. Per the veterinarian's order, an E-collar was immediately placed on the animal. The dog urinated and defecated when brought to the exercise area. Meloxicam was administered; food and water were offered.

10 AM

A brief exam revealed no change in physical status. The dog was released to the owner, and meloxicam was dispensed. Prior to discharge, the owner was educated regarding E-collar usage, incision line care, exercise, and medication administration. An appointment was scheduled for suture removal in 10 days, and a notation was made to contact the owner once pathology results became available.

Tumor Removal Documentation

Patient ID: Gizmo Jones

Progress Notes

10/1/08 4 PM
S- Firm, nonpainful, nonadhered mass approximately 0.2" x 0.25" noted on dorsal surface of right radius. Overlying dermis normal. Owner states mass was first noticed 6 months ago and has been enlarging over that time. Remainder of PE WNL.
O- T 100.2°F, P 110 bpm, panting. Weight 6.8 lb.
A- 1. Preoperative Compliance
P- 1. Initiate MAOR: NPO
2. Provided cost estimate; owner consent for surgery and preoperative blood work obtained.
3. Schedule surgery.
C. Smith, CVT

10/1/08 5 PM
Notation: Preoperative blood work completed; all results WNL.
C. Smith, CVT

10/2/08 8 AM
S- PE WNL. Mucous membranes pink, CRT 1 second. Patient urinated and defecated this AM.
O- T 99.9°F, P 110 bpm, R 30 bpm.
A- 1. Preoperative Compliance
P- 1. Prepare surgical suite.
C. Smith, CVT

10/02/08 10 AM
Notation: Tissue sample sent out for pathology.
C. Smith, CVT

10/02/08 12 PM
S- Bandage intact; no swelling or hyperemia above or below bandage. Pain response 1/5.
O- No data
A- 1. Acute Pain 2. Risk of Infection
P- 1. Modify MAOR: add meloxicam.
2. Monitor pain response.
3. Monitor for signs of infection.
C. Smith, CVT

10/2/08 4 PM
Notation: Intact gag reflex noted. Patient offered water.
C. Smith, CVT

Figure 7-11a

Patient ID: Gizmo Jones

Progress Notes

10/2/08 6 PM
S- BAR, mucus membranes pink, CRT 1 second. 1/3 of water consumed; patient urinated when brought to exercise area. Slight limp during ambulation; pain response 1/5. Bandage intact, no swelling noted.
O- T 100.5°F, P 133 bpm, R 28 bpm.
A- Continue TE 1, 2. Add 3. Reduced Mobility
P- 1. Modify MAOR: Diet: Provide water and moist Hill's p/d
2. Provide controlled exercise.
3. Continue existing plan.
C. Smith, CVT

10/3/08 8 AM
S- Bandage removed: incision site clean and dry; no discharge or hyperemia. Patient began licking incision site. All food and half of water consumed. Patient urinated and defecated this AM when brought to exercise area.
O- T 100.1°F, P 120 bpm, panting.
A- 1. Acute Pain 2. Risk of Infection 3. Risk of Self-Inflicted Injury
P- 1. Modify MAOR: add E-collar; continue diet and meloxicam.
2. Provide controlled exercise.
3. Monitor pain status.
4. Monitor for signs of infection.
C. Smith, CVT

10/3/08 10 AM
S- PE WNL. BAR
O- No data
A. continue TE 1, 2, 3. Add 4. Client Knowledge Deficit.
P- 1. Dispense meloxicam.
2. Client education:
   a. Purpose and administration of medication
   b. Signs of infection
   c. Incision line care
   d. E-collar usage
3. Schedule suture removal in 10 days (10-13-08).
4. Call in 48 hours for progress report (10-5-08).
5. Notify DVM when pathology report received.
C. Smith, CVT

Figure 7-11b

Medication Administration/Order Record

Patient ID	Allergies	Initials	Signature	Title
Gizmo Jones	none	CS	Cathy Smith	CVT

Medication Administration

Date of Order	Medication	Time	10/1	10/2	10/3	10/4	10/5	10/6	10/7
10/2	Meloxicam 0.1 mg/kg p.o. s.i.d. for 3 days	8A	—	12P CS	CS			—	

Fluids and IV Drips

Figure 7-12a

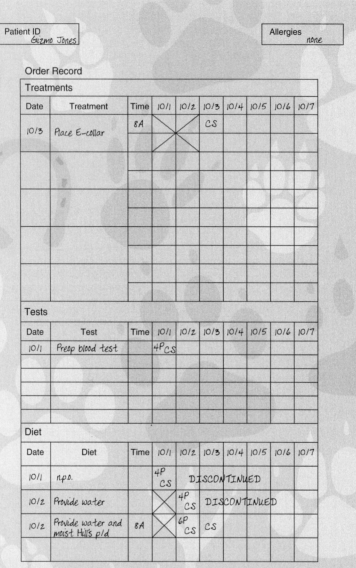

Figure 7-12b

Case 10: Urethral Obstruction

Scenario
Day 1
4 PM

A five-year-old, 8-lb, indoor-only castrated DLH cat was brought to the clinic in the late afternoon. The owner reported that the cat had been "acting funny" since morning and that the odd behavior had become more pronounced as the day progressed. The owner described the cat getting in and out of its litter box without urinating, meowing frequently, and continually grooming the perineal area.

A physical examination was performed with the following results: TPR 101°F, 80 bpm, 20 bpm. Mucous membranes dark/tacky, abdomen painful, bladder palpated large and turgid, penis a dark red color. Abdominal palpation elicited a 3/5 pain response.

The veterinarian diagnosed urethral obstruction with imminent risk of bladder perforation. The owner was informed of the medical condition and was provided a cost estimate and treatment consent forms.

Upon receiving consent, the attending veterinarian placed a urinary catheter, hydropulsed the urethral calculi, obtained a urine sample, and flushed the bladder with 0.9 percent NaCl. An indwelling urinary catheter was secured.

Per order, the veterinary technicians placed an IV catheter; obtained a PCV, TP, BUN, and creatinine and electrolyte panel; performed an ECG; and initiated administration of 0.9 percent NaCl at a rate of 200 cc/h. The attending veterinarian prescribed amoxicillin 10 mg/kg p.o. b.i.d. for 7 days, buprenorphine 0.01 mg/kg p.o. t.i.d. for 3 days, and phenoxybenzamine 5 mg p.o. per day for 3 days, which were administered by the technician. The cat was made n.p.o. An E-collar was placed on the cat prior to placing it in a cage equipped with an elevated grate.

Results of the urinalysis, blood work, and ECG obtained by the technician included the following:

- Urinalysis: pH 7.8, SG 1.040. Sediment: blood +4, occasional WBC, no bacteria, numerous triple phosphate crystals.
- Blood work: PCV 50, TP 7.5, K 6.6 mEq/L, BUN 70 mg/dl, Cr 2.4 mg/dl, Na 157 mEq/L, Cl 129 mEq/L
- The ECG indicated bradycardia (85 bpm) but was otherwise normal.

8 PM

Recheck of select blood work revealed the following results: K 5.5 mEq/L, PCV 46, TP 7.2, BUN 41 mg/dl.

The urinary catheter was flowing well, urine was blood-tinged, and approximate urine output was 100 cc. Fluid rate was changed to 100 cc/h.

Day 2

12 AM

Physical examination revealed no changes in status. No tests were conducted. The IV fluid administration rate was changed to 25 cc/h.

6 AM

Physical examination revealed TPR 101.2°F, 140 bpm, 24 bpm; mucous membranes were pink, moist; thoracic auscultation was normal; cat did not resist abdominal palpation. The urinary catheter was flowing well, urine was blood-tinged, and urine output was approximately 200 cc.

The veterinarian ordered an IV fluid change to Normosol-R at 25 cc/h. amoxicillin, buprenorphine, and phenoxybenzamine were administered. Water and Hill's s/d were ordered.

Blood was collected and results were as follows: PVC 42, TP 7.2, BUN 28 mg/dl, K 5 mEq/L.

12 PM

The urinary catheter was flowing well; urine was less blood-tinged, and urine output was approximately 150 cc. The cat had not eaten or consumed water.

2 PM

The buprenorphine was administered by the technician.

6 PM

Physical examination revealed TPR 101.4°F, 148 bpm, 24 bpm; mucous membranes pink, moist; thoracic auscultation normal. The cat did not resist abdominal palpation, and pain response was 1/5. The urinary catheter was flowing well, urine was faintly blood-tinged, and the urine output was approximately 75 cc. The cat had not eaten.

Amoxicillin was administered by the technician. Normosol-R was continued at a rate of 25 cc/h. The cat was offered fresh water and moist s/d.

8 PM

Buprenorphine was administered by the technician.

Day 3

12 AM

Physical examination revealed no changes in status. No tests were conducted.

6 AM

A physical exam was performed revealing the following results: TPR 101.4°F, 142 bpm, 22 bpm. Mucous membranes were pink and moist; thoracic auscultation was normal. Abdominal palpation did not elicit pain. The urine was almost clear and pale yellow with an estimated volume of 200 cc. The cat had eaten half the Hill's s/d offered.

Upon veterinary order, the urinary catheter and E-collar were removed by the technician, who also administered amoxicillin and the final dose of phenoxybenzamine. The buprenorphine was discontinued. Metacam was ordered at 0.1 mg/kg p.o. s.i.d. for 2 days, and the first dose was administered by the technician. Fluids were continued at a rate of 25 cc/h, and fresh water and moist s/d were provided.

12 PM

The cat had urinated several times in the litter box and had eaten 40 percent of the s/d.

6 PM

A brief physical revealed TPR 101°F, 140 bpm, 24 bpm. No pain upon palpation and normal auscultation. Amoxicillin was administered. The IV fluids and phenoxybenzamine were discontinued. The cat had eaten 50 percent of the food offered and had urinated 2 times since 12 PM. The cat was offered fresh water and moist s/d.

Day 4

6 AM

A physical exam was performed revealing the following results: TPR 101.1°F, 150 bpm, 26 bpm. Mucous membranes were pink and moist; thoracic auscultation was normal. Abdominal palpation did not elicit pain. The urine was clear pale yellow. The cat had eaten all of the Hill's s/d offered. Fresh water and moist Hill's s/d were provided.

The IV catheter was removed. Amoxicillin and metacam were administered by the technician.

12 PM

No changes in physical status were noted. The cat maintained normal appearance and posture during urination, and two large urine spots were noted in the litter box.

4 PM

The cat was released to the owner, amoxicillin and Hill's s/d were dispensed, and the owner was provided with home-care instructions.

Urethral Obstruction Documentation

Patient ID: Zepplin Jones

Progress Notes

10/1/08 4 PM

S- Five-year-old, castrated DLH. Owner states cat "acting funny," getting in and out of litter box, grooming perineal area, and vocalizing. PE: Mucous membranes dark and tacky, abdomen pain response 3/5, bladder large and turgid, penis dark red. Other body systems WNL.

O- TPR 101°F, 80 bpm, 20 bpm. Weight 8 lb.
See urinalysis results (>pH, blood+4, triple phosphate crystals)
See results of blood work (>PCV, TP, K, BUN, Cr)
See EKG strip (bradycardia)

A- 1. Altered Urinary Output 2. Hypovolemia 3. Electrolyte Imbalance 4. Acute Pain 5. Risk of Infection 6. Client Knowledge Deficit

P- 1. Initiate MAOR to include:
Medication: amoxicillin, buprenorphine, phenoxybenzamine
Fluids: Normal saline
Treatments: IV catheter, E-collar, cage grate
Tests: urinalysis, blood work: PCV, TP, serum chemistry, ECG, recheck blood work at 8 PM and in AM
2. Monitor quantity and quality of urine output.
3. Monitor pain response.
4. Monitor vital signs.
5. Provide owner education prior to discharge.
C. Smith, CVT

10/1/08 8 PM

S- Urinary catheter flowing well; urine blood-tinged (dark red); output estimated at 100 cc

O- TPR 101°F, 110 bpm, 24 bpm
See results of blood work (> PCV, TP, K, BUN, Cr)

A- Continue TE 1, 2, 3, 4, 5, 6. Blood work improving.

P- 1. Modify MAOR: change fluid rate.
2. Continue plan.
C. Smith, CVT

Figure 7-13a

Patient ID: Zepplin Jones

Progress Notes

10/2/08 6 AM
S- Catheter flowing well, urine blood-tinged, output estimated at 200 cc. Mucous membranes pink and moist. Thoracic auscultation normal. Cat does not resist abdominal palpation, pain response 1/5.

O- TPR 101.2°F, 140 bpm, 24 bpm.
See results of blood work: All parameters within normal limits.

A- TE: Hypovolemia and electrolyte imbalance resolved.
Continue TE: 1. Altered Urinary Output 2. Acute Pain 3. Risk of Infection 4. Client Knowledge Deficit 5. SWAP
Cat improving: urine quality improving, not yet normal.

P- 1. Modify MAOR fluid type, rate, and diet. Continue medications, treatments as ordered.
2. Monitor quantity and quality of urine output.
3. Monitor pain response.
4. Monitor vital signs.
5. Provide owner education prior to discharge.
C. Smith, CVT

10/2/08 12 PM
Notation: Catheter flowing well, urinary output 150 cc, less blood-tinged, IV catheter patent. Hasn't eaten.
C. Smith, CVT

10/2/08 6 PM
S- Catheter flowing well. Urine faintly pink, output estimated at 75 cc. Mucous membranes pink and moist; thoracic auscultation normal. Pain response 1/5; does not resist palpation. Anorexic.

O- TPR 101.4°F, 148 bpm, 24 bpm

A- Continue TE 1, 2, 3, 4, 5. Urine quality continues to improve.

P- Continue plan.
B. Bell, CVT

Figure 7-13b

Patient ID: Zepplin Jones

Progress Notes

10/3/08 6 AM
S- Urine clear; output estimated at 200 cc. Mucous membranes pink and moist. Thoracic auscultation normal. No pain with abdominal palpation. Ate half of s/d
O- TPR 101.4°F, 142 bpm, 22 bpm.
A- 1. Altered Urinary Output 2. Acute Pain 3. Risk of Infection 4. Client Knowledge Deficit 5. SWAP

P- 1. Modify MAOR; remove urinary catheter, E-collar. Discontinue buprenorphine; add metacam, continue amoxicillin, phenoxybenzamine, fluids, diet.
2. Monitor quantity and quality of urine output.
3. Monitor pain response.
4. Monitor vital signs.
5. Provide owner education prior to discharge.
C. Smith, CVT

10/3/08 12 PM
Notation: Drinking water and eating, urinated multiple times in litter box.
C. Smith, CVT

10/3/08 6 PM
S- Drinking well, ate 50% of s/d. No pain on palpation, normal auscultation.

O- TPR 101°F, 140 bpm, 24 bpm

A- Continue TE: 1, 2, 3, 4, 5

P- 1. Modify MAOR: IV fluids discontinued, phenoxybenzamine discontinued. Continue amoxicillin, diet.
2. Continue plan.
B. Bell, CVT

10/4/08 6 AM
S- Cat is urinating normally, eating and drinking well.
PE: All systems WNL.

O- TPR 101.1°F, 150 bpm, 26 bpm

Figure 7-13c

Patient ID: Zepplin Jones

Progress Notes

A- Altered Urinary Output resolved
1. SWAP 2. Acute Pain 3. Risk of Infection 4. Client Knowledge Deficit

P- Modify MAOR: remove IV catheter, continue amoxicillin, metacam.
2. Monitor quantity and quality of urine output.
3. Monitor pain response.
4. Provide owner education prior to discharge.

C. Smith, CVT

10/4/08 4 PM (postdischarge)
S- BAR, eating, drinking, and urinating normally
O- No data
A- TE: Client Knowledge Deficit resolved
TE: 1. Risk of Altered Urinary Output 2. Acute Pain 3. Risk of Infection

P- 1. Owner education:
Purpose and administration of medications
Home-care goals:
   a. Decrease urine pH < 6 and < magnesium concentration via diet modification.
   b. Ensure adequate fluid intake.
2. Dispense amoxicillin.
3. Set appointment 10/11/08 for recheck.
4. Call owner in 48 hours for update on progress.
5. Discharge cat.
C. Smith, CVT

Figure 7-13d

Medication Administration/Order Record

Patient ID	Allergies
Zepplin Jones	none

Initials	Signature	Title
CS	Cathy Smith	CVT
SO	Sheila Olgrady	DVM
JP	John Peters	CVT

Medication Administration

Date of Order	Medication	Time	10/1	10/2	10/3	10/4	10/5	10/6	10/7
10/1	Amoxicillin 10 mg/kg p.o. b.i.d. 7 days	6A	✕	CS	CS	CS			
		6P	4P CS	CS	CS				
10/1	Buprenorphine 0.01 mg/kg p.o. t.i.d. 3 days	6A	✕	CS					
		2P	✕	CS					
		8P	4P CS	JP		✕	✕	✕	✕
10/1	Phenoxybenzamine 5 mg p.o. s.i.d. 3 days	6A	4P CS	CS	CS		✕	✕	✕
10/3	Meticom 0.1 mg/kg p.o. s.i.d. 2 days	6A	✕	✕	CS	CS	✕	✕	✕

Fluids and IV Drips			10/1	10/2	10/3	10/4	10/5	10/6	10/7
10/1	0.9% NaCl @ 200 cc/hr	4P	N#1 CS	DISCONTINUED					
10/1	0.9% NaCl @ 100 cc/hr	8P	CS	DISCONTINUED					
10/2	0.9% NaCl @ 25 cc/hr	12A	✕	N#2 SO	DISCONTINUED				
10/2	Normosol R @ 25 cc/hr	6A	✕	CS N#1	DISCONTINUED				

Figure 7-14a

Patient ID	Allergies
Zipplin Jones	none

Order Record

Treatments

Date	Treatment	Time	10/1	10/2	10/3	10/4	10/5	10/6	10/7
10/1	Place urinary catheter	4P	SO	CS	DISCONTINUED				
10/1	Place IV catheter	4P	CS	CS		DISCONTINUED			
10/1	Place E-collar	4P	CS	CS	DISCONTINUED				

Tests

Date	Test	Time							
10/1	Urinalysis	4P	SO						
10/1	Blood test	4P	CS						
10/1	ECG	4P	CS						
10/1	Blood test	8P	JP						
10/2	Blood test	6A	CS						

Diet

Date	Diet	Time							
10/1	n.p.o.	4P	CS	DISCONTINUED					
10/2	Provide water and Hill's s/d	6A / 6P		CS	CS	CS			

Figure 7-14b

BIBLIOGRAPHY

Aiello, S. (Ed.). (1998). *The Merck veterinary manual* (8th ed.). Whitehouse Station, NJ: Merck & Co.

Cote, E. (2007). *Clinical veterinary advisor: Dogs and cats.* St. Louis, MO: Mosby.

Jack, C. M., Watson, P. M., & Donovan, M. S. (2003). *Veterinary technician's daily reference guide. Canine and feline.* Baltimore: Lippincott Williams & Wilkins.

Nelson, R. W., & Couto, C. G. (1999). *Manual of small animal internal medicine* (2nd ed.). St. Louis, MO: Mosby.

Plumb, D. C. (2005). *Plumb's veterinary drug handbook* (5th ed.). Stockholm, WI: Blackwell Publishing.

Appendices

Appendix Table 1 Canine and Feline Vaccinations

CORE VACCINES		NONCORE VACCINES	
Canine	Feline	Canine	Feline
Distemper	Panleukopenia	Parainfluenza	Chlamydiosis
Parvovirus	Viral Rhinotracheitis	Bordetella	FIP
Adenovirus Type 2	Calicivirus	Giardia	Bordetella
Rabies	Coronavirus	Lyme	FIV
		Leptospirosis	

Common Abbreviations/Acronyms Used in Documentation

amp	ampule
bar	bright alert and responsive
b.i.d.	twice a day
c.	with
cap.	capsule
cc	cubic centimeter
CVT	certified veterinary technician
d.h.l.	canine distemper hepatitis leptospirosis vaccine
D.I.C.	disseminated intravascular coagulation
ECG	electrocardiogram
ELISA	enzyme-linked immunosorbent assay
FeLV	feline leukemia virus
F.I.V.	feline immunodeficiency virus
g	gram
GDV	gastric dilation volvulus
gm	gram
gtt.	drop
h	hour
IM	intramuscular
IP	intraperitoneal
IV	intravenous

lb	pound
MAOR	Medication Administration/Order Record
mL	milliliter
O.D.	right eye
O.S.	left eye
O.U.	both eyes
p.o.	by mouth
prn.	as needed
q	every
Q.A.R.	quite alert and responsive
q4h	every 4 hours
q.i.d.	four times a day
q.o.d.	every other day
q.s.	a sufficient quantity
RVT	registered veterinary technician
SC	subcutaneous
s.i.d.	once a day
SOAP	subjective, objective, assessment, plan
SQ	subcutaneous
stat	immediately
susp.	suspension
tab	tablet
t.i.d.	three times a day

Weight and Liquid Conversions Commonly Used in Practice

1 pound (lb) = 0.454 kilograms (kg) = 454 grams (g) = 16 ounces (oz)

1 kilogram (kg) = 2.2 pounds (lb) = 1,000 grams (g)

1 grain (gr) = 64.8 milligrams (mg)

1 ounce (oz) = 28.4 grams (g)

1 gram = 1,000 milligrams (mg)

1 milligram (mg) = 1,000 micrograms (mgc or µg)

1 microgram (mcg or µg) = 1,000 nanograms (ng)

1 gallon (gal) = 4 quarts (qts) = 8 pints (pts) = 128 ounces (oz) = 3.785 liters (l) = 3,785 milliliters (mL)

1 quart (qt) = 2 pints (pts) = 32 ounces (oz) = 946 milliliters (mL)

1 pint (pt) = 2 cups (c) = 16 ounces (oz) = 473 milliliters (ml)
1 cup (c) = 8 ounces (oz) = 237 milliliters (ml) = 16 tablespoons (T)
1 tablespoon (T) = 15 milliliters (ml) = 3 teaspoons (tsp)
1 teaspoon (tsp) = 5 milliliters (ml)
4 liters (L) = 1.57 gallons
1 liter = 1,000 milliliters (ml) = 10 deciliters (dl)
1 deciliter (dl) = 100 milliliters (ml)
1 milliliter (ml) = 1 cubic centimeter (cc) = 1,000 microliters (µl or mcl)

Temperature Conversions

°C to °F = (°C × 1.8) + 32 = °F
°F to °C = (°F − 32) × 0.555 = °C

Appendix Table 2 Normal TPR Values of the Dog and Cat

	FELINE	CANINE
Temperature (F)	100–103.1	99.5–102.5
Pulse Rate (Young)	130–140	110–120
Pulse Rate (Old)	100–120	80–120
Pulse Rate (Large Breed)	–	80–120
Respiratory Rate	20–30	15–30

Appendix Table 3 Normal Hematology Values of the Dog and Cat

	CANINE	FELINE
PCV (%)	37–55	24.0–45.0
Hct	29.8–57.5	25.8–41.8
Hb (g/dl)	12.4–19.1	8.5–14.4
RBC (× 10^6/µl)	5.2–8.06	4.95–10.53
WBC (× 10^3)	5.4–15.3	3.8–19
Total Protein (TPP) (g/dl)	5.8–7.2	5.7–7.5
MCV (fl)	62.7–72	36–50
MCH (pg)	22.2–25.4	12.2–16.8

(continued)

	CANINE	FELINE
MCHC (g/dl)	34–36.6	32.4–35.2
Reticulocytes (%)	0–1.5	0.2–1.6
RBC Diameter (microns)	6.7–7.2	5.5–6.3
RBC Life (days)	100–120	66–78
M:E Ratio	0.75–2.5:10	0.6–3.9:10
Platelets ($\times 10^3/\mu l$)	160–525	160–660
Icterus Index	2–5	2–5
Fibrinogen (mg/dl)	200–400	150–300
RDW	12.2–14.9	14.1–18.4
PCT	0.182–0.416	0.179–0.916
MPV	6.6–10.9	10.0–15.5
PDW	14.5–16	14.4–17
WBC Diff Absolute Count/µl (% of total)		
Stabs	0–150 (0–10)	0–190 (0–10)
Segs	2,750–12,850 (51–84)	1,290–15,950 (34–84)
Lymphs	430–5,800 (8–38)	260–11,400 (7–60)
Monos	50–14,000 (1–9)	0–950 (0–5)
Eos	0–14,000 (0–9)	0–2,300 (0–12)
Basos	Rare (0–1)	0–400 (0–2)
Coagulation (seconds)		
PT	6–8.4	8.7–10.5
PTT	11.0–17.4	12.3–16.7
TT	4.3–7.1	5.6–9.0

Appendix Table 4 Normal Blood Chemistry Values of the Dog and Cat

	CANINE	FELINE
ALB—Albumin (g/dl)	2.6–4.0	2.6–4.3
ALP—Alkaline Phosphatase (U/I)	3–60	3–61
ALT—Alanine Transaminase (U/I)	4–91	13–75

	CANINE	FELINE
AMYL—Amylase (U/l)	220–1,070	400–15,900
T-Bili—Bilirubin Total (mg/dl)	0–0.7	0–0.6
BUN—Blood Urea Nitrogen (mg/dl)	7–26	10–30
Calcium Total (mg/dl)	9.6–11.6	9.3–11.7
Creatinine (mg/dl)	0.6–1.4	0.8–2.0
CK—Creatine Kinase (IU/l)	36–155	21–275
Glucose (g/dl)	79–126	63–132
PHOS—Phosphorus (mg/dl)	2.5–6.2	2.9–7.7
TP—Total Protein (g/dl)	5.8–7.9	6.1–8.8
Uric Acid (mg/dl)	0–0.6	0–0.2
Na^+—Sodium (mEq/L)	146–156	151–161
K^+—Potassium (mEq/L)	3.8–5.1	3.5–5.1
Cl^-—Chloride	109–122	117–129
T CO_2—CO_2 Total (mM/L)	17–27	13–25
Anion Gap	8–19	9–21
Osmolality (mOsm/L)—Calculated	289–313	299–327
Additional Serum Chemistries		
Ammonia (µg/dl)	19–120	-
Ammonia Tolerance (µg/dl)	<200 at 30 min.	<300
AST (SGOT) (IU/l)	<105	<51
Bilirubin Direct	0–0.4	0–0.2
BSP Retention at 30 minutes	0–5%	-
Cholesterol (mg/dl)	125–300	95–130
GGT—Gamma-Glutamyl Transpeptidase	0–2.26	-
Lactate (mg/dl)	4–12	-
Methemalbumin (mg/dl)	0–5	-
Lipase (U/l)	0–600	0–600
Free Plasma Hgb (mg/dl)	<10	<10
Xylose Tolerance (mg/dl) at 60–90 min.	70–90	-

(continued)

	CANINE	FELINE
PABA Tolerance (plasma) (µg/dl)	670+/−140	386+/−134
Additional Acid-Base/Electrolytes		
pH	7.31–7.53	7.32–7.44
pO_2 (mm Hg) Arterial	85–95	-
Venous	35–40	35–40
pCO_2 (mm Hg) Arterial	29–36	-
Venous	35–44	38–46
Bicarb (HCO_3^-) (mEq/L)	25–35	24–34
Base Excess (mEq/L)	+6 to 0	+2 to −5
Magnesium (mg/dl)	1.7–2.9	2–3

Appendix Table 5 Normal Urinalysis Values of the Dog and Cat

	CANINE	FELINE
Specific Gravity	1.001–1.070	1.001–1.080
pH	5.5–7.5	5.5–7.5
Color	Straw yellow	Straw yellow
Clarity	Clear	Clear
Calcium	2.1	3.0
Bilirubin	0–trace	0
Blood	0	0
Urobilinogen	0	0
Glucose	0	0
Ketones	0	0
Protein	0	0
Sediment		
WBC #/HPF	0–5	0–5
RBC #/HPF	0–5	0–5
Casts #LPF	0	0

Appendix Table 6 Comparison of Common Crystalloid Fluids

FLUID	USE	ROUTE	TONICITY	Na (mEq/L)	K (mEq/L)	Ca (mEq/L)	OSMOLARITY (mOsm/L)
Dextrose 5%	Supply free water and calories to patient (1 g dextrose = 4.1 calories) as a replacement solution	IV	Hypotonic	77	-	-	280 (pH 4.5)
Lactated Ringer's	Replacement fluid Maintenance fluid Shock	IV SC	Isotonic	130	4	2.7	273 (pH 6.7)
Ringer's Solution Normal Saline (0.9%)	Replacement fluid Corrects metabolic alkalosis	IV SC IP	Isotonic	154	-	-	308 (pH 5.7)
Normosol-R	Balanced, multiple electrolyte solution	IV	Isotonic	140	5	-	295 (pH 7.4)

Conversion Formula for mm Hg to cm H₂O

mm Hg × 1.36 = cm H$_2$O

Normal Blood Pressure Values

Dogs: Systolic/Diastolic = 147/83 mm Hg
Cats: Systolic/Diastolic = 160/100 mm Hg

Appendix Table 7 Normal Blood Gas Values of the Dog and Cat

	UNITS	CANINE	FELINE
Arterial Blood Gas			
pH	-	7.36–7.44	7.36–7.44
pCO$_2$	mm Hg	36–44	28–32
pO$_2$	mm Hg	90–100	90–100
TCO$_2$	mEq/L	25–27	21–23
HCO$_3$	mEq/L	24–26	20–22
Venous Blood Gas			
pH	-	7.34–7.46	7.33–7.41
pCO$_2$	mm Hg	32–49	34–38
pO$_2$	mm Hg	24–48	35–45
TCO$_2$	mEq/L	21–31	37–31
HCO$_3$	mEq/L	20–29	22–24

Directory of Pet Loss Hotlines

1-800-565-1526: University of California Davis School of Veterinary Medicine
1-877-394-2273: (CARE)—University of Illinois College of Veterinary Medicine
1-888-478-7574: Iowa State University College of Veterinary Medicine
630-325-1600: Chicago Veterinary Medical Association
607-253-3932: Cornell University College of Veterinary Medicine
517-432-2696: Michigan State University College of Veterinary Medicine
614-292-1823: Ohio State University College of Veterinary Medicine
508-839-7966: Cummings School of Veterinary Medicine at Tufts University
540-231-8038: Virginia-Maryland Regional College of Veterinary Medicine
509-335-5704: Washington State University College of Veterinary Medicine

Directory of Pet Loss Web Sites

http://www.animalclergy.com   Animal Chaplains
http://www.aplb.org   Association for Pet Loss and Bereavement
http://www.petvets.com/petloss   Harmony Animal Hospital
http://vet.osu.edu   Ohio State University College of Veterinary Medicine
http://www.petcaring.com   PetCaring
http://www.griefonline.com/candleceremony.htm   Pet Grief Support
http://www.pet-loss.net   Pet Loss Support Page

Bibliography

Aiello, S. (Ed.). (1998). *The Merck veterinary manual* (8th ed.). Whitehouse Station, NJ: Merck & Co.

Bistner, S. I., Ford, R. B., & Raffe, M. R. (2000). *Kirk and Bistner's handbook of veterinary procedures & emergency treatment* (7th ed.). Philadelphia: W. B. Saunders.

Jack, C. M., Watson, P. M., & Donovan, M. S. (2003). *Veterinary technician's daily reference guide. Canine and feline.* Philadelphia: Lippincott Williams & Wilkins.

Plumb, D. C. (2005). *Plumb's veterinary drug handbook* (5th ed.). Stockholm, WI: Blackwell Publishing.

Pugh, S. (2006, Winter). Educating pet owners about vaccines and vaccine reactions. *The NAVTA Journal.*

Sirois, M., & McBride, D. F. (Eds.). (1995). *Veterinary clinical laboratory procedures.* St. Louis, MO: Mosby.

Wingfield, W. (1997). *Veterinary emergency medicine secrets.* Philadelphia: Hanley & Belfus.

Glossary

Abduction Movement of an extremity away from the body.

Abscess Accumulation of pus within a tissue cavity.

Adventitious Externally acquired; not normal for location. With reference to lung sounds, refers to sounds that are not normally heard with auscultation.

Agalactia Condition in which milk is not secreted by the dam after having delivered offspring.

Alopecia Loss of hair.

Anorexia Decrease in or complete loss of appetite.

Anuria Absence of urination.

Ascites Accumulation of fluid in the peritoneal cavity.

Ataxia Uncoordinated movement; staggering gait.

Atelectasis Collapse of lung tissue or failure of lung to expand in a neonate.

Atrophy Decrease in size of a body part or organ.

Bacteremia Presence of bacteria in the blood.

Blepharospasm Spasm of the eyelid.

Bradycardia Heart rate lower than normal.

Buccal Relating to the cheek.

Cachexia Extreme malnutrition.

Calcinosis cutis Deposits of calcium salts in the skin.

Catabolism Metabolic process that breaks down complex molecules into simpler molecules, usually resulting in the release of energy.

Cellulitis Inflammation of connective tissue below the skin.

Cirrhosis Condition of the liver in which fat, fibrous tissue, and nodules replace normal tissue, reducing functional capacity of the organ.

Coprophagy Ingestion of feces.

Crackles Bubbling or crackling sound in the lungs. Also called rales.

Crepitus Grating sound in the lungs or joints.

Cupage Massage therapy treatment. The palm of the hand is cupped as if holding water, then gently tapped over the body.

Cyanosis Bluish discoloration of the skin and mucous membranes caused by ischemia.

Cystitis Inflammation of the urinary bladder and/or tract.

Cytotoxic Agent that is destructive or poisonous to cells.

Debridement Removal of foreign material and necrotic or nonviable tissue from a wound.

Dialysate Solution used in dialysis therapy.

Dialysis Process of using a semipermeable membrane to remove wastes from the blood.

Diaphoresis Profuse perspiration.

Dysphagia Difficulty swallowing.

Dyspnea Shortness of breath or labored, difficult breathing.

Dystocia Difficult or abnormal labor.

Dysuria Painful or difficult urination.

Ectopic Not located in normal anatomical position.

Embolism Blockage or obstruction of a blood vessel.

Endotoxemia Presence of endotoxins in the blood.

Endotoxin Toxin produced by gram negative bacteria that is released when the bacterial cell dies.

Enteric Pertaining to the intestines.

Epiphora Tearing or watering of the eye.

Epistaxis Bleeding from the nose.

Erythema Abnormal redness of the skin resulting from dilatation of the capillaries.

Erythropoetin Hormone produced by the kidneys that stimulates production of red blood cells.

Exophthalmos Protrusion of the eyeball.

Exudate Fluid containing proteins and cellular debris that is produced in response to inflammation.

Flatulence Excessive amount of gas expelled from the anus.

Fomite Object that can support and transmit pathogens.

Glossitis Inflammation of the tongue.

Glucosuria Presence of glucose in the urine.

Halitosis Offensive or fetid breath.

Hematemesis Vomiting of blood.

Hematochezia Passage of feces containing visible blood.

Hematuria Presence of blood in the urine.

Hemoptysis Coughing up blood from the respiratory tract.

Hyper- Prefix meaning excessive, above, or beyond.

Hyperesthesia Extreme sensitivity to stimuli.

Glossary

Hyperglycemia Higher-than-normal glucose in the blood.
Hypermetria Overreaching form of ataxia.
Hyperpigmentation Increase in pigmentation.
Hypo- Prefix meaning under, below, beneath, or deficient.
Hypometria Underreaching ataxia.
Hypovolemic Low circulating blood volume.
Icterus Yellowing of the skin and sclerae of the eyes caused by an accumulation of bilirubin.
Idiopathic Of unknown cause.
Ischemia Decreased blood supply.
Keratitis Inflammation of the cornea.
Ketosis Abnormal accumulation of ketones in the body.
Leukopenia Decreased number of leukocytes in the blood.
Lingual Pertaining to the tongue.
Lochia Discharge from the vagina after having given birth.
Melena Stools containing occult blood; usually dark and tarry colored.
Metabolic acidosis Acid-base imbalance produced by loss of bicarbonate or excess production of acid, as in ketoacidosis.
Micturition Urination.
Miosis Constriction of the pupil.
Neoplasia Abnormal and uncontrolled cellular growth; cancer.
Nocturia Excessive urination at night.
Mydriasis Dilatation of the pupil.
Nosocomial Of or pertaining to a hospital.
Nosocomial infection Infection acquired during a hospital stay.
Oliguria Decreased urine production.
Orthopnea Condition in which animal can breath comfortably only when sitting, standing, or extending the neck.
Pallor Abnormal paleness of the skin.
Parenchyma Functional tissue of an organ apart from supporting or connective tissue.
Patent Open as a tube or passageway.
Petechiae Tiny reddish or purplish flat spots appearing as a result of pinpoint hemorrhages.
Phlebitis Inflammation of the wall of a vein.
Photophobia Abnormal intolerance to light.
Pica Eating of nonfood substances.

Piloerection Hair standing on end or fluffed out.

Polydipsia Excessive thirst.

Polyphagia Overeating.

Polyuria Production of large volumes of usually dilute, pale urine.

Proptosis Forward displacement of the eyeball.

Ptyalism Excessive salivation.

Pudendal Pertaining to the external genitalia.

Purulent Producing or containing pus.

Pustules Small pus-containing elevations on the skin.

Rales Abnormal lung sound, usually a bubbling or crackling noise.

Regurgitation Return of previously swallowed food into the mouth.

Reticulocytopenia Reduced reticulocytes in blood.

Rhonchi Abnormal chest sound of rumbling or scraping.

Septicemia Infection in which pathogens are present in the circulating blood.

Stranguria Slow, frequent, difficult, and/or painful urinary flow.

Syncope Brief loss of consciousness; fainting.

Tachycardia Abnormally rapid heart rate.

Tenesmus Straining to defecate.

Thrombosis Condition in which a blood clot forms within a blood vessel.

Trocar Sharp, pointed rod that fits inside a tube.

Turgor Normal resiliency of the skin caused by the outward pressure of the cells and interstitial fluid.

Urticaria Raised, itchy patch of skin usually associated with an allergic reaction. Also called hives.

Zoonotic Transmissible between humans and animals.

Index

A

Abbreviations, common, 321–322
Abdomen palpation, 43–44
Abdominal quadrants, 43–44
Abdominocentesis, 50
ABGs (arterial blood gases), 62
Abnormal eating behavior, 54–57
Abortion, 168
Abscess, anal glands, 284–285
Acronyms. *See* Abbreviations, common
ACT (activated clotting time), 49, 159
ACTH (adrenocorticotropic hormone), 48, 169
Activated clotting time (ACT), 49, 159
Activated partial thromboplastin time (APTT), 49, 159
Activity, 8
Acute pain, 6–7, 58–60
Acute renal failure (ARF), 154, 185–186
ADA (American Dental Association), 65
Addisonian crisis, 169
Addison's disease, 169–170
Adenovirus, 228–229
ADH (antidiuretic hormone), 171
Adrenal glands, 169
Adrenocorticotropic hormone (ACTH), 48, 169
Agalactia, 170
Aggression, 60–62
Airways, obstructed, 100–101
Alanine aminotransferase (ALT), 48
Alimentary lymphoma, 213
Alkaline phosphatase (ALP), 48
Allergic inhalant dermatitis, 158
Allergic reactions, 156–157, 204–206
Allergy testing (serum), 47
Alopecia, 304
ALP (alkaline phosphatase), 48
ALT (alanine aminotransferase), 48
Altered gas diffusion, 62–63
Altered mentation, 63–64
Altered oral health, 64–66
Altered sensory perception, 66
Altered urinary production, 67–68
Altered ventilation, 68
Ambulation, 107
American Dental Association (ADA), 65
American Veterinary Medical Association (AVMA), 2, 219, 276
Amaroid rings, 142–143
Ammonia, 47, 136–137, 143
Amp. (ampule), 321
Amplification. *See* Interventions
Ampule (amp.), 321
Amputation, 241–242
ANA (antinuclear antibody) tests, 47, 164
Anal gland abscess, 284–285
Anal gland removal, 240–241
Anastomosis, intestinal, 255–256
Anemia, 157–158, 226
Anthropomorphisizing, 73–74
Antibodies, 166
Antidiuretic hormone (ADH), 171
Antifreeze intoxication, 183–184
Antinuclear antibody (ANA) tests, 47, 164
Antisedan, 301
Anxiety, 69
Applanation tonometry, 50
APTT (activated partial thromboplastin time), 49, 159
ARF (acute renal failure), 154, 185–186
Arrhythmia, cardiac, 144
Arterial blood gases (ABGs), 62
Arterial catheters, 280
Arthritis, 164–165, 189–191
Arthrocentesis, 50
Arthroscopy, 242
As needed (p.r.n.), 322
ASDs (atrial septal defects), 152
Aspartate aminotransferase (AST), 48
Aspiration, risk of, 108–109
AST (aspartate aminotransferase), 48
Ataxia, 191–192
Atopy, 158–159
Atrial septal defects (ASDs), 152
Attending behaviors, 74
Attraction therapies, 94
Auditory interventions, 66
Aural hematoma, 242–243
Auricular vessels, 242
Auscultate thoracic cavity, 40–42
Aversion therapies, 95
AVMA (American Veterinary Medical Association), 2, 219, 276
Avulsion fractures, 251

B

Bacterial diseases, 233
BAR (Bright, Alert, Responsive), 40
Bathing, 112

335

BCS (body condition score), 38–39, 102, 115
Begging, 57
Behavioral counseling, 266–267
Behavioral inappropriate elimination, 93–96
Benign tumors, 212, 262
Beta cells, 213
Bile, 135, 143
Bile acid measurement, 47
Bilirubin, 139
Biopsies, 51
Bladder, 117–118, 187, 243–244, 311
Bland diets, 236
Bloat, 128, 253
Blood backwashing, 155
Blood chemistry values, 324–326
Blood clotting, 159, 167
Blood conditions
  anemia, 157–158
  coagulopathies, 159–160
  disseminated intravascular coagulation, 160–161
  hemorrhage, 161–162
  septicemia, 165
  thrombosis, 167–168
Blood gas analysis, 47
Blood gas values, 328
Blood glucose curves, 47
Blood pressure, 50, 148–149, 328
Blood tests, 47–49
Blood transfusions, 267–268
Blood types, 267
Blood urea nitrogen (BUN), 48
Blood values, 323–326
Body condition score (BCS), 38–39, 102, 115
Body system evaluations, 56
Bone
  infections, 200
  tumors, 216–217
Bordetella bronchiseptica, 229–230
Both eyes (O.U.), 322
Bowel habits, 76
Bowel incontinence, 69–71
Brachial plexus avulsion, 192–193
Bright, Alert, Responsive (BAR), 40
Bronchoalveolar lavage, 50
Bronchopneumonia, 167
Brucella, 183
Brushing, 112
Bubonic plague, 205
BUN (blood urea nitrogen), 48
Burns, 268–270
By mouth (po), 322

C

C. (with), 321
Calcium levels, 176–177
Calculi, 187
Caloric requirements, 102, 115
Cancers. *See* Neoplastic conditions
Canine adenovirus type 2 (CAV-2), 228
Canine bloat, 253
Canines
  behavioral inappropriate elimination, 96
  blood chemistry values, 324–326
  blood gas values, 328
  castration, 98
  hematology values, 323–324
  housetraining, 97
  infectious diseases, 228–237
  neutering, 285–290
  pulse rate ranges, 323
  respiratory rate ranges, 323
  temperature ranges, 323
  territorial inappropriate elimination, 97–98
  TPR values, 323
  urinalysis values, 326
  vaccinations, 321
Cannula, 243
Cap. (capsule), 321
Capillary refill time (CRT), 38–39, 71
Capsule (cap.), 321
Carcinoma, squamous cell, 217
Cardiac arrhythmia, 144
Cardiac compression, 271
Cardiac insufficiency, 71–72
Cardiomyopathy, 145
Cardiopulmonary arrest (CPA), 270–272
Cardiopulmonary auscultation terms, 41–42
Cardiopulmonary cerebrovascular resuscitation (CPCR), 270–272
Cardiopulmonary perfusion, 77–78
Cardiovascular diets, 72
Cardiovascular system conditions
  cardiac arrhythmia, 144
  cardiomyopathy, 145
  congestive heart failure, 146–147
  heartworm disease, 147–148
  hypertension, 148–149
  pericarditis, 149
  septal defects, 152–153
  shock, 153–155
  valvular insufficiency, 155–156

Case scenarios
  documenting, 284–319
  SOAP notes, notations, and MAOR, 27–33
Castration, 96, 98, 285–290
Casts, 272–273
Catheters, 106, 110, 118, 280–281
Cats. *See* Felines
Cauliflower ear, 243
CAV-2 (canine adenovirus type 2), 228
CBC (complete blood count), 48, 64
Cc (cubic centimeter), 321
Cell-mediated deficiency, 162
Cellophane bands, 142–143
Centesis, 50
Central nervous system (CNS), 191
Central venous pressure (CVP), 154
Cerebral perfusion, 77–78
Cerebrospinal fluid (CSF), 165
Certified veterinary technician (CVT), 321
Cesarean sections, 244–245, 291–297
Chemical spills, 184
Chemoreceptor trigger zone (CRTZ), 188, 274
Chemotherapy, 273–274
Chest tubes, 151
Chews, 65
CHF (congestive heart failure), 145–147
Cholangiohepatitis, 134–135
Chronic pain, 7, 8, 72–73
Chronic renal failure (CRF), 185–186
Client coping deficits, 73–74
Client knowledge deficits, 75
Client privacy, 73
Clinic questionnaires, 37
Clipping, 112
Clotting factors, 160
CNS (central nervous system), 191
Coagulation blood tests, 49
Coagulopathies, 159–160
Colitis, 124–125
Colloids, 90
Colostrum, 98–99
Coma, 193–194
Comatose animals, 193–194
Comminuted fractures, 251
Complete blood count (CBC), 48, 64
Compliance, 104–106
Computed tomography (CT), 49
Conditions, medical. *See body systems by name; individual conditions by name*
Congenital disorders, 159
Congestive heart failure (CHF), 145–147
Conjunctivitis, 202–203
Consent forms, 106
Constipation, 75–77
Conversions, 322–323
Convulsions, 201
Coombs' tests, 48
Coping deficits, 73–74
Coprophagy, 57
Core vaccinations, 321
Corneal ulceration, 203–204, 298
Coronavirus, 219–220
Corticotropin-releasing hormone (CRH), 169
Coughing spasms, 229
Counseling, behavioral, 266–267
CPA (cardiopulmonary arrest), 270–272
CPCR (cardiopulmonary cerebrovascular resuscitation), 270–272
CRF (chronic renal failure), 185–186
CRH (corticotropin-releasing hormone), 169
Critical safety, Needs Ladder, 6–7
CRT (capillary refill time), 38–39, 71
CRTZ (chemoreceptor trigger zone), 188, 274
Cryptorchidism, 218
Crystalloids, 90, 327
C-sections. *See* Cesarean sections
CSF (cerebrospinal fluid), 165
CT (computed tomography), 49
Cubic centimeter (cc), 321
Culture tests, 51
Cushing's disease, 170–171
Cutaneous, 171, 217
CVP (central venous pressure), 154
CVT (certified veterinary technician), 321
Cystitis, 184–185, 198
Cystotomies, 243–244
Cytology, 51
Cytotoxic drugs, 273

D

Databases
  computerized, 4
  diagnostic tests and procedures, 46–51
  versus flow sheets, 10
  overview, 35–36
  physical examination, 36–46
DEA (Drug Enforcement Administration), 276
Declawing, 245–246
Decompression techniques, 258–259
Decreased perfusion, 77–80
Decreasing stress therapies, 94–95
Degenerative joint disease (DJD), 190, 194–196, 197
Dehydration, 90–91

Dental abnormalities, 125
Dental cleaning, 64
Dental disease, 125–126
Dental prophylaxis, 246–247
Dental-specific diets, 65–66
Dermal ulcerations, 164
Dermal wounds, 151
Dermatitis, flea allergy, 204–206
Dermatophyte test medium (DTM), 211
Dermatophytoses, 211
Dermis, 268–269
Dexamethasone suppression tests, 48
D.h.l. (distemper hepatitis leptospirosis) vaccine, 321
Diabetes insipidus, 171–172
Diabetes mellitus, 172–174
Diabetic animals, 173–174
Diabetic ketoacidosis, 172
Diagnostic tests and procedures
    blood tests, 47–49
    fecal/urine, 51
    fluid collection/evaluation, 50–51
    overview, 46–47
    pathology/culture, 51
    physical tests, 49–50
Dialysis, peritoneal, 277–279
Diaphragmatic hernias, 247–248, 254
Diarrhea, 80–82
DIC (disseminated intravascular coagulation), 139–140, 160–161, 321
Diets
    bland, 236
    calorically dense, 115
    cardiovascular, 72
    dental-specific, 65–66
    growth, 196
    high-calorie, 131
    high-energy, 169
    high-fiber, 70, 76
    highly digestible, 132, 141
    high-protein, 135, 138
    hypoallergenic, 131, 157
    limited-antigen, 70
    low-allergen, 129
    low-calorie, 102–104
    low-fat, 81, 140
    low-fiber, 141
    low-residue, 70
    low-sodium, 88
    nutrient-dense, 221
    nutritionally dense, 233
    osteoarthritis, 190
    postsurgery, 104–105
    protein-restricted, 136
    recovery, 120, 236
    restricted salt, 145–147, 156
    syringeable, 109
    tubeable, 109
    weight-reduction, 190
Dilute urine, 172
Dinamap, 50
Dipsticks, 117
Dirofilaria immitis, 147
Discharge, 298–300
Diseases. *See individual medical conditions by name*
Disinfection, 111
Dislocations, 248–249
Disseminated intravascular coagulation (DIC), 139–140, 160–161, 321
Distemper, 223, 230–233
Distemper hepatitis leptospirosis (d.h.l.) vaccine, 321
Diuretics, 88
DJD (degenerative joint disease), 190, 194–196, 197
Documentation
    case scenarios, 284–310
    MAORs, 13–27
    notations and incorporations, 13
    overview, 9
    physical examination findings, 44–46
    SOAP notes, 9–12
Dogs. *See* Canines
Dominance aggression, 61
Domitor, 301
Doppler, 49
Dorsal laminectomy, 259
Drop (gtt.), 321
Drug Enforcement Administration (DEA), 276
Dry eye, 207
DTM (dermatophyte test medium), 211
Dumb form, rabies, 218
Dysfunction, reproductive, 108
Dysphagia, 126
Dystocia, 175–176, 291

E

Ears, 208–210, 242, 249
Eating behavior, abnormal, 54–57
ECG (electrocardiogram), 50, 71–72, 321
Echocardiography, 49
Eclampsia, 176–177
E-collars, 92, 105, 208, 241, 284–285
Ectropion, 204
Edema, 72, 79, 88
EG (ethylene glycol) intoxication, 183–184
Ehmer slings, 248–251
Electrocardiogram (ECG), 50, 71–72, 321
Electrolyte imbalances, 63, 81–83, 224, 269
Elimination, 7, 93–98

ELISA (enzyme-linked immunosorbent assay), 321
ELISA Snap test, 221–222
Emboli, 167
Emesis, 119
Encephalitis, 199
Endocrine system conditions
   Addison's disease, 169–170
   agalactia, 170
   Cushing's disease, 170–171
   diabetes insipidus, 171–172
   diabetes mellitus, 172–174
   hyperthyroidism, 177–178
   hypothyroidism, 178–179
Endodontics, 125
Endogenous ACTH, 48
Endoscopy, 50, 250
Enteral feeding tubes, 274–276
Enteritis, 129
Entropion, 204
Enzyme-linked immunosorbent assay (ELISA), 321
Enzymes, 141
EPI (exocrine pancreatic insufficiency), 131
Epidermis, 268–269
Epilepsy, 201–202
Esophageal feeding tubes, 274
Esophagostomy tubes, 274
Estrogens, 168
Ethylene glycol (EG) intoxication, 183–184
Euthanasia, 74, 276
Evaluations, 8. *See also* Technician evaluations
Every (q), 322
Every 4 hours (q4h), 322
Every other day (qod), 322
Examinations, physical. *See* Physical examinations
Exercise intolerance, 83–84
Exocrine pancreatic insufficiency (EPI), 131
External genitalia, 44
External stimuli, 114
Extrahepatic shunts, 142–143
Extranodal lymphoma, 213–214
Extremities, 43
Eye conditions. *See* Ocular conditions
Eye injuries, 256–257, 298

F

Facial fractures, 252–253
False pregnancy, 181
Fatty liver disease, 137
FCV (feline calicivirus), 224, 299–300
Fear, 84–85
Fear-induced aggression, 61
Fecal cultures, 51
Fecal flotation, 51, 81
Fecal output, 76
Fecal scalding, 70–71, 81
Fecal tests, 51
Feeding, 111–112, 274–276
Feline calicivirus (FCV), 224, 299–300
Feline immunodeficiency virus infection (FIV), 220–222, 321
Feline infectious anemia, 226
Feline infectious peritonitis (FIP), 135, 219–220
Feline leukemia (FeLV), 111, 222–223, 321
Feline ovariohysterectomies, 96
Feline panleukopenia, 223–224
Feline upper respiratory disease (FURD), 224–226, 299–300
Feline viral rhinotracheitis (FVR), 224, 299–300
Felines
   behavioral inappropriate elimination, 93–95
   castration, 96
   infectious diseases, 111, 135, 219–228, 299–300, 321
   territorial inappropriate elimination, 95–96
   vaccinations, 321
FeLV (feline leukemia), 111, 222–223, 321
Fenestration, 259
Fertility, 108
Fevers, 85–86
Fibrinogen, 49
Fine needle aspiration, 51
FIP (feline infectious peritonitis), 135, 219–220
First-degree burns, 268
FIV (feline immunodeficiency virus infection), 220–222, 321
Flea allergy dermatitis, 204–206
Fleas, 226–227
Flowsheets, 10
Fluid
   collection and evaluation, 50–51
   intake, 72
   requirement formula, 91
   retention, 87
Fluorescein stain, 50, 298
Forced oral feeding, 137–138
Foreign body obstructions, 126–127
Forelimb amputation, 241
Fractures, 250–253, 272
Fungal infections, 211
FURD (feline upper respiratory disease), 224–226, 299–300
FVR (feline viral rhinotracheitis), 224, 299–300

G

G or gm (gram), 321
Gag reflex, 109
Gallbladder conditions. *See* Liver conditions
Gas diffusion, altered, 62–63
Gastric bleeding, 129
Gastric dilatation volvulus (GDV), 127–128, 253–254
Gastric torsion, 253
Gastritis, 128–129
Gastroenteritis, 235
Gastrointestinal (GI) perfusion, 79
Gastrointestinal (GI) system conditions
  colitis, 124–125
  dental disease, 125–126
  dysphagia, 126
  foreign body obstruction, 126–127
  gastric dilatation volvulus, 127–128
  gastritis, 128–129
  ileus, 130
  intussusception, 129–130
  malabsorption, 130–131
  maldigestion, 131–132
  megaesophagus, 132–133
  oral trauma, 133
  salivary mucocele, 134
Gastrostomy tubes (G-tube), 275
GDV (gastric dilatation volvulus), 127–128, 253–254
Genitalia, 44
Germinal epithelium, 246
GI (gastrointestinal) perfusion, 79
GI (gastrointestinal) system conditions. *See* Gastrointestinal (GI) system conditions
Glaucoma, 206–207
Glucose, 47, 172
Gram (g or gm), 321
Granulocytopathy syndrome, 162
Grass eating, 57
Greenstick fractures, 251
Grooming, self-care deficit, 112
Growth diets, 196
Growths, removal of, 305–310
Gtt. (drop), 321
G-tubes, 275

H

H (hour), 321
H&H (hemoglobin and hematocrit), 81
Hand washing techniques, 110–111
Head, physical examinations of, 42–43
Heart, 40–42. *See also* index entries beginning with "cardio"
Heartworm disease, 147–148
Heartworm tests, 48
Heimlich valves, 151
Hemangioma, 212
Hemangiosarcoma, 212–213
Hematology values, 323–324
Hematoma, aural, 242–243
Hemilaminectomy, 258
Haemobartonella, 226–227
Hemoglobin and hematocrit (H&H), 81
Hemolytic disorders, 163
Hemorrhages, 161–162, 245
Hemostasis, 159
Hepatic encephalopathy, 136–137, 143
Hepatic lipidosis, 137–139
Hepatitis, 135–136, 232–233
Hernias, 247–248, 254–255
Herpesviruses, 183
Hiatal hernias, 254
High blood pressure, 148–149
High-calorie diets, 131
High-energy diets, 169
High-fiber diets, 70, 76
Highly digestible diets, 132, 141
High-protein diets, 135, 138
Hind limb amputation, 241
Hip dysplasia, 196–197
Histology, 51
History, obtaining, 37–38
Hormonal blood tests, 48–49
Hormone-responsive incontinence, 117
Hospital physical examination forms, 36
Hour (h), 321
Humidified oxygen, 89
Hydration, 7–8
Hyperadrenocorticism, 170
Hyperbilirubinemia, 139
Hyperkalemia, 82–83
Hypernatremia, 82–83
Hyperphosphatemia, 82–83
Hypertension, 148–149
Hyperthermia, 85–87
Hyperthyroidism, 177–178, 279–280
Hypervolemia, 87–88
Hypoadrenocorticism, 169
Hypoallergenic diets, 131, 157
Hypocalcemia, 83
Hypoglycemia, 63
Hypokalemia, 82–83, 128
Hypomagnesemia, 82
Hyponatremia, 82–83
Hypothermia, 88–90
Hypothyroidism, 178–179
Hypoventilation, 248
Hypovolemia, 90–91

I

IBD (inflammatory bowel disease), 70, 130, 135
Icterus, 139
Ictus seizures, 202
IDDM (insulin-dependent diabetes mellitus), 172
Identification, patient, 37
IgA (immunoglobulin A) deficiency, 162
Ileus, 130
IM (intramuscular), 321
Imaging physical tests, 49–50
Imbalance, electrolyte, 63, 81–83, 224, 269
IMHA (immune-mediated hemolytic anemia), 163
Immediately (stat), 322
Immune system conditions
  allergic reactions, 156–157
  atopy, 158–159
  IMHA, 163
  immunodeficiencies, 162–163
  pemphigus, 164
  rheumatoid arthritis, 164–165
  systemic lupus erythematosus, 166–167
Immune-mediated hemolytic anemia (IMHA), 163
Immunoglobulin A (IgA) deficiency, 162
Immunosuppressive viruses, 235
Impaired tissue integrity, 91–92
Inappropriate elimination, 93–98
Incontinence
  behavioral, 118
  bowel, 69–70
  colitis-associated, 124
  hormone-responsive, 117
  management techniques, 70–71
  submissive, 118
  urge/inflammatory, 117
  urinary, 116–119
Incorporations, 13
Indentation tonometry, 50
Ineffective nursing, 98–99
Infections
  bone, 200
  disinfection, 111
  fungal, 211
  risk of, 109–110
  skin, 210
  transmission of, 110–111
  urinary tract, 93, 117, 171, 184–185
  yeast, 209
Infectious diseases
  canine, 228–237
  feline, 219–228
  rabies, 218–219
Infectious enteritis, 223
Infectious tracheobronchitis (IT), 228–229, 234
Inflammatory bowel disease (IBD), 70, 130, 135
Inguinal hernias, 254
Inhalation burns, 269–270
Injuries
  eye, 256–257, 298
  lung, 147
  self-inflicted, 112–113
Insulin, 172, 213
Insulin-dependent diabetes mellitus (IDDM), 172
Insulinoma, 213
Integument conditions, 204–206, 210, 211
Interventions
  abnormal eating behavior, 54–57
  acute pain, 58–60
  aggression, 60–62
  altered gas diffusion, 62–63
  altered mentation, 63–64
  altered oral health, 64–66
  altered sensory perception, 66
  altered urinary production, 67–68
  altered ventilation, 68
  anxiety, 69
  bowel incontinence, 70–71
  cardiac insufficiency, 71–72
  chronic pain, 72–73
  client coping deficit, 73–74
  client knowledge deficit, 75
  constipation, 76–77
  decreased perfusion, 78–80
  diarrhea, 80–82
  electrolyte imbalance, 82–83
  exercise intolerance, 83–84
  fear, 84–85
  hyperthermia, 86–87
  hypervolemia, 87–88
  hypothermia, 89–90
  hypovolemia, 90–91
  impaired tissue integrity, 92
  implementing, 8
  inappropriate elimination, 93–98
  ineffective nursing, 98–99
  noncompliant owners, 100
  obstructed airway, 100–101
  overweight animals, 102–104
  postoperative compliance, 104–105
  preoperative compliance, 105–106
  reduced mobility, 107
  reproductive dysfunction, 108
  risk of aspiration, 109
  risk of infection, 109–110
  risk of infection transmission, 110–111
  self-care deficit, 111–112

Interventions (*continued*)
   self-inflicted injuries, 113
   sleep disturbance, 113–114
   status within acceptable parameters, 114
   underweight animals, 115–116
   urinary incontinence, 116–119
   vomiting/nausea, 119–120
Intervertebral disc disease (IVDD), 197–199, 260
Intestinal obstructions, 130
Intestinal resection and anastomosis, 255–256
Intestinal telescoping, 129
Intoxication, ethylene glycol, 183–184
Intracardiac shunts, 152
Intradermal testing, 50
Intrahepatic shunts, 142
Intramuscular (IM), 321
Intraocular pressure (IOP), 50, 206
Intraperitoneal (IP), 321
Intravenous (IV), 321
Intussusception, 129–130
Iodine therapy, radioactive, 279–280
IOP (intraocular pressure), 50, 206
IP (intraperitoneal), 321
IT (infectious tracheobronchitis), 228–229, 234
IV (intravenous), 321
IVDD (intervertebral disc disease), 197–199, 260

J
Jaundice, 139
Joint diseases, 164–165
Joint displacement, 248
Joint immobilization, 272
Jugular vein distention (JVD), 71

K
KCS (keratoconjunctivitis sicca), 207
Kennel cough, 229
Kennel/barn workers, 2
Keratoconjunctivitis sicca (KCS), 207
Ketones, 172
Kidneys. *See also* Renal failure
Knott's tests, 48

L
Labor, 175
Laboratory results, 47
Lacerations, 256, 301–302
Lactate dehydrogenase (LDH), 48
Lactated Ringer's solution (LRS), 90
Lactation, 116, 176

Lactose intolerance, 131
Lactulose treatments, 137, 143
Lateral intercostal thoracotomy, 261
Lb (pound), 322
LDH (lactate dehydrogenase), 48
LE cell preparation, 48
Left eye (O.S.), 322
Lentivirus, 220
Leptospirosis, 233–234
Leukemia, feline, 111, 222–223
Level of consciousness (LOC), 38–40
Leydig cell tumors, 217
Limbs, 192, 241, 251–252
Limited-antigen diets, 70
Liquid conversions, 322–323
Litter box aversion, 93–94
Liver conditions, 134–139, 142–144, 232
LOC (level of consciousness), 38–40
Lochia, 175
Low-allergen diets, 129
Low-calorie diets, 102–104
Low-fat diets, 81, 140
Low-fiber diets, 141
Low-residue diets, 70
Low-sodium diets, 88
LRS (lactated Ringer's solution), 90
Lubricants, ocular, 208
Lungs, 40–42, 68, 87, 147, 150
Lymphatic conditions. *See* Immune system conditions
Lymphoma, 213–214

M
Magnetic resonance imaging (MRI), 49
Malabsorption, 130–131
Maldigestion, 131–132
Malignant tumors, 262
Malnutrition, 115–116
Mammary disease, 170
Mammary glands, 179, 214–215
Mammary tumors, 214–215
MAORs. *See* Medical Administration/Order Records
Maslow's Hierarchy of Needs, 5
Masses, removal of, 305–310
Mast cell tumors, 215–216
Mastitis, 179–180
MCHC (mean corpuscular hemoglobin concentration), 48
MCV (mean corpuscular volume), 48
Mean corpuscular hemoglobin concentration (MCHC), 48
Mean corpuscular volume (MCV), 48
Mediastinal lymphoma, 213
Medical Administration/Order Records (MAORs)

case scenario, 27–33
examples, 18–27
overview, 13–17
Medical conditions. *See also* Infectious diseases
Medical records, 9
Megaesophagus, 132–133
Meningitis, 199–200
Meningoencephalitis, 199
Mentation, altered, 63–64
Metastasis, 216
Microchips, 37
Microsporum, 304
Mild pain, 7
Milk production, 170, 179–180, 245
Milliliter (mL), 322
Mitral valves, 155
ML (milliliter), 322
Mobility, reduced, 106–107
Moderate pain, 7
Morphine, 140
Mosquitoes, 147–148
MRI (magnetic resonance imaging), 49
Mucous membrane color, 39
Multicentric lymphoma, 213
Musculoskeletal system conditions
  arthritis, 189–191
  ataxia, 191–192
  degenerative joint disease, 194–196
  hip dysplasia, 196–197
  intervertebral disc disease, 197–199
  osteomyelitis, 200
  panosteitis, 201
Muzzles, 60, 72, 85
Myelogram, 49

N

Narcotics, 59
Nasogastric (NG) intubation, 274
National Association of State Public Health Veterinarians (NASPHV), 219
Nausea, 119–120
Neck, physical examination of, 42–43
Needs Ladder, 6–8, 54–55
Neonatal rejection, 292
Neoplastic conditions, 212–218
Nephrectomy, 258
Nephrolithotomy, 258
Neurologic system conditions
  brachial plexus avulsion, 192–193
  coma, 193–194
  meningitis, 199–200
  seizures, 201–202
Neutering, 258
New puppy exam, 302–303
NG (nasogastric) intubation, 274

Noncompliant owners, 99–100
Noncore vaccinations, 321
Noncritical safety, Needs Ladder, 7–8
Nonprofessional staff, 2
Nonsteroidal anti-inflammatory drugs (NSAIDs), 59
Notations, 13, 27–33
N.p.o. status, 119–120
NSAIDs (nonsteroidal anti-inflammatory drugs), 59
Nuclear scintigraphy, 49
Nursing, ineffective, 98–99
Nutrient-dense diets, 221, 233
Nutrition, 7–8, 64, 137, 276–277

O

O_2 (oxygen) measurements, 50
Obese animals, 103–104
Objective data, 3–4, 36
Oblique fractures, 251
Obstructed airways, 100–101
Ocular conditions
  conjunctivitis, 202–203
  corneal ulceration, 203–204
  entropion and ectropion, 204
  glaucoma, 206–207
  injuries and surgery, 256–257
  keratoconjunctivitis sicca (KCS), 207
  ocular discharge, 298
  ocular proptosis, 208
  retinal detachment, 210
Ocular emergencies, 210
Ocular lubricants, 208
Ocular tests, 50
Ocular trauma, 208
O.D. (right eye), 322
OHE (ovariohysterectomy), 96, 168
Once a day (sid), 322
Onychectomies, 245–246
Open fractures, 251
Operations. *See* Surgical procedures
Ophthalmic fluorescent test strips, 95
Opioids, 59
Oral carcinoma, 217
Oral health, 64–66
Oral trauma, 133
Oral ulcerations, 167
Orthodontics, 125
Orthopedic Foundation Association evaluation, 49
O.S. (left eye), 322
Osteoarthritis, 190, 196
Osteomyelitis, 200
Osteosarcoma, 216–217
Otitis, 208–210
O.U. (both eyes), 322

Ovariohysterectomy (OHE), 96, 168
Overweight animals, 101–104
Owners, 99–100, 112
Oxygen (O_2) measurements, 50
Oxygen saturation levels, 62, 68, 71, 151
Oxygenation, 6, 77, 101, 154–155, 245
Oxytocin, 244–245

P

P3 (third phalanx), 245–246
Packed cell volume (PCV), 48, 267
Pain
  acute, 6–7, 58–60
  chronic, 7, 72–73
Pain scale, 40, 58–59, 80
Pain-induced aggression, 61
Palpation, abdomen, 43–44
Pancreas, 172, 213
Pancreatic duct spasms, 140
Pancreatic enzymes, 132
Pancreatic insufficiency, 141–142
Pancreatitis, 135, 139–141
Panleukopenia, feline, 223–224
Panosteitis, 201
Parainfluenza virus (PIV), 229, 234–235
Paralysis, 198
Paraphimosis, 180
Parasites, 147, 227
Parental aggression, 61
Parenteral nutrition (PN), 276–277
Partial parenteral nutrition (PPN), 137, 276–277
Parvovirus, 223, 235–237
Pathology tests, 51
Patient identification, 37
PCV (packed cell volume), 48, 267
Pelvic fractures, 250–251
Pemphigus, 164
Penis engorgement, 180
Pennsylvania Hip Improvement Program (PennHIP), 49
Penrose drains, 243
Perforation, urinary bladder, 187
Perfusion, decreased, 77–80
Pericarditis, 149
Perineal area, 284–285
Periocular mucoid discharge, 207
Periodontal disease, 246–247
Periodontics, 125
Periparturient hypocalcemia, 176
Peripheral edema, 146
Peripheral parenteral nutrition (PPN), 270
Peripheral perfusion, 79–80
Peritoneal dialysis, 277–279
Peritoneum, 44
Peritonitis, 129–130, 187, 219–220

Pet loss hotlines, 328
Pet loss Web sites, 329
Phantom pregnancy, 181
Pharyngeal walls, 134
Physeal fractures, 251
Physical examinations
  abdomen palpation, 43–44
  auscultate thoracic cavity, 40–42
  head and neck, 42–43
  obtaining history, 37–38
  overview, 36
  patient identification, 37
  peritoneum and external genitalia, 44
  recording findings, 44–46
  trunk and extremities, 43
  vital signs, 38–40
  weight and body condition scores, 38
Physical tests, 49–50
Pica, 54, 57, 127
Pinna removal, 249
Pituitary glands, 169
PIV (parainfluenza virus), 229, 234–235
Plan of care, developing, 8
Plaque, 65
Play aggression, 61
PN (parenteral nutrition), 276–277
Pneumonia, 151–152
Pneumothorax, 150–151
P.o. (by mouth), 322
Polyuria/Polydipsia (PU/PD), 148
Portosystemic shunts (PSSs), 136, 142–144
Possessive aggression, 61
Postictal seizures, 202
Postoperative compliance, 104–105
Postsurgery diets, 104–105
Pound (lb), 322
PPN (partial parenteral nutrition), 137, 276–277
PPN (peripheral parenteral nutrition), 270
Predatory aggression, 61
Pregnancy, 181
Preictal seizures, 201
Preoperative compliance, 105–106
Prepuce, 180
Pressure transducers, 50
P.r.n. (as needed), 322
Procedures. *See* Diagnostic tests and procedures
Prolapses, 257
Prophylaxis, dental, 246–247
Proptosis, ocular, 208
Prostaglandins, 168
Prostate glands, 180–181
Prostate wash, 51
Prostatitis, 180–181
Protein-restricted diets, 136
Prothrombin time (PT), 49, 159

Pruritus, 158–159, 304
Pseudopregnancy, 181
PSSs (portosystemic shunts), 136, 142–144
PT (prothrombin time), 49, 159
Puerperal tetany, 176
Pulmonary edema, 144, 146
Pulse oximetry, 50
Pulse rate, 323
Pulse strength scale, 79
PU/PD (polyuria/polydipsia), 148
Puppy recovery stations, 291
Pus retention, 182
Pyoderma, 210
Pyometra, 182

Q

Q (every), 322
Q4h (every 4 hours), 322
QAR (quite alert and responsive), 40, 322
Q.i.d. (4 times a day), 322
Q.o.d. (every other day), 322
Qs (a sufficient quantity), 322
Quite alert and responsive (QAR), 40, 322

R

Rabies, 218–219
Radioactive iodine therapy (I-131), 279–280
Radiographs, 50
Range of motion (ROM), 107, 190, 242
Rapid bone growth, 201
Rationales. *See* Interventions
RBCs (red blood cells), 90, 157–158, 163
Reagent strips, 117
Recovery diets, 120, 236
Recovery stations, 244, 291
Rectal prolapse, 257
Red blood cells (RBCs), 90, 157–158, 163
Reduced mobility, 106–107
Reevaluation, 8
Registered veterinary technician (RVT), 322
Regurgitation, 132–133
Renal failure, 154, 185–186
Renal perfusion, 77–79
Renal surgery, 258
Renal system conditions, 154, 183–186. *See also* Urinary system conditions
Reproductive system conditions. *See also* Endocrine system conditions
  abortion, 168
  dysfunction, 108
  dystocia, 175–176
  eclampsia, 176–177
  mastitis, 179–180
  paraphimosis, 180
  prostatitis, 180–181

  pseudopregnancy, 181
  pyometra, 182
  vaginitis, 182–183
Resection, intestinal, 255–256
Resolution, desired. *See* Desired resolution
Respiratory distress, 68
Respiratory patterns, 41
Respiratory rates, 323
Respiratory secretions, 101
Respiratory system conditions, 150–152
Restraint techniques, 60, 85
Restricted salt diets, 145–147, 156
Retinal detachment, 148, 210
RF (rheumatoid factor) test, 164
Rheumatoid arthritis, 164–165
Rheumatoid factor (RF) test, 164
Right eye (O.D.), 322
Ringworm, 211, 304
"Risk for" assessments, 5
Risks, 108–111
ROM (range of motion), 107, 190, 242
Rule of One, 267
RVT (registered veterinary technician), 322

S

Safety, 6–7
Salivary glands, 134
Salivary mucocele, 134
Sample cases. *See* Case scenarios
Sc (subcutaneous), 322
Schiotz tonometer, 50
Schirmer's test, 50, 298
Secondary glaucoma, 206
Second-degree burns, 268–269
Seizures, 201–202
Self-care deficits, 111–112
Self-inflicted injuries, 112–113
Self-mutilation, 113
Seminomas, 217
Sensory perception, altered, 66
Sepsis, 166
Septal defects, 152–153
Septic shock, 165
Septicemia, 165, 167
Sertoli cell tumors, 217
Serum (allergy testing), 47
Serum antibody titers, 48
Serum chemistry, 48
Serum cortisol, 48
Severe pain, 6–7
Shock, 153–155
S.i.d. (once a day), 322
Skin, 76, 210, 215–216
Skin breakdown, 107
Skin conditions. *See* Integument conditions

SLE (systemic lupus erythematosus), 47, 166–167
Sleep disturbances, 113–114
SOAP (subjective, objective, assessment, plan) notes, 9–12, 27–33
Sodium bicarbonate, 184
Sodium pentobarbital, 276
Soft tissue tumors, 212–213
Spaying, 258
Spinal cord surgeries, 258–260
Spiral fractures, 251
Spleen, 127, 165–166
Splenomegaly, 165–166
Spontaneous pneumothorax, 150
SQ (subcutaneous), 322
Squamous cell carcinoma, 217
Standardized forms, 36–37
Stat (immediately), 322
Sterilization, 258
Subcutaneous (SC/SQ), 322
Subjective data, 3–4
Subjective, objective, assessment, plan (SOAP) notes, 9–12, 27–33
Sublingual glands, 134
Superhydration, 68
Surgical procedures
   amputation, 241–242
   anal gland removal, 240–241
   arthroscopy, 242
   aural hematoma, 242–243
   bladder surgery, 243–244
   cesarean section, 244–245
   declawing, 245–246
   dental prophylaxis, 246–247
   diaphragmatic hernia, 247–248
   dislocations, 248–249
   ear cropping, 249
   endoscopy, 250
   fractures, 250–253
   gastric dilatation volvulus, 253–254
   hernias, 254–255
   intestinal resection and anastomosis, 255–256
   lacerations, 256
   ocular injury/surgery, 256–257
   overview, 239–240
   prolapses, 257
   renal surgery, 258
   spaying/neutering, 258
   spinal cord/vertebrae surgery, 258–260
   tail docking, 260–261
   thoracic surgery, 261–262
   tumor removal, 262
   wounds, 262–263
Suspension (susp.), 322
Swab cultures, 51
Swallow reflex, 109

Swallowing, 126
Syringeable diets, 109
Systemic lupus erythematosus (SLE), 47, 166–167

T

T4 (thyroxine), 49
Tablet (tab), 322
Tail docking, 260–261
Tapeworms, 205
Tattoos, 37
Tear production, 207
Technician evaluations. *See also* Medical conditions; Surgical procedures; Therapeutic procedures
   abnormal eating behavior, 54–57
   acute pain, 58–60
   aggression, 60–62
   altered gas diffusion, 62–63
   altered mentation, 63–64
   altered oral health, 64–66
   altered sensory perception, 66
   altered urinary production, 67–68
   altered ventilation, 68
   anxiety, 69
   bowel incontinence, 69–71
   cardiac insufficiency, 71–72
   chronic pain, 72–73
   client coping deficit, 73–74
   client knowledge deficit, 75
   constipation, 75–77
   decreased perfusion, 77–80
   diarrhea, 80–82
   electrolyte imbalance, 82–83
   exercise intolerance, 83–84
   fear, 84–85
   hyperthermia, 85–87
   hypervolemia, 87–88
   hypothermia, 88–90
   hypovolemia, 90–91
   identifying and prioritizing, 4
   impaired tissue integrity, 91–92
   inappropriate elimination, 93–98
   ineffective nursing, 98–99
   noncompliant owners, 99–100
   obstructed airway, 100–101
   overview, 53–54
   overweight, 101–104
   postoperative compliance, 104–105
   preoperative compliance, 105–106
   reduced mobility, 106–107
   reproductive dysfunction, 108
   risk of aspiration, 108–109
   risk of infection, 109–110
   risk of infection transmission, 110–111

self-care deficit, 111–112
self-inflicted injuries, 112–113
sleep disturbances, 113–114
status within acceptable parameters, 114
underweight animals, 114–116
urinary incontinence, 116–119
vomiting/nausea, 119–120
Teeth cleaning, 246–247
Telescoping, 129
Temperature conversions, 323
Temperature values, 86, 89, 323
Tenesmus, 76, 81
Territorial aggression, 61
Territorial inappropriate elimination, 95–98
Testicular tumors, 217–218
Tests. *See* Diagnostic tests and procedures
Therapeutic procedures
  behavioral counseling, 266–267
  blood transfusion, 267–268
  burn therapy, 268–270
  cardiopulmonary cerebrovascular resuscitation, 270–272
  casts, 272–273
  chemotherapy, 273–274
  enteral feeding tubes, 274–276
  euthanasia, 276
  overview, 265–266
  parenteral nutrition, 276–277
  peritoneal dialysis, 277–279
  radioactive iodine therapy, 279–280
  vascular access via catheterization, 280–281
Third phalanx (P3), 245–246
Third-degree burns, 268–269
Thoracic cavity, 40–42
Thoracic limbs, 192
Thoracic surgery, 261–262
Thoracocentesis, 50
Thromboembolism, 149
Thrombophlebitis, 277
Thrombosis, 167–168
Thyroglobulin antibody tests, 48
Thyroid conditions, 177–179, 279
Thyroid-stimulating hormone (TSH), 49
Thyroxine (T4), 49
T.i.d. (3 times a day), 322
Tissue integrity, impaired, 91–92
TLI (trypsin-like immunoreactivity), 131
Toileting, self-care deficit, 112
Tono-Pen, 50
Tooth brushing, 65
Total parenteral nutrition (TPN), 64, 137, 276–277
Toxicology panels, 48
Toxocara canis, 303
Toxoplasmosis, 227–228
Toys, 65
TPN (total parenteral nutrition), 64, 137, 276–277
TPR values, 323
Training, 266–267
Transfusions, blood, 267–268
Transverse fractures, 251
Trauma, 133, 208
Traumatic pneumothorax, 150
Triglyceride measurements, 48
Trunk, physical examinations of, 43
Trypsin-like immunoreactivity (TLI), 131
TSH (thyroid-stimulating hormone), 49
Tube feeding, 137–138
Tubeable diets, 109
Tumor removal, 262, 305–310. *See also* Neoplastic conditions
Tumors
  benign, 212, 262
  bone, 216–217
  Leydig cell, 217
  malignant, 262
  mammary, 214–215
  mast cell, 215–216
  removal of, 262, 305–310
  Sertoli cell, 217
  skin, 215–216
  soft tissue, 212–213
  testicular, 217–218
Twice a day (b.i.d.), 321

U

Ulcerations
  corneal, 203–204
  dermal, 164
  oral, 167
Ultrasound examinations, 50
Umbilical hernias, 254
Underweight animals, 114–116
Urethral obstructions, 310–319
Urge/inflammatory incontinence, 117
Urinalysis values, 51, 326
Urinary catheters, 68, 311–313
Urinary system conditions
  cystitis, 184–185
  incontinence, 116–119
  production, altered, 67
  urinary bladder perforation, 187
  urinary tract infections (UTIs), 93, 117, 171
  uroliths, 187–189
Urine, 51, 67, 117, 188, 310–311
Uterine hemorrhages, 245
Utility, Needs Ladder, 8

V

Vaccinations, 62, 321
Vaginal prolapse, 257
Vaginitis, 182–183
Valvular insufficiency, 155–156
Vascular access, 280–281
Vascular system conditions. *See* Blood conditions
Venous catheters, 280
Ventilation, altered, 68
Ventral slot decompression, 259
Ventricular premature complex (VPC), 188
Ventricular septal defects (VSDs), 152
Vertebrae surgeries, 258–260
Veterinarian assistants, 2
Veterinarians, 2
Veterinary Oral Health Council (VOHC), 65
Veterinary technician practice model, 2–8
Veterinary technicians, 2–3
Viral diseases, 218–222
Viruses
   CAV-2, 228
   coronavirus, 219–220
   FCV, 224
   FIV, 220–222, 321
   herpesviruses, 183
   immunosuppressive, 235
   lentivirus, 220
   parvovirus, 223, 235–237

PIV, 229, 234–235
Vision loss, 210
Visual impairment, 66
Vital signs, 38–40
VOHC (Veterinary Oral Health Council), 65
Volume overload, 161
Vomiting, 119–120
von Willebrand factor (vWF) antigen assay, 49, 159
VPC (ventricular premature complex), 188
VSDs (ventricular septal defects), 152
vWF (von Willebrand factor) antigen assay, 49, 159

W

WBC (white blood cells), 48
Weight, 38, 101–104
Weight and liquid conversions, 322–323
Weight-reduction diets, 190
White blood cells (WBC), 48
With (c.), 321
Within normal limits (WNL), 4, 44–46
Wobbler syndrome, 259
Wolfing food, 57
Wounds, 92, 256, 262–263

Y

Yeast infections, 209